System Architecture

Wolfgang J. Paul · Christoph Baumann
Petro Lutsyk · Sabine Schmaltz

System Architecture

An Ordinary Engineering Discipline

 Springer

Wolfgang J. Paul
FR 6.1 Informatik
Universität des Saarlandes
Saarbrücken, Saarland
Germany

Petro Lutsyk
FR 6.1 Informatik
Universität des Saarlandes
Saarbrücken, Saarland
Germany

Christoph Baumann
School of Computer Science
  and Communication
KTH Royal Institute of Technology
Stockholm
Sweden

Sabine Schmaltz
FR 6.1 Informatik
Universität des Saarlandes
Saarbrücken, Saarland
Germany

ISBN 978-3-319-82729-2   ISBN 978-3-319-43065-2   (eBook)
DOI 10.1007/978-3-319-43065-2

Printed on acid-free paper

This Springer imprint is published by Springer Nature
The registered company is Springer International Publishing AG
The registered company address is: Gewerbestrasse 11, 6330 Cham, Switzerland

Contents

1

Introduction

This text contains the lecture notes of a class we teach in Saarbrücken to first-year students within a single semester. The purpose of the class is simple: to exhibit constructions of

- a simple MIPS processor
- a simple compiler for a C dialect
- a small operating system kernel

and to give detailed explanations for why they work.

We are able to cover all this material within a single lecture course, because we treat computer science as an engineering discipline like any other: for any topic there is appropriate math which is intuitive and adequate to deal with it, both for specifications and for explanations for why constructions work. High school mathematics happens to suffice to deal with all subjects in this text.

As a warm-up exercise we study in Chap. 2 how to prove some very basic properties of decimal addition. Chapter 3 contains reference material about basic mathematical concepts which are used throughout the book. Elementary properties of binary numbers, two's complement numbers, and Boolean algebra are treated in Chap. 4. In particular the correctness of algorithms for binary addition and subtraction is shown. Of course we also include a proof of the fact that every switching function can be computed by a Boolean formula.

A digital hardware model consisting of circuits and registers is introduced in Chap. 5 and some simple circuits are constructed. Chapter 6 contains various constructions of random access memory (RAM) that are needed later in the text. Several adder constructions (including conditional-sum adders and carry-look-ahead adders), an arithmetic logic unit (ALU), and a simple shifter are covered in Chap. 7. In Chap. 8 a subset of the MIPS instruction set architecture (ISA) is introduced and a simple sequential (i.e., not pipelined) MIPS processor is constructed. Although the correctness proof of the basic processor is straightforward, it exhibits an important concept: processor hardware usually does not interpret all ISA programs; only programs which satisfy certain software conditions are correctly executed. In our case we require so-called 'alignment' and the absence of writes into the ROM (read-only

© Springer International Publishing Switzerland 2016
W.J. Paul et al., *System Architecture*, DOI 10.1007/978-3-319-43065-2_1

memory) region. Because we have a correctness proof we know that this list of conditions is *exhaustive* for our simple processor. In the text we contrast this with the situation for commercial multi-core processors and arrive at slightly worrisome conclusions.

Example programs written in assembly code are presented in Chap. 9. In particular we present algorithms for multiplication and division and show that they work. These algorithms are used later in the compiler for the translation of multiplications and divisions in expressions.

Chapter 10 contains a fairly detailed treatment of context-free grammars and their derivation trees. The reason why we invest in this level of detail is simple: compilers are algorithms operating on derivation trees, and the theory developed in this chapter will permit us to argue about such algorithms in a concise and — hopefully — clear way. In particular we study grammars for arithmetic expressions. In expressions one uses priorities between operators to save brackets. For instance, we abbreviate $1 + (2 \cdot x)$ by $1 + 2 \cdot x$ because multiplication takes priority over addition. Fortunately with this priority rule (and a few more) any expression can be interpreted in a unique way, both by readers of mathematical text and by compilers. We feel that this is fundamental and have therefore included the proof of this result: in a certain grammar, which reflects the priorities of the operators, the derivation trees for expressions are unique.

Next, syntax and semantics of the programming language $C0$ are presented in Chap. 11. In a nutshell $C0$ is Pascal with C syntax. Although we avoid address arithmetic and type casting, $C0$ is quite a rich language with recursive data types, pointers, and a heap. Its complete grammar fits on a single page (Table 12). We begin with an informal description and elaborate somewhat on the implementation of tree algorithms in $C0$. The semantics of expressions is defined via the unique derivation trees studied in Chap. 10. The effect of statement execution is defined by a so-called small-step semantics hinging on $C0$ configurations with a program rest. A big-step semantics would be simpler, but it would not work in the context of kernel programming, where the $C0$ computations of a kernel must be interleaved with the ISA computations of user processes.

A non-optimizing compiler translating $C0$ programs into MIPS programs is described in Chap. 12. We assume that the derivation tree T of an arbitrary $C0$ program p is given, and we describe by induction over the subtrees of T a target program p' in MIPS ISA which simulates source program p. Having formal semantics both for $C0$ and MIPS ISA makes life remarkably simple. We invest some pages in the development of a consistency relation $consis(c,d)$, which formalizes how $C0$ configurations c are encoded in ISA configurations d. Then we use this relation as a guide for code generation by induction over the subtrees of derivation tree T: arguing locally we always generate ISA code such that in runs of the source program and the target program consistency is maintained. Once the consistency relation is understood this is remarkably straightforward with two exceptions. First, for expression evaluation we use the elegant and powerful Sethi-Ullman algorithm. Second, there is a 'natural mismatch' between the way program execution is controlled in ISA (with a program counter) and in $C0$ (with a program rest). For the correctness proof of the compiler we bridge this gap with the help of the very useful technical Lemma 90. There we identify for every node n in the derivation tree, which generates a statement s in the program rest, the unique node $succ(n)$ which generates the next statement after s in the program rest, unless s is a return statement.

At this place in the text we have laid the foundations which permit us to deal with system programming — here the programming of a kernel — in a precise way. Kernels perform so-called process switches, where they save or restore processor registers in $C0$ variables, which are called process control blocks. In higher-level programming languages such as $C0$ one can only access C variables; the processor registers are not directly accessible. Thus in order to program 'process save' and 'restore' one has to be able to switch from $C0$ to assembly code and back. We call the resulting language '$C0$ + assembly' or for short '$C+A$'. A semantics for this language is developed in Sect. 13.1. We begin by studying an apparently quite academic question: given an ISA configuration d, what are all possible well-formed $C0$ configurations c satisfying $consis(c,d)$? We call every such configuration a possible C *abstraction* of d. It turns out that the construction of such C abstractions is unique except for the heap. During the reconstruction of possible C abstractions Lemma 90 is reused for the reconstruction of the program rest, and reachable portions of the heap are found essentially by a graph search starting from pointer variables in the global memory and on the stack. In general, heaps cannot be uniquely reconstructed, but an easy argument shows that different reconstructed heaps are basically isomorphic in the graph theoretic sense. We call $C0$ configurations c and c' equivalent if their heaps are isomorphic and their other components are identical and show: i) computations starting in equivalent configurations continue in equivalent configurations and ii) expression evaluation in equivalent computations leads to the same result, unless the expression has pointer type.

This result opens the door to define the semantics of C+A in a quite intuitive way. As long as abstracted C code is running we run the ISA computation and C abstractions in lock step. If the computation switches to inline assembly or jumps completely outside of the C program (this happens if a kernel starts a user process), we simply drop the C abstraction and continue with ISA semantics. When we return to translated C code we reconstruct a C abstraction at the earliest possible ISA step. Which C abstraction we choose does not matter: the value of expressions, except those of pointer type, does not depend on the (nondeterministic) choice.

Finally, there is a situation where a user process whose program lies outside of the kernel returns control to the kernel, and the kernel performs the aforementioned 'process save'. The return to the code of the kernel is very similar to a 'goto', a construct absent in the original definition of the structured programming language $C0$. As a warm-up exercise for the later treatment of this situation we augment $C0$ with labels and gotos. Compilation is trivial. Providing small-step semantics for it is not, but is again easily solved with Lemma 90.

The pointer chasing on the heap for the reconstruction of $C0$ configurations is also known from garbage collection. For interested readers we have therefore included a short treatment of garbage collection, although we will not make use of it later.

Chapter 14 defines and justifies extensions of the programming models encountered so far for system programmers. On the hardware side MIPS ISA is extended by mechanisms for interrupt processing and address translation. Extending the processor construction and proving its correctness is fairly straightforward. There is, however, a small new kind of argument: the external interrupt signals eev_{isa}^i seen by ISA instructions i have to be constructed from the external hardware interrupt signals eev_h^t observed during the hardware cycles t.

Next we define a hardware model of a hard disk, integrate it into the processor construction, and try to abstract an ISA model of the system obtained in this way. This becomes surprisingly interesting for a simple reason: the hardware of the sequential processor together with the disk forms already a parallel system, and it is basically impossible to predict how many ISA instructions a disk access needs until its completion after it has started. Nondeterminism appears now for good. Processor steps and steps when disk accesses complete are interleaved sequentially in a nondeterministic order. In order to justify such a model one has to construct this order from the hardware cycles in which processor instructions and disk accesses are completed; the interesting case is when both kinds of steps complete in the same cycle. Abstracting the external interrupt event signals eev_{isa}^i observed at the ISA level also becomes slightly more involved, because processor steps consume external interrupt signals, whereas disk steps ignore them. This difficulty is solved by keeping track of completed processor steps in the configurations of the resulting 'ISA + disk' model, much in the style of a head position on an input tape.

It turns out that not all orders of interleaving have to be considered. As long as the processor does not observe the disk (either by polling or reacting to an interrupt signal generated by the disk), the completion of a disk step has no influence on the computation of a processor. This allows us to restrict the possible occurrences of disk steps: it suffices to consider disk steps which immediately precede instructions when the processor observes the disk. A result of this nature is called an *order reduction theorem*. Order reduction is amazingly useful when it comes to integrating the disk into the C+A model: if we disable external interrupts while translated C code is running and if we do not allocate C variables on I/O-ports of the disk, then translated C code cannot observe the disk; thus we can ignore the disk and external interrupts other than reset while translated C code is running.

Disk liveness, i.e., the termination of every disk access that was started, holds in the ISA model with order reduction only if certain natural software conditions holds. After this fact is established in a somewhat nontrivial proof, one can construct disk drivers which obey these conditions, and show their correctness in the 'C+A + disk + interrupt' model in a straightforward way.

In Chap. 15 we finally specify and implement what we call a generic operating system kernel. In a nutshell a kernel has two tasks:

- to simulate on a single physical processor multiple virtual processors of almost the same ISA, and
- to provide services to these virtual processors via system calls.

This suggests to split the kernel implementation into two layers, one for each of the above tasks: i) a lower virtualization layer and ii) an upper layer which includes the scheduler and the handlers for system calls. In this text we treat the lower layer and leave the programmer the freedom to program the upper layer in any way she or he wishes. With all the computational models developed so far in place and justified, we can afford the luxury of following our usual pattern of specification, implementation, and correctness proof. We have to explain the differences between physical and virtual processors; roughly speaking virtual processors cannot use address translation and not all interrupts are visible. In particular the virtual machine sees page faults only if it tries to access pages that are not allocated by it. Page faults due to invalid page table entries are invisible and must be handled transparently by the implementation of the kernel.

Then we formally specify a computational model, which is realized by the virtualization layer. The user of the virtualization layer simply sees a number of virtual user machines communicating with a so-called *abstract kernel*. The latter is an arbitrary (!) $C0$ program which is allowed to call a very small number of special functions that we call *CVM primitives*. The entire specification of the CVM model including the semantics of all special functions takes only four pages. No reference to inline assembly is necessary in the specification. Only the semantics of ISA and $C0$, and the parallelism of CVM are used.

CVM is implemented by what we call a concrete kernel written in 'C+A + disk + interrupts'. Again we identify the simulation relation we have to maintain and then hope that this will guide the implementation in a straightforward way. Now system programming has a reputation for being somewhat more tricky than ordinary programming, and even with completely clean computational models we cannot make the complexity involved completely disappear. It turns out that we have to maintain three simulation relations:

- between the physical machine configuration d and the virtual user machines vm. This involves address translation and process control blocks.
- between the C abstraction k of the concrete kernel configuration (if present) and the physical machine d. This is handled basically by $consis(k,d)$, but if users are running we must maintain a substitute for this C abstraction.
- between the C abstraction k of the concrete kernel configuration and the configuration c of the abstract kernel. Because the code of the abstract kernel is a part of the concrete kernel this requires a very rudimentary theory of linking and gives rise to a simulation relation *k-consis*.

Moreover, each of the involved relations changes slightly when the simulated machine (user, concrete kernel, or abstract kernel) starts or stops running. Identifying these relations takes six pages (Sects.15.3.3 to 15.3.5) and is the heart of the matter. From then on things become easier. A top-level flow chart for the CVM primitive *runvm*, which starts user processes, handles page faults transparently and saves user processes when they return to the kernel, fits on a single page (Fig. 235). We have to annotate it with a nontrivial number of invariants, but identifying these invariants is not very hard: they are just the invariants needed to maintain the three simulation relations. Programming the building blocks of the flow chart such that the invariants are maintained is mostly done in a straightforward way. Showing that it is done correctly would of course be impossible, if we would lack a formal semantics for a single one of the instructions that we use. But this is not our problem any more.

Throughout the book we give in the introduction of each chapter comments on how we teach this material, what we cover in full detail and where we refer students to the lecture notes. At the end of each chapter we give a summary of what we think were the key ideas in the chapter. This is what we advise you to memorize i) for examinations and ii) — if you can do it — for life. This is not as immodest as it may appear: as a lecturer of a basic course one has a very large responsibility. If not even the lecturer believes that some things in the lecture are worth memorizing for life, then he or she should by all means change the content of the lecture. Explaining why implementations of systems meet their specification by mathematical proofs not only speeds up delivery in the classroom. It also opens the way to what is called *formal verification*. There, one uses so-called *computer-aided verification systems* (CAV systems) to i) formulate the correctness theorems in machine readable form and then ii) verify by computer that the proofs have no gaps. In this way one can check that even complex systems work according to their specification for *all* inputs, something that cannot be done with conventional testing. For very simple systems this was done as early as 1987 [BJMY89], for complex processors in 2003 [BJK+03], for non-optimizing compilers in 2008 [LP08], and for optimizing compilers in 2009 [Ler09], for a complete kernel written in C including all lines of assembly code in 2008 [IdRT08], and for the C portions of a small industrial kernel in 2010 [KAE+10]. Throughout the text we will refer the reader to related work in formal verification.

Developing a theory of system architecture like the one in this book for *multi-core systems* is an ongoing research effort. Again, throughout the text we will inform the reader about the state of this research at the time of writing.

2

Understanding Decimal Addition

2.1 Experience Versus Understanding

This book is about understanding system architecture in a quick and clean way: no black art, nothing you can only get a feeling for after years of programming experience. While experience is good, it does not replace understanding. For illustration, consider the basic method for decimal addition of one-digit numbers as taught in the first weeks of elementary school: everybody has experience with it, almost nobody understands it; *very* few of the people who do not understand it realize that they don't.

Recall that, in mathematics, there are definitions and statements. Statements that we can prove are called theorems. Some true statements, however, are so basic that there are no even more basic statements that we can derive them from; these are called axioms. A person who understands decimal addition will clearly be able to answer the following simple

Questions: Which of the following equations are definitions? Which ones are theorems? If an equation is a theorem, what is the proof?

$$2 + 1 = 3,$$
$$1 + 2 = 3,$$
$$9 + 1 = 10.$$

We just stated that these questions are simple; we did not say that answering them is easy. Should you care? Indeed you should, for at least three reasons: i) In case you don't even understand the school method for decimal addition, how can you hope to understand computer systems? ii) The reason why the school method works has *very* much to do with the reason why binary addition in the fixed-point adders of processors works. iii) You should learn to distinguish between having experience with something that has not gone wrong (yet) and having an explanation of why it always works. The authors of this text consider iii) the most important.

© Springer International Publishing Switzerland 2016
W.J. Paul et al., *System Architecture*, DOI 10.1007/978-3-319-43065-2_2

2.2 The Natural Numbers

In order to answer the above questions, we first consider counting. Since we count by repeatedly adding 1, this should be a step in the right direction.

The set of *natural numbers* $\mathbb{N}$ and the properties of counting are not based on ordinary mathematical definitions. In fact, they are so basic that we use five axioms due to Peano to simultaneously lay down all properties about the natural numbers and of counting we will ever use without proof. The axioms talk about

- a special number 0,
- the set $\mathbb{N}$ of all natural numbers (with zero),
- counting formalized by a *successor* function $S : \mathbb{N} \to \mathbb{N}$, and
- subsets $A \subset \mathbb{N}$ of the natural numbers.

Peano's axioms are

1. $0 \in \mathbb{N}$. Zero is a natural number. Note that this is a modern view of counting, because zero counts something that could be there but isn't.
2. $x \in \mathbb{N} \to S(x) \in \mathbb{N}$. You can always count to the next number.
3. $x \neq y \to S(x) \neq S(y)$. Different numbers have different successors.
4. $\nexists y.\ 0 = S(y)$. By counting you cannot arrive at 0. Note that this isn't true for computer arithmetic, where you **can arrive** at zero by an overflow of modulo arithmetic (see Sect. 3.2).
5. $A \subseteq \mathbb{N} \land 0 \in A \land (n \in A \to S(n) \in A) \to A = \mathbb{N}$. This is the famous induction scheme for proofs by induction. We give plenty of examples later.

In a proof by induction, one usually considers a set A consisting of all numbers n satisfying a certain property $P(n)$:

$$A = \{n \in \mathbb{N} \mid P(n)\}.$$

Then,

$$n \in A \leftrightarrow P(n),$$
$$A = \mathbb{N} \leftrightarrow \forall n \in \mathbb{N} : P(n),$$

and the induction axiom translates into a proof scheme you might or might not know from high school:

- Start of the induction: show P(0).
- Induction step: show that $P(n)$ implies $P(S(n))$.
- Conclude $\forall n \in \mathbb{N} : P(n)$. Property P holds for all natural numbers.

With the rules of counting laid down by the Peano axioms, we are able to make two 'ordinary' definitions. We define 1 to be the next number after 0 if you count. We also define that addition of 1 is counting.

Definition 1 (Adding 1 by Counting).

$$1 = S(0),$$
$$x + 1 = S(x).$$

With this, the induction step of proofs by induction can be reformulated to the more familiar form

- Induction step: show that $P(n)$ implies $P(n+1)$.

2.2.1 $2+1=3$ is a Definition

One can now give meaning to the other digits of decimal numbers with the following mathematical definition.

Definition 2 (The Digits 2 to 9).

$$2 = 1+1 = S(1),$$
$$3 = 2+1 = S(2),$$
$$4 = 3+1 = S(3),$$
$$\vdots$$
$$9 = 8+1 = S(8).$$

Thus, $2+1=3$ is the definition of 3.

2.2.2 $1+2=3$ is a Theorem

Expanding definitions, we would like to prove it by

$$
\begin{aligned}
1+2 &= 1+(1+1) \quad \text{(Definition of 2)} \\
&= (1+1)+1 \\
&= 2+1 \quad \text{(Definition of 2)} \\
&= 3 \quad \text{(Definition of 3).}
\end{aligned}
$$

With the axioms and definitions we have so far we cannot prove the second equation yet. This is due to the fact that we have not defined addition completely. We fix this by the following inductive definition.

Definition 3 (Addition).

$$x+0 = x,$$
$$x+S(y) = S(x+y).$$

In words: adding 0 does nothing. In order to add $y+1$, first add y, then add 1 (by counting to the next number). From this we can derive the usual laws of addition.

Lemma 1 (Associativity of Addition).

$$(x+y)+z = x+(y+z).$$

Proof. by induction on z. For $z = 0$ we have

$$(x+y)+0 = x+y = x+(y+0)$$

by definition of addition (adding 0).

For the induction step we assume the induction hypothesis $x + (y+z) = (x+y) + z$. By repeatedly applying the definition of addition we conclude

$$\begin{aligned}
(x+y)+S(z) &= S((x+y)+z) \quad \text{(by definition of addition)}\\
&= S(x+(y+z)) \quad \text{(by induction hypothesis)}\\
&= x+S(y+z) \quad \text{(by definition of addition)}\\
&= x+(y+S(z)).
\end{aligned}$$

Substituting $x = y = z = 1$ in Lemma 1, we get

$$(1+1)+1 = 1+(1+1)$$

which completes the missing step in the proof of $1+2 = 3$.

Showing the commutativity of addition is surprisingly tricky. We first have to show two special cases.

Lemma 2. $0+x = x$.

Proof. by induction on x. For $x = 0$ we have

$$0+0 = 0$$

by the definition of addition.

For the induction step we can assume the induction hypothesis $0+x = x$ and use this to show

$$\begin{aligned}
0+S(x) &= S(0+x) \quad \text{(definition of addition)}\\
&= S(x) \quad \text{(induction hypothesis).}
\end{aligned}$$

Lemma 3. $x+1 = 1+x$.

Proof. by induction on x. For $x = 0$ we have

$$0+1 = 1 = 1+0$$

by the previous lemma and the definition of addition.

For the induction step we can assume the induction hypothesis $x+1 = 1+x$ and show

$$\begin{aligned}
1+S(x) &= S(1+x) \quad \text{(definition of addition)}\\
&= S(x+1) \quad \text{(induction hypothesis)}\\
&= S(x)+1 \quad \text{(definition of counting by adding 1).}
\end{aligned}$$

Lemma 4 (Commutativity of Addition).

$$x+y = y+x.$$

Proof. by induction on y. For $y = 0$ we have

$$x+0 = x \quad \text{(definition of addition)}$$
$$= 0+x \quad \text{(Lemma 2)}.$$

For the induction step we can assume the induction hypothesis $x+y = y+x$ and show

$$x+S(y) = S(x+y) \quad \text{(definition of addition)}$$
$$= S(y+x) \quad \text{(induction hypothesis)}$$
$$= y+S(x) \quad \text{(definition of addition)}$$
$$= y+(x+1) \quad \text{(definition of counting by adding 1)}$$
$$= y+(1+x) \quad \text{(Lemma 3)}$$
$$= (y+1)+x \quad \text{(associativity of addition)}$$
$$= S(y)+x \quad \text{(definition of counting by adding 1)}.$$

By induction one shows the following in the same way.

Lemma 5.

$$(x+y)\cdot z = x\cdot z+y\cdot z \quad \textit{(distributivity)},$$
$$(x\cdot y)\cdot z = x\cdot(y\cdot z) \quad \textit{(associativity of multiplication)},$$
$$x\cdot y = y\cdot x \quad \textit{(commutativity of multiplication)}.$$

The proof is left as an exercise.

2.2.3 $9+1 = 10$ is a Brilliant Theorem

The proof of $9+1 = 10$ is much more involved. It uses a special biological constant defined as

$$Z = 9+1$$

which denotes the number of our fingers or, respectively, toes[1]. Moreover it uses a definition attributed to the brilliant mathematician al-Khwarizmi, which defines the decimal number system.

Definition 4 (Decimal Numbers). *An n-digit decimal number $a_{n-1}\ldots a_0$ with digits $a_i \in \{0,\ldots,9\}$ is interpreted as*

$$a_{n-1}\ldots a_0 = \sum_{i=0}^{n-1} a_i \cdot Z^i.$$

---

[1] We use the letter Z here because in German, our native language, the word for 'ten' is 'zehn' and the word for 'toes' is 'Zehen', thus it's almost the same.

Substituting $n = 2$, $a_1 = 1$, and $a_0 = 0$ we can derive the proof of $9 + 1 = 10$. We must however evaluate the formula obtained from the definition of decimal numbers; in doing so, we need properties of exponentiation and multiplication:

$$\begin{aligned} 10 &= 1 \cdot Z^1 + 0 \cdot Z^0 \quad \text{(definition of decimal number)} \\ &= 1 \cdot Z + 0 \cdot 1 \quad \text{(properties of exponentiation)} \\ &= Z + 0 \quad \text{(properties of multiplication)} \\ &= Z \quad \text{(definition of addition)} \\ &= 9 + 1 \quad \text{(definition of } Z \text{)}. \end{aligned}$$

Observe that addition and multiplication are taught in elementary school, whereas exponentiation is only treated much later in high school. In this order we cannot possibly fill the gaps in the proof above. Instead, one defines multiplication and exponentiation without relying on decimal numbers, as below.

Definition 5 (Multiplication).

$$x \cdot 0 = 0,$$
$$x \cdot S(y) = x \cdot y + x.$$

Definition 6 (Exponentiation).

$$x^0 = 1,$$
$$x^{S(y)} = x^y \cdot x.$$

By Lemma 2 we get

$$x \cdot 1 = x \cdot (S(0)) = x \cdot 0 + x = 0 + x = x.$$

i.e., multiplication of x by 1 from the right results in x. In order to progress in the proof of $9 + 1 = 10$, we show the following.

Lemma 6. $1 \cdot x = x.$

Proof. by induction on x. For $x = 0$ we have

$$1 \cdot 0 = 0$$

by the definition of multiplication.

For the induction step we can assume the induction hypothesis $1 \cdot x = x$ and show

$$\begin{aligned} 1 \cdot S(x) &= 1 \cdot x + 1 \quad \text{(definition of multiplication)} \\ &= x + 1 \quad \text{(induction hypothesis)}. \end{aligned}$$

Using Lemma 6, we get

$$x^1 = x^{S(0)} = x^0 \cdot x = 1 \cdot x = x. \tag{1}$$

We finish the section by showing a classical identity for exponentiation.

Lemma 7. $x^{y+z} = x^y \cdot x^z$.

Proof. by induction on z. For $z = 0$ we have (leaving the justification of the steps as an exercise)

$$x^{y+0} = x^y = x^y \cdot 1 = x^y \cdot x^0.$$

For the induction step we assume the induction hypothesis $x^{y+z} = x^y \cdot x^z$ and show

$$
\begin{aligned}
x^{y+S(z)} &= x^{S(y+z)} && \text{(definition of addition)} \\
&= x^{(y+z)} \cdot x && \text{(definition of exponentiation)} \\
&= (x^y \cdot x^z) \cdot x && \text{(induction hypothesis)} \\
&= x^y \cdot (x^z \cdot x) && \text{(associativity of multiplication, Lemma 5)} \\
&= x^y \cdot (x^{S(z)}) && \text{(definition of exponentiation).}
\end{aligned}
$$

2.3 Final Remarks

Using decimal addition as an example, we have tried to convince the reader that being used to something that has not gone wrong yet and understanding it are very different things. We have reviewed Peano's axioms and have warned the reader that computer arithmetic does not satisfy them. We have formally defined the value of decimal numbers; this will turn out to be helpful, when we study binary arithmetic and construct adders later. We have practiced proofs by induction and we have shown how to derive laws of computation without referring to decimal representations.

We recommend to remember as key technical points:

- Peano's axioms.
- the inductive definition of addition as iterated counting

$$
\begin{aligned}
x + 0 &= x, \\
x + (y+1) &= (x+y) + 1.
\end{aligned}
$$

- for parties: $2+1 = 3, 1+2 = 3, 9+1 = 10$. Which are theorems?

Everything else is easy. Rules of computation are shown by induction; what else can you do with an inductive definition of addition? And of course multiplication is defined by iterated addition and exponentiation is defined by iterated multiplication.

A proof of the usual rules of computation governing addition, subtraction, multiplication, and division of the rational numbers and/or the real numbers can be found in [Lan30]. The recursive definitions of addition, multiplication, and exponentiation that we have presented also play a central role in the theory of computability [Rog67, Pau78].

2.4 Exercises

1. For each of the following statements, point out which ones are definitions and which ones are theorems for $x, y \in \mathbb{N}$. Prove the theorems using only definitions and statements proven above.
    a) $x = x + 0$,
    b) $x = 0 + x$,
    c) $x + (y + 1) = (x + y) + 1$,
    d) $(x + 1) + y = (y + x) + 1$,
    e) $x \cdot 0 = 0$,
    f) $0 \cdot x = 0$,
    g) $5 + 1 = 6$,
    h) $7 + 4 = 11$.
2. Prove Lemma 5.
3. Prove the following properties of exponentiation for $a, b, c \in \mathbb{N}$.

$$(a \cdot b)^c = a^c \cdot b^c,$$
$$(a^b)^c = a^{b \cdot c}.$$

4. We define a modified subtraction of natural numbers

$$-' : \mathbb{N} \to \mathbb{N}$$

by

$$x -' y = \begin{cases} x - y, & x \geq y, \\ 0, & \text{otherwise.} \end{cases}$$

Give an inductive definition of this function, starting with the successor function $S(x)$.

Hint: clearly you want to define a modified decrement function $D(x) = x -' 1$ first. Do this by an inductive definition.

3

Basic Mathematical Concepts

After a few short comments on implicit quantification, in Sect. 3.1 we treat very basic definitions like intervals $[i : j]$ of natural numbers. We consider different ways to number sequence elements; usually sequence elements are numbered from left to right starting with index 1, but in number representations it is much nicer to number them from right to left starting with index 0. We elaborate very briefly on the formal consequences of this. We also introduce vector operations on bit strings.

Section 3.2 on modulo arithmetic was included for several reasons. i) The notation mod k is overloaded: it is used to denote both the congruence *relation* modulo a number k or the *operation* of taking remainder of integer division by k. We prefer our readers to clearly understand this. ii) Fixed-point arithmetic is modulo arithmetic, so we will clearly have to make use of it. The most important reason however is iii) to give with Lemma 12 a tiny twist to a trivial result about the solution of congruences mod k by applying it not only with the standard set $[0 : k-1]$ of representatives. In Sect. 7.2 this will yield a very quick proof of the remarkable fact that addition/subtraction of binary numbers and of two's complement numbers is done by exactly the same hardware[1].

The very short Sect. 3.3 on geometric and arithmetic sums is simply there to remind the reader of the proof of the classical formulas for the computation of geometric and arithmetic sums, which are much easier to memorize than the formula itself. The formula for geometric sums is later used to bound the range of numbers represented by bit strings, and the formula for arithmetic sums is used in the correctness proof of an example program to show the absence of overflows.

Section 3.4 contains some very basic graph theory. Lemma 14 contains the result that path length in directed acyclic graphs is bounded. This little result is remarkably important for several reasons: i) its short proof is a great illustration of the pigeonhole principle; ii) it justifies the definition of depth of nodes in directed acyclic graphs. This is later used to show — by an induction on the depth of nodes — that the behavior of switching circuits is well defined; iii) it is used to show the equivalence of a recursive definition of rooted trees and the classical graph theoretic definition of

---

[1] Except for the computation of overflow and negative signals.

© Springer International Publishing Switzerland 2016
W.J. Paul et al., *System Architecture*, DOI 10.1007/978-3-319-43065-2_3

rooted trees. This result is later used to relate our formalization of derivation trees for context-free grammars with classical graph theory.

Except for Sect. 3.2 the chapter contains fairly simple reference material. We recommend the reader should glance at it, notice what is there, but worry later about detailed understanding. In lectures we reorder the topics; initially we just sketch the material of this chapter, and elaborate details when it is used.

3.1 Basics

3.1.1 Implicit Quantification

One uses the quantifiers $\exists$ and $\forall$ as abbreviations of 'there exists' and 'for all'. Quantifiers range over sets. So one writes

$$\forall x, y \in \mathbb{N} : x + y = y + x.$$

If the set over which one quantifies is clear from the context, one usually does not specify this set in the quantifier again. Thus one could also write

$$\forall x, y : x + y = y + x.$$

Rules of computation are often formulated in an even shorter way, for instance

$$x + y = y + x.$$

This is an example of a general principle. Formulas without quantifiers are often treated like statements. The convention is that one assumes all variables in the formula to be universally quantified. So, in a context where we speak about real numbers

$$x \cdot x \geq 0$$

is interpreted as

$$\forall x \in \mathbb{R} : x \cdot x \geq 0,$$

which is true, whereas in a context where we speak about the complex numbers, it is interpreted as

$$\forall x \in C : x \cdot x \geq 0,$$

which is false because $i^2 = -1$.

Implicit quantification is very frequently used. It keeps notation uncluttered and saves time at the blackboard. We will return to implicit quantification in Sect. 4.3.

3.1.2 Numbers and Sets

We denote by

$$\mathbb{N} = \{0, 1, 2, \ldots\}$$

the set of natural numbers including zero, by

$$\mathbb{Z} = \{\ldots, -2, -1, 0, 1, 2, \ldots\}$$

the set of integers. Unless explicitly stated otherwise we identify natural numbers with their decimal representation, as we always have done since elementary school.

We denote by

$$\mathbb{B} = \{0, 1\}$$

the set of Boolean values. We also use the term *bit* to refer to a Boolean value.

For integers i, j with $i < j$ we define the interval of integers from i to j by

$$[i : j] = \{i, i+1, \ldots, j\}.$$

Strictly speaking, definitions using three dots are never precise; they resemble intelligence tests, where the author hopes that all readers who are forced to take the test arrive at the same solution. Usually, one can easily find a corresponding and completely precise recursive definition (without three dots) in such a situation. Thus, we define $[i : j]$ in a rigorous way as

$$[i : i] = \{i\},$$
$$[i : j+1] = [i : j] \cup \{j+1\}.$$

The Hilbert $\in$-Operator $\in A$ picks an element from a set A. Applied to a singleton set, it returns the unique element of the set:

$$\in \{x\} = x.$$

For finite sets A, we denote by $\#A$ the *cardinality*, i.e., the number of elements in A.

Given a function f operating on a set A and a set $A_1 \subseteq A$, we denote by $f(A_1)$ the image of set A_1 under function f, i.e.,

$$f(A_1) = \{f(a) \mid a \in A_1\}.$$

For a few complexity arguments in this book we use big O notation for comparing the asymptotic growth of functions $f, g : \mathbb{N} \to \mathbb{N}$.

$$f = O(g) \iff \exists N_0 \in \mathbb{N}, c > 0 \forall n \geq N_0 : f(n) \leq c \cdot g(n).$$

In computer science logarithms are to the base two unless explicitly stated otherwise. This text is no exception.

3.1.3 Sequences, Their Indexing and Overloading

We start this section with a remark on overloading. A mathematical symbol is *overloaded* if it has different meanings in different contexts. A standard example is the addition symbol + in arithmetic expressions, which is for instance interpreted as integer addition in $1 + 2$ and as addition of fractions in $\frac{2}{3} + \frac{3}{4}$. Overloading is helpful to keep notation uncluttered, but it should of course only be used where the meaning can be inferred from the context. Some situations — like $1 + \frac{2}{3}$ — suggest several meanings of a symbol. In such cases one has to resolve the conflict by recoding one or more operands; in our example

$$1 + \frac{2}{3} = \frac{1}{1} + \frac{2}{3}.$$

Finite sequences (resp. words or strings) a of n elements a_i or $a[i]$ from a set A are a basic mathematical concept that is intuitively completely clear. If we try to formalize it, we find that there are several completely natural ways to do this. Let $n \geq 1$.

1. If we start indices with 1 and index from left to right we write

$$a = (a_1, \ldots, a_n) = a[1 : n]$$

   which is formalized without three dots as a mapping

$$a : [1 : n] \to A.$$

   This is convenient for numbering symbols in a program text or statements in a statement sequence. With this formalization the set A^n of sequences of length n with elements from A is defined as

$$A^n = \{a \mid a : [1 : n] \to A\}.$$

2. If we start indices at 0 and index from left to right we write

$$a = (a_0, \ldots, a_{n-1}) = a[0 : n-1]$$

   which is coded as

$$a : [0 : n-1] \to A.$$

   This is convenient, e.g., for indexing nodes of paths in graphs. With this formalization the set A^n of sequences of length n with elements from A is defined as

$$A^n = \{a \mid a : [0 : n-1] \to A\}.$$

   As usual we refer by $hd(a) \equiv a_0$ to the head, i.e., the first element, of such sequences and by $tail(a) \equiv (a_1, \ldots, a_{n-1})$ to the remaining part.
3. When dealing with number representations it turns out to be most convenient by far to start counting from 0 from right to left. We write

$$a = (a_{n-1}, \ldots, a_0) = a[n-1 : 0]$$

which is also formalized as

$$a : [0 : n-1] \to A$$

and get again the formalization

$$A^n = \{a \mid a : [0 : n-1] \to A\}.$$

Thus the direction of ordering does not show in the formalization *yet*. The reason is that the interval $[0 : n-1]$ is a set, and elements of sets are unordered.

This is changed when we define concatenation $a \circ b = c$ of sequences a and b. For the three cases we get

1.

$$a[1 : n] \circ b[1 : m] = c[1 : n+m]$$

with

$$c[i] = \begin{cases} a[i], & i \leq n, \\ b[i-n], & i \geq n+1. \end{cases}$$

2.

$$a[0 : n-1] \circ b[0 : m-1] = c[0 : n+m-1]$$

with

$$c[i] = \begin{cases} a[i], & i \leq n-1, \\ b[i-n], & i \geq n. \end{cases}$$

3.

$$a[n-1 : 0] \circ b[m-1 : 0] = c[n+m-1 : 0]$$

with

$$c[i] = \begin{cases} b[i], & i \leq m-1, \\ a[i-m], & i \geq m. \end{cases}$$

Concatenation of sequences a with single symbols $b \in A$ is handled by treating elements b as sequences with one element $b = b[1]$ if counting starts with 1 and $b = b[0]$ if counting starts with 0. The *empty sequence* ε is the unique sequence of length 0; thus one defines

$$A^0 = \{\varepsilon\}$$

and it satisfies

$$a \circ \varepsilon = \varepsilon \circ a = a$$

in all formalizations. For any variant of indexing we define a to be a prefix of b if b has the form $a \circ c$:

$$prefix(a,b) \equiv \exists c : b = a \circ c.$$

The set A^+ of nonempty finite sequences with elements from A is defined as

$$A^+ = \bigcup_{n \in \mathbb{N} \setminus \{0\}} A^n$$

and the set A^* of all finite sequences with elements from A as

$$A^* = \bigcup_{n \in \mathbb{N}} A^n = A^+ \cup \{\varepsilon\}.$$

Both definitions use A^n. Because A^n is overloaded, A^+ and A^* are overloaded too. Given two sets A and B, then $A \circ B$ denotes the set of all possible concatenations of one element from B to one element from A:

$$A \circ B = \{a \circ b \mid a \in A \wedge b \in B\}.$$

When forming subsequences, we get for the three cases

1. for $a[1 : n]$ and $i \leq j$

$$a[i : j] = c[1 : j - i + 1] \quad \text{with} \quad c[k] = a[i + k - 1].$$

2. for $a[0 : n - 1]$ and $i \leq j$

$$a[i : j] = c[0 : j - i] \quad \text{with} \quad c[k] = a[i + k].$$

3. for $a[0 : n - 1]$ and $j \geq i$

$$a[j : i] = c[j - i : 0] \quad \text{with} \quad c[k] = a[i + k].$$

In very few places we will have different indexing and direction in the same place, for instance if we identify in a program text $w[1 : n]$ a substring $w[i : j]$ that we wish to interpret as a number representation $c[j - i : 0]$ we have to do the trivial conversion

$$w[j : i] = c[j - i : 0] \quad \text{with} \quad c[k] = w[i + k].$$

3.1.4 Bytes, Logical Connectives, and Vector Notation

A *byte* is a string $x \in \mathbb{B}^8$. Let $n = 8 \cdot k$ be a multiple of 8, let $a \in \mathbb{B}^n$ be a string consisting of k bytes. For $i \in [k - 1 : 0]$ we define byte i of string a as

$$byte(i, a) = a[8 \cdot (i + 1) - 1 : 8 \cdot i].$$

For bits $x \in \mathbb{B}$ and natural numbers $n \in \mathbb{N}$ we denote the string obtained by repeating x exactly n times by x^n. In the form of an intelligence test:

$$x^n = \underbrace{x \ldots x}_{n \text{ times}}$$

and in rigorous form

$$x^1 = x,$$
$$x^{n+1} = x \circ x^n,$$

$\wedge$	and
$\vee$	or
$\neg$	not
$\oplus$	exclusive or, + modulo 2
$\rightarrow$	implies
$\leftrightarrow$	if and only if
$\forall$	for all
$\exists$	exists

Table 1. Logical connectives and quantifiers

where $\circ$ denotes the concatenation of bit strings.

Examples:

$$1^2 = 11 \quad \text{and} \quad 0^4 = 0000.$$

For the concatenation of bit strings x_1 and x_2 we often omit $\circ$ and write

$$x_1 x_2 = x_1 \circ x_2.$$

In statements and predicates, we use the logical connectives and quantifiers from Table 1. For $\neg x$ we also write $\bar{x}$.

For $\bullet \in \{\wedge, \vee, \oplus\}$, $a, b \in \mathbb{B}^n$ and a bit $c \in \mathbb{B}$, we borrow notation from vector calculus to define the corresponding bit-operations on bit-vectors:

$$\bar{a} = (\overline{a_{n-1}}, \ldots, \overline{a_0}),$$
$$a[n-1:0] \bullet b[n-1:0] = (a_{n-1} \bullet b_{n-1}, \ldots, a_0 \bullet b_0),$$
$$c \bullet b[n-1:0] = (c \bullet b_{n-1}, \ldots, c \bullet b_0).$$

3.2 Modulo Computation

There are infinitely many integers and every computer can only store finitely many numbers. Thus, computer arithmetic cannot possibly work like ordinary arithmetic. Fixed-point arithmetic[2] is usually performed modulo 2^n for some n. We review the basics of modulo computation.

Definition 7 (Congruence Modulo). *For integers $a, b \in \mathbb{Z}$ and natural numbers $k \in \mathbb{N}$ one defines a and b to be* congruent mod k *or* equivalent mod k *iff they differ by an integer multiple of k:*

$$a \equiv b \bmod k \leftrightarrow \exists z \in \mathbb{Z} : a - b = z \cdot k.$$

---

[2] The only arithmetic considered in this book. For the construction of floating-point units see [MP00].

Definition 8 (Equivalence Relation). *Let R be a relation between elements of a set A. We say that R is* reflexive *if we have aRa for all $a \in A$. We say that R is* symmetric *if aRb implies bRa. We say that R is* transitive *if aRb and bRc imply aRc. If all three properties hold, R is called an* equivalence relation *on A.*

An easy exercise shows the following.

Lemma 8. *Congruence mod k is an equivalence relation.*

Proof. We show that the properties of an equivalence relation are satisfied:

- Reflexivity: For all $a \in \mathbb{Z}$ we have

$$a - a = 0 \cdot k.$$

Thus

$$a \equiv a \bmod k$$

and congruence mod k is reflexive.

- Symmetry: Let

$$a \equiv b \bmod k \quad \text{with} \quad a - b = z \cdot k.$$

Then

$$b - a = -z \cdot k,$$

thus

$$b \equiv a \bmod k.$$

- Transitivity: Let

$$a \equiv b \bmod k \quad \text{with} \quad a - b = z_1 \cdot k \quad \text{and} \quad b \equiv c \bmod k \quad \text{with} \quad b - c = z_2 \cdot k.$$

Then

$$a - c = (z_1 + z_2) \cdot k,$$

thus

$$a \equiv c \bmod k.$$

Lemma 9. *Let $a, b \in \mathbb{Z}$ and $k \in \mathbb{N}$ with $a \equiv a' \bmod k$ and $b \equiv b' \bmod k$. Then,*

$$a + b \equiv a' + b' \bmod k,$$
$$a - b \equiv a' - b' \bmod k,$$
$$a \cdot b \equiv a' \cdot b' \bmod k.$$

Proof. Let $a - a' = u \cdot k$ and $b - b' = v \cdot k$, then we have

$$
\begin{aligned}
a + b - (a' + b') &= a - a' + b - b' \\
&= (u + v) \cdot k, \\
a - b - (a' - b') &= a - a' - (b - b') \\
&= (u - v) \cdot k, \\
a \cdot b &= (a' + u \cdot k) \cdot (b' + v \cdot k) \\
&= a' \cdot b' + k \cdot (a' \cdot v + b' \cdot u + k \cdot u \cdot v),
\end{aligned}
$$

which imply the desired congruences.

Two numbers r and s in an interval of the form $[i : i+k-1]$ that are both equivalent to $a \bmod k$ are identical:

Lemma 10. *Let* $i \in \mathbb{Z}$, $k \in \mathbb{N}$, *and let* $r,s \in [i : i+k-1]$, *then*

$$a \equiv r \bmod k \wedge a \equiv s \bmod k \rightarrow r = s.$$

Proof. By symmetry we have

$$s \equiv a \bmod k$$

and by transitivity we get

$$s \equiv r \bmod k.$$

Thus

$$r - s = z \cdot k$$

for an integer z. We conclude $z = 0$ because $|r - s| < k$.

Definition 9 (System of Representatives). *Let R be an equivalence relation on A. A subset $B \subset A$ is called a* system of representatives *if and only if for every $a \in A$ there is exactly one $r \in B$ with aRr. The unique $r \in B$ satisfying aRr is called the* representative *of a in B.*

Lemma 11. *For $i \in \mathbb{Z}$ and $k \in \mathbb{N}$, the interval of integers $[i : i+k-1]$ is a system of representatives for equivalence mod k.*

Proof. Let $a \in \mathbb{Z}$. We define the representative $r(a)$ by

$$f(a) = max\{j \in \mathbb{Z} \mid a - k \cdot j \geq i\},$$
$$r(a) = a - k \cdot f(a).$$

Then

$$r(a) \equiv a \bmod k$$

and $r(a) \in [i : i+k-1]$. Uniqueness follows from Lemma 10.

Note that in case $i = 0$

$$f(a) = \lfloor a/k \rfloor$$

is the result of the integer division of a by k, and

$$r(a) = a - \lfloor a/k \rfloor \cdot k$$

is the remainder of this division.

We have to point out that in mathematics the three letter word 'mod' is not only used for the *relation* defined above. It is also used as a *binary operator* in which case $a \bmod k$ denotes the representative of a in $[0 : k-1]$.

Definition 10 (Modulo Operator). *For $a,b \in \mathbb{Z}$ and $k \in \mathbb{N}$,*

$$(a \bmod k) = \in\{b \mid a \equiv b \bmod k \wedge b \in [0 : k-1]\}.$$

Thus, $(a \bmod k)$ is the remainder of the integer division of a by k for $a \geq 0$. In order to stress when mod is used as a binary operator, we *always* write $(a \bmod k)$ in brackets. For later use in the theory of two's complement numbers we define another modulo operator.

Definition 11 (Two's Complement Modulo Operator). *For $a, b \in \mathbb{Z}$ and an* even *number $k = 2 \cdot k'$ with $k' \in \mathbb{N}$,*

$$(a \operatorname{tmod} k) = \in \{b \mid a \equiv b \bmod k \wedge b \in [-k/2 : k/2 - 1]\}.$$

From Lemma 10 we infer a simple but useful lemma about the solution of equivalences mod k:

Lemma 12. *Let k be even and $x \equiv y \bmod k$, then*

1. $$x \in [0 : k - 1] \to x = (y \bmod k),$$
2. $$x \in [-k/2 : k/2 - 1] \to x = (y \operatorname{tmod} k).$$

3.3 Sums

3.3.1 Geometric Sums

For $q \neq 1$ we consider

$$S = \sum_{i=0}^{n-1} q^i$$

the geometric sum over q. Then,

$$q \cdot S = \sum_{i=1}^{n} q^i,$$
$$q \cdot S - S = q^n - 1,$$
$$S = \frac{q^n - 1}{q - 1}.$$

For $q = 2$ we get the following.

Lemma 13. *For $n \in \mathbb{N}$,*

$$\sum_{i=0}^{n-1} 2^i = 2^n - 1$$

which we will use in the next chapter.

Fig. 1. Drawing an edge (u,v) from u to v

3.3.2 Arithmetic Sums

For $n \in \mathbb{N}$ let

$$S_n = \sum_{i=0}^{n} i$$

be the sum of the natural numbers from 0 to n. We recall that Gauss wrote $2 \cdot S_n$ as

$$
\begin{aligned}
2 \cdot S_n &= 0 + \quad 1 \quad + \ldots + (n-1) + n + \\
&\quad\; n + (n-1) + \ldots + \quad 1 \quad + 0 \\
&= n \cdot (n+1)
\end{aligned}
$$

which gives

$$S_n = n \cdot (n+1)/2.$$

If you are suspicious about proofs involving three dots (which you should be), use the equation as an induction hypothesis and prove it by induction. While doing this do not define

$$\sum_{i=0}^{n} f_i = f_0 + \ldots + f_n$$

because that would involve three dots too. Instead define

$$\sum_{i=0}^{0} f_i = f_0,$$

$$\sum_{i=0}^{n} f_i = \left(\sum_{i=0}^{n-1} f_i \right) + f_n.$$

3.4 Graphs

3.4.1 Directed Graphs

In graph theory a *directed graph G* is specified by

- a set $G.V$ of *nodes*. Here we consider only finite graphs, thus $G.V$ is finite.
- a set $G.E \subseteq G.V \times G.V$ of *edges*. Edges $(u,v) \in G.E$ are depicted as arrows from node u to node v as shown in Fig. 1. For $(u,v) \in G.E$ one says that v is a *successor* of u and that u is a *predecessor* of v.

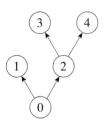

Fig. 2. Graph G **Fig. 3.** Graph G'

If it is clear which graph G is meant, one abbreviates

$$V = G.V,$$
$$E = G.E.$$

The graph G in Fig. 2 is formally described by

$$G.V = \{0,1,2,3\},$$
$$G.E = \{(0,1),(1,2),(2,3),(3,0)\}.$$

The graph G' in Fig. 3 is formally described by

$$G'.V = \{0,1,2,3,4\},$$
$$G'.E = \{(0,1),(0,2),(2,3),(2,4)\}.$$

Let $G = (V,E)$ be a directed graph. For nodes $u,v \in V$ a *path* from u to v in G is a sequence $p[0:k]$ of nodes $p[i] \in V$, that begins with u, ends with v, and in which subsequent elements are connected by edges:

$$p_0 = u,$$
$$p_k = v,$$
$$i > 0 \rightarrow (p_{i-1}, p_i) \in E.$$

The number k is called the *length* of the path. We denote it by $le(p)$. A path from u to u with length at least 1 is called a *cycle*. In Fig. 2 the sequence of nodes $(0,1,2,3,0,1)$ is a path of length 5 from node 0 to node 1, and the sequence $(1,2,3,0,1)$ is a cycle. In Fig. 3 the sequence of nodes $(0,2,4)$ is a path of length 2 from node 0 to node 4.

For graphs G and nodes $v \in G.V$ the indegree $indeg(v,G)$ of node v in graph G is defined as the number of edges ending in v and the outdegree $outdeg(v,G)$ of node v in graph G is defined as the number of edges starting in v:

$$indeg(v,G) = \#\{u \mid (u,v) \in G.E\},$$
$$outdeg(v,G) = \#\{x \mid (v,x) \in G.E\}.$$

In the graph of Fig. 2 we have

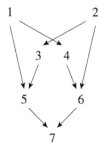

Fig. 4. Example of a directed acyclic graph

$$indeg(v, G) = outdeg(v, G) = 1$$

for all nodes v. In the graph G' of Fig. 3 we have

$$indeg(0, G') = 0,$$
$$v \neq 0 \to indeg(v, G') = 1,$$
$$v \in \{1, 3, 4\} \to outdeg(v, G') = 0,$$
$$v \in \{0, 2\} \to outdeg(v, G') = 2.$$

A source of a graph G is a node v with $indeg(v, G) = 0$ and a sink of a graph is a node v with $outdeg(v, G) = 0$. The graph in Fig. 2 has neither sources nor sinks. The graph in Fig. 3 has a single source $v = 0$ and sinks 1, 3, and 4. If it is clear which graph is meant one abbreviates

$$indeg(v) = indeg(v, G),$$
$$outdeg(v) = outdeg(v, G).$$

3.4.2 Directed Acyclic Graphs and the Depth of Nodes

A *directed acyclic graph* or DAG is simply a directed graph without cycles. For the remainder of this subsection we will use the graph from Fig. 4 as a running example.

Lemma 14. *In directed acyclic graphs $G = (V, E)$ the length $le(p)$ of paths p is bounded by the number of nodes minus 1:*

$$le(p) \leq \#V - 1.$$

Proof. Assume you have n pigeons sitting in $n - 1$ holes. Then there must be two pigeons which are sitting in the same hole, because otherwise there would be at most $n - 1$ pigeons. We use this so-called *pigeonhole argument* to prove the lemma by contradiction.

Let $n = \#V$ be the number of nodes and assume that $p[0 : n]$ is a path in G. We treat the indices $i \in [0 : n]$ as pigeons, the vertices $v \in V$ as pigeonholes, and pigeon i sits in hole $p(i)$. We conclude that there must be two different indices i and j such that $i < j$ and $p(i) = p(j)$. But then $p[i : j]$ is a cycle and G is not acyclic.

In the sequel this simple lemma will turn out to be amazingly useful.

Lemma 15. *Let $G = (V, E)$ be a directed acyclic graph. Then G has at least one source and at least one sink.*

Proof. Let $n = \#V$. Pick any node $v \in V$ in the graph. Starting from v follow edges forward repeatedly as long as they are present. By Lemma 14 one hits a sink after at most $n - 1$ edges. Following edges backward one finds a source.

In the graph in Fig. 4 we always end in sink 7. Following edges backward we arrive in source 1 if we start in nodes 4 or 5, in source 2 if we start in nodes 3 or 6, and in source 1 or 2 if we start in nodes 5, 6, or 7, depending on which of the edges leading into nodes 5 and 6 we follow.

The previous two lemmas allow us to define the *depth* $d(v, G)$ of a node v in a DAG G as the length of a longest path in G from a source to v. If it is clear what graph is meant one drops the argument G and abbreviates

$$d(v) = d(v, G).$$

A simple argument shows that the depth of sources is 0 and that the depth of node v is one greater than the depth of one of its predecessors

$$d(v, G) = \begin{cases} 0, & indeg(v, G) = 0, \\ \max\{d(u, G) \mid (u, v) \in G.E\} + 1, & \text{otherwise.} \end{cases}$$

Thus in Fig. 4 we have

$$d(v) = \begin{cases} 1, & v \in \{3, 4\}, \\ 2, & v \in \{5, 6\}, \\ 3, & v \in \{7\}. \end{cases}$$

The depth of a DAG is the maximum of the depth of its nodes

$$d(G) = \max\{d(v, G) \mid v \in G.V\}.$$

Thus the graph in Fig. 4 has depth 3.

3.4.3 Rooted Trees

We define rooted trees in the following way:

1. a graph with a single node and no edges is a rooted tree.
2. if G is a rooted tree, u is a sink of G and $v_0, \dots, v_{k-1} \notin G.V$ are new nodes, then the graph G' defined by

$$G'.V = G.V \cup \{v_0, \dots, v_{k-1}\},$$
$$G'.E = G.E \cup \{(u, v_0), \dots, (u, v_{k-1})\}$$

is a rooted trees (see Fig. 5).

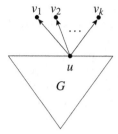

Fig. 5. Generating a rooted tree G' by adding new nodes $v_1, \ldots, v_k$ as successors of a sink u of rooted tree G

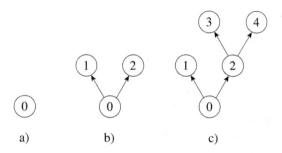

Fig. 6. Steps taken to generate the graph from Fig. 3

3. all rooted trees can be generated in finitely many steps by the above two rules.

By this definition the graph in Fig. 3 is a rooted tree, because it can be generated in three steps as shown in Fig. 6 a) to c). Rooted trees have a single source which is called the *root* of the tree. The sinks of trees are called *leaves* of the tree. If (u, v) is an edge of a tree one calls u the *father* of v and one calls v a *son* of u.

3.5 Final Remarks

We recommend to remember as key technical points:

- about modulo computation
  - the definition of equivalence modulo k
  - that we consider *two* systems R of representatives, namely

  $$R = [0 : k-1] \quad \text{and} \quad R = [-n/2 : n/2 - 1]$$

  and that the representatives $rep(n)$ for n in these systems are named

  $$(n \bmod k) \quad \text{and} \quad (n \bmod k)$$

  - the trivial fact 'if x and y are equivalent mod k and x lies in a system R of representatives, then x is the representative of y in R' is applicable to any system of representatives.

- about sums we strictly advise against remembering the formulas for arithmetic and geometric sums. Instead, remember the proofs.
- about graph theory remember by all means the proof that in a directed acyclic graph with n nodes no path can have a length greater than $n - 1$, because otherwise — by the pigeonhole principle — a node would have to appear repeatedly on the path and this would constitute a cycle.
- for parties: Sequences $a[n : 1] \in A^*$ and $a[1 : n] \in A^*$ are both formalized as $f : [1 : n] \to A$. Where is the formal difference? You remember: in the definition of concatenation.

3.6 Exercises

1. Consider the set $\{1, 2, 3, \ldots, 8, 9, 10\}$ and the following equivalence classes: $\{1, 4, 6\}, \{2, 8\}, \{3\}, \{5, 7\}, \{9\}, \{10\}$.
   a) Give a system of representatives.
   b) Give the equivalence classes of 1, 2, 3, and 4.
2. Informally explain the difference between mod and tmod, then give an example where they differ.
3. Informally explain the difference between

$$a = (b \bmod k) \quad \text{and} \quad a \equiv b \bmod k.$$

Give an example (i.e., an a, b, and k) where they differ, e.g.,

$$a \neq (b \bmod k) \quad \text{but} \quad a \equiv b \bmod k.$$

4. Show

$$\#(A^n) = (\#A)^n.$$

5. Refer to the inductive definition of sums (without three dots) and prove by induction on m

$$\sum_{i=m}^{n-1} s_i + \sum_{i=0}^{m-1} s_i = \sum_{i=0}^{n-1} s_i.$$

Hint: use associativity of addition.

6. Prove by induction on n

   a)
   $$\sum_{i=0}^{n} i = n \cdot (n+1)/2,$$

   b)
   $$\sum_{i=0}^{n-1} q^i = (q^n - 1)/(q - 1),$$

   c)
   $$\sum_{i=0}^{n} 6 \cdot i^2 = n \cdot (n+1) \cdot (2n+1),$$

   d)
   $$\sum_{i=0}^{n} i^3 = \left(\sum_{i=0}^{n} i \right)^2,$$

e)
$$x \cdot \sum_{i=0}^{n} f_i = \sum_{i=0}^{n} (x \cdot f_i).$$

The next exercises relate the recursive definition of rooted trees of Sect. 3.4.3 to a common definition of rooted trees in graph theory.

7. Show that every rooted tree has exactly one source (called the root) and that for every node u of a rooted tree there is exactly one path from the root to u.

8. Now assume that G is a finite directed graph with a single source r (called the root) and such that for every node $u \in G.V$ there is exactly one path from r to u in G. Let $n = \#G.V$ be the number of nodes in G.

   a) Show that every path in G has length at most n.

      Hint: construct a path $p[0:y]$ in G' in the following way:

       i) Initially $y = 0$ and $p(y) = r$.

       ii) If all successors v_i of $p(y)$ are sinks, set $u = p(y)$ and stop. Otherwise, extend the path by choosing $p(y+1)$ among the successors of $p(y)$ that are not sinks. Show that we stop with some $y \leq n$.

   b) Show that there is a node u in G such that all successors of u are sinks.

   c) Show that G is a rooted tree.

4

Number Formats and Boolean Algebra

Section 4.1 introduces the binary number format, presents the school method for binary addition, and proves that it works. Although this will look completely familiar and the correctness proof of the addition algorithm is only a few lines long, the reader should treat this result with deep respect: it is probably the first time that he or she sees a proof of the fact that the addition algorithm he or she learned at school always works. The ancient Romans, who were fabulous engineers in spite of their clumsy number system, would have *loved* to see this proof.

Integers are represented in computers as two's complement numbers. In Sect. 4.2 we introduce this number format and derive several basic identities for such numbers. From this we derive a subtraction algorithm for binary numbers which is quite different from the school method, and show that it works. Sections 3.2, 3.3 and 4.2 are the basis of our construction of an arithmetic unit later.

Finally, in Sect. 4.3 on Boolean algebra, we provide a very short proof of the fundamental result that Boolean functions can be computed using Boolean expressions in disjunctive normal form. This result can serve to construct all small circuits — e.g., in control logic — where we only specify their functionality and do not bother to specify a concrete realization. The proof is intuitive and looks simple, but it will give us the occasion to explain formally the difference between what are often called 'two kinds of equations': i) identities[1] $e(x) = e'(x)$ which hold for all x, and ii) equations $e(x) = e'(x)$ that we want to solve by determining the set of all x such that the equation holds[2]. The reader will notice that this might be slightly subtle, because the two kinds of equations appear to have exactly the same form.

4.1 Binary Numbers

Definition 12 (Binary Number Interpretation). *For bit strings $a = a[n-1:0] \in \mathbb{B}^n$ we denote by*

---

[1] In German: Identitäten.
[2] In German: Bestimmungsgleichung.

© Springer International Publishing Switzerland 2016
W.J. Paul et al., *System Architecture*, DOI 10.1007/978-3-319-43065-2_4

$$\langle a \rangle = \sum_{i=0}^{n-1} a_i \cdot 2^i$$

the interpretation of bit string a as a binary number. *We call a the* binary representation *of length n of the natural number $\langle a \rangle$.*

Examples:

$$\langle 100 \rangle = 4,$$
$$\langle 111 \rangle = 7,$$
$$\langle 10^n \rangle = 2^n.$$

Applying Lemma 13, we get

$$\langle 1^n \rangle = \sum_{i=0}^{n-1} 2^i = 2^n - 1,$$

i.e., the largest binary number representable with n bits corresponds to the natural number $2^n - 1$.

Note that binary number interpretation is an injective function.

Lemma 16. *Let $a, b \in \mathbb{B}^n$. Then*

$$a \neq b \rightarrow \langle a \rangle \neq \langle b \rangle.$$

Proof. Let

$$j = \max\{i \mid a_i \neq b_i\}$$

be the largest index where strings a and b differ. Without loss of generality assume $a_j = 1$ and $b_j = 0$. Then

$$\langle a \rangle - \langle b \rangle = \sum_{i=0}^{j} a_i \cdot 2^i - \sum_{i=0}^{j} b_i \cdot 2^i$$
$$\geq 2^j - \sum_{i=0}^{j-1} 2^i$$
$$= 1$$

by Lemma 13.

Definition 13. *Let $n \in \mathbb{N}$. We denote by*

$$B_n = \{\langle a \rangle \mid a \in \mathbb{B}^n\}$$

the set of natural numbers that have a binary representation of length n.

Since

$$0 \leq \langle a \rangle \leq \sum_{i=0}^{n-1} 2^i = 2^n - 1$$

we deduce

$$B_n \subseteq [0 : 2^n - 1].$$

Since $\langle \cdot \rangle$ is injective we have $\#B_n = \#\mathbb{B}^n$, thus we observe that $\langle \cdot \rangle$ is bijective. Then by

$$\#B_n = \#\mathbb{B}^n = 2^n = \#[0 : 2^n - 1]$$

we obtain the following.

Lemma 17. *For $n \in \mathbb{N}$, we have*

$$B_n = [0 : 2^n - 1].$$

Definition 14 (Binary Representation). *For $x \in B_n$ we denote the binary representation of x of length n by $bin_n(x)$:*

$$bin_n(x) = \in\{a \mid a \in \mathbb{B}^n \wedge \langle a \rangle = x\}.$$

To shorten notation even further, we write x_n instead of $bin_n(x)$:

$$x_n = bin_n(x).$$

It is often useful to decompose n-bit binary representations $a[n-1:0]$ into an upper part $a[n-1:m]$ and a lower part $a[m-1:0]$.

Lemma 18 (Decomposition Lemma). *Let $a \in \mathbb{B}^n$ and $n > m > 0$. Then*

$$\langle a[n-1:0] \rangle = \langle a[n-1:m] \rangle \cdot 2^m + \langle a[m-1:0] \rangle.$$

Proof.

$$
\begin{aligned}
\langle a[n-1:0] \rangle &= \sum_{i=m}^{n-1} a_i \cdot 2^i + \sum_{i=0}^{m-1} a_i \cdot 2^i \\
&= \sum_{j=0}^{n-1-m} a_{m+j} \cdot 2^{m+j} + \langle a[m-1:0] \rangle \\
&= 2^m \cdot \sum_{j=0}^{n-1-m} a_{m+j} \cdot 2^j + \langle a[m-1:0] \rangle \\
&= 2^m \cdot \langle a[n-1:m] \rangle + \langle a[m-1:0] \rangle.
\end{aligned}
$$

We obviously have

$$\langle a[n-1:0] \rangle \equiv \langle a[m-1:0] \rangle \bmod 2^m.$$

Using Lemma 12, we infer the following.

a b c	c' s
0 0 0	0 0
0 0 1	0 1
0 1 0	0 1
0 1 1	1 0
1 0 0	0 1
1 0 1	1 0
1 1 0	1 0
1 1 1	1 1

Table 2. Binary addition of 1-bit numbers a, b with carry c

Lemma 19. *For $a \in \mathbb{B}^n$ and $n \geq m > 0$,*

$$\langle a[m-1:0]\rangle = (\langle a[n-1:0]\rangle \bmod 2^m).$$

Intuitively, taking a binary number modulo 2^m means 'throwing away' the bits with position m or higher.

Table 2 specifies the addition algorithm for binary numbers a, b of length 1 and a carry bit c. The binary representation $(c', s) \in \mathbb{B}^2$ of the sum of bits $a, b, c \in \mathbb{B}$ is computed as

$$\langle c', s\rangle = a + b + c.$$

For the addition of n-bit numbers $a[n-1:0]$ and $b[n-1:0]$ with carry in c_0 we first observe for the sum S:

$$\begin{aligned}
S &= \langle a[n-1:0]\rangle + \langle b[n-1:0]\rangle + c_0 \\
&\leq 2^n - 1 + 2^n - 1 + 1 \\
&= 2^{n+1} - 1.
\end{aligned}$$

Thus, the sum $S \in \mathbb{B}^{n+1}$ can be represented as a binary number $\langle s[n:0]\rangle$ with $n+1$ bits. For the computation of the sum bits we use the method of long addition that we learned in elementary school for adding decimal numbers. Using the addition algorithm for 1-bit numbers from Table 2 we compute (c_{i+1}, s_i), where by c_{i+1} we denote the carry to position $i+1$ from position $i \in [0:n-1]$:

$$\begin{aligned}
\langle c_{i+1}, s_i\rangle &= a_i + b_i + c_i, \\
s_n &= c_n.
\end{aligned} \tag{2}$$

That this algorithm indeed computes the sum of the input numbers is asserted as follows.

Lemma 20 (Correctness of the Binary Addition Algorithm). *Let $a, b \in \mathbb{B}^n$ and let $c \in \mathbb{B}$. Further, let $c_n \in \mathbb{B}$ and $s \in \mathbb{B}^n$ be computed according to the addition algorithm described above. Then*

$$\langle c_n, s[n-1:0]\rangle = \langle a[n-1:0]\rangle + \langle b[n-1:0]\rangle + c_0.$$

Proof. by induction on n. For $n = 1$ this follows directly from Equation 2. For the induction step from $n - 1$ to n we conclude:

$$\langle a[n-1:0] \rangle + \langle b[n-1:0] \rangle + c_0$$
$$= (a_{n-1} + b_{n-1}) \cdot 2^{n-1} + \langle a[n-2:0] \rangle + \langle b[n-2:0] \rangle + c_0$$
$$= (a_{n-1} + b_{n-1}) \cdot 2^{n-1} + \langle c_{n-1}, s[n-2:0] \rangle \qquad \text{(induction hypothesis)}$$
$$= (a_{n-1} + b_{n-1} + c_{n-1}) \cdot 2^{n-1} + \langle s[n-2:0] \rangle$$
$$= \langle c_n, s_{n-1} \rangle \cdot 2^{n-1} + \langle s[n-2:0] \rangle \qquad \text{(Equation 2)}$$
$$= \langle c_n, s[n-1:0] \rangle \qquad \text{(Lemma 18)}.$$

The following simple lemma allows us to break the addition of two long numbers into two additions of shorter numbers. It is amazingly useful. In this text we will use it in Sect. 7.1.2 for showing the correctness of recursive addition algorithms as well as in Sect. 8.3.8 for justifying the addition of 32-bit numbers with 30-bit adders if the binary representation of one of the numbers ends with 00. In the hardware-oriented text [KMP14] it is also used to show the correctness of the shifter constructions supporting aligned byte and half-word accesses in cache lines.

Lemma 21 (Decomposition of Binary Number Addition). *For $a, b \in \mathbb{B}^n$, for $d, e \in \mathbb{B}^m$ and for $c_0, c', c'' \in \mathbb{B}$ let*

$$\langle d \rangle + \langle e \rangle + c_0 = \langle c't[m-1:0] \rangle,$$
$$\langle a \rangle + \langle b \rangle + c' = \langle c''s[n-1:0] \rangle,$$

then

$$\langle ad \rangle + \langle be \rangle + c_0 = \langle c''st \rangle.$$

Proof. Repeatedly using Lemma 18, we have

$$\langle ad \rangle + \langle be \rangle + c_0$$
$$= \langle a \rangle \cdot 2^m + \langle d \rangle + \langle b \rangle \cdot 2^m + \langle e \rangle + c_0$$
$$= (\langle a \rangle + \langle b \rangle) \cdot 2^m + \langle c't \rangle$$
$$= (\langle a \rangle + \langle b \rangle + c') \cdot 2^m + \langle t \rangle$$
$$= \langle c''s \rangle \cdot 2^m + \langle t \rangle$$
$$= \langle c''st \rangle.$$

4.2 Two's Complement Numbers

Definition 15 (Two's Complement Number Interpretation). *For bit strings $a[n-1:0] \in \mathbb{B}^n$ we denote by*

$$[a] = -a_{n-1} \cdot 2^{n-1} + \langle a[n-2:0] \rangle$$

the interpretation of a as a two's complement number. *We refer to a as the* two's complement representation *of $[a]$.*

Definition 16. *For $n \in \mathbb{N}$, we denote by*

$$T_n = \{[a] \mid a \in \mathbb{B}^n\}$$

the set of integers that have a two's complement representation of length n.

Since

$$\begin{aligned} T_n &= \{[0b] \mid b \in \mathbb{B}^{n-1}\} \cup \{[1b] \mid b \in \mathbb{B}^{n-1}\} \\ &= B_{n-1} \cup \{-2^{n-1} + x \mid x \in B_{n-1}\} \\ &= [0 : 2^{n-1} - 1] \cup \{-2^{n-1} + x \mid x \in [0 : 2^{n-1} - 1]\} \qquad \text{(Lemma 17)} \end{aligned}$$

we have the following.

Lemma 22. *Let $n \in \mathbb{N}$. Then*

$$T_n = [-2^{n-1} : 2^{n-1} - 1].$$

Definition 17 (Two's Complement Representation). *By $twoc_n(x)$ we denote the two's complement representation of $x \in T_n$:*

$$twoc_n(x) = \in \{a \mid a \in \mathbb{B}^n \wedge [a] = x\}.$$

Basic properties of two's complement numbers are summarized as follows.

Lemma 23. *Let $a \in \mathbb{B}^n$. Then the following holds:*

$$\begin{aligned} [0a] &= \langle a \rangle \qquad \text{(embedding)}, \\ [a] &\equiv \langle a \rangle \bmod 2^n, \\ [a] < 0 &\leftrightarrow a_{n-1} = 1 \qquad \text{(sign bit)}, \\ [a_{n-1}a] &= [a] \qquad \text{(sign extension)}, \\ -[a] &= [\bar{a}] + 1. \end{aligned}$$

Proof. The first line is trivial. The second line follows from

$$\begin{aligned} [a] - \langle a \rangle &= -a_{n-1} \cdot 2^{n-1} + \langle a[n-2:0] \rangle - (a_{n-1} \cdot 2^{n-1} + \langle a[n-2:0] \rangle) \\ &= -a_{n-1} \cdot 2^n. \end{aligned}$$

If $a_{n-1} = 0$ we have $[a] = \langle a[n-2:0] \rangle \geq 0$. If $a_{n-1} = 1$ we have

$$\begin{aligned} [a] &= -2^{n-1} + \langle a[n-2:0] \rangle \\ &\leq -2^{n-1} + 2^{n-1} - 1 \qquad \text{(Lemma 17)} \\ &= -1. \end{aligned}$$

This shows the third line. The fourth line follows from

$$[a_{n-1}a] = -a_{n-1} \cdot 2^n + \langle a[n-1:0] \rangle$$
$$= -a_{n-1} \cdot 2^n + a_{n-1} \cdot 2^{n-1} + \langle a[n-2:0] \rangle$$
$$= -a_{n-1} \cdot 2^{n-1} + \langle a[n-2:0] \rangle$$
$$= [a].$$

For the last line we observe that $x + \bar{x} = 1$ for $x \in \mathbb{B}$. Then

$$[a] + [\bar{a}] = -a_{n-1} \cdot 2^{n-1} - \overline{a_{n-1}} \cdot 2^{n-1} + \sum_{i=0}^{n-2} a_i \cdot 2^i + \sum_{i=0}^{n-2} \overline{a_i} \cdot 2^i$$
$$= -2^{n-1} + \sum_{i=0}^{n-2} 2^i$$
$$= -1 \qquad \text{(Lemma 13)}.$$

We conclude the discussion of binary numbers and two's complement numbers with a lemma that provides a subtraction algorithm for binary numbers.

Lemma 24. *Let $a, b \in \mathbb{B}^n$. Then*

$$\langle a \rangle - \langle b \rangle \equiv \langle a \rangle + \langle \bar{b} \rangle + 1 \mod 2^n.$$

If additionally $\langle a \rangle - \langle b \rangle \geq 0$, we have

$$\langle a \rangle - \langle b \rangle = (\langle a \rangle + \langle \bar{b} \rangle + 1 \mod 2^n).$$

Proof. By Lemma 23 we have

$$\langle a \rangle - \langle b \rangle = \langle a \rangle - [0b]$$
$$= \langle a \rangle + [1\bar{b}] + 1$$
$$= \langle a \rangle - 2^n + \langle \bar{b} \rangle + 1$$
$$\equiv \langle a \rangle + \langle \bar{b} \rangle + 1 \mod 2^n \qquad \text{(Lemma 9)}.$$

The extra hypothesis $\langle a \rangle - \langle b \rangle \geq 0$ implies

$$\langle a \rangle - \langle b \rangle \in B_n.$$

The second claim now follows from Lemma 12.

4.3 Boolean Algebra

We consider Boolean expressions with constants 0 and 1, variables $x_0, x_1, \ldots, a, b, \ldots$, and function symbols $^-$, $\wedge$, $\vee$, $\oplus$, $f(\ldots), g(\ldots), \ldots$. Four of the function symbols have predefined semantics as specified in Table 3. For a more formal definition one collects the constants, variables, and function symbols allowed into sets

x y	$\bar{x}$	$x \wedge y$	$x \vee y$	$x \oplus y$
0 0	1	0	0	0
0 1	1	0	1	1
1 0	0	0	1	1
1 1	0	1	1	0

Table 3. Boolean operators

$$C = \{0,1\},$$
$$V = \{x_0, x_1, \ldots\},$$
$$F = \{f_0, f_1, \ldots\}.$$

and denotes the number of arguments for function f_i with n_i. Now we can define the set BE of Boolean expressions by the following rules:

1. constants and variables are Boolean expressions

$$C \cup V \subseteq BE.$$

2. if e is a Boolean expression, then so is $(\bar{e})$

$$e \in BE \rightarrow (\bar{e}) \in BE.$$

3. if e and e' are Boolean expressions, then also $(e \circ e')$ is a Boolean expression for any of the predefined binary connectives $\circ$

$$e, e' \in BE \wedge \circ \in \{\wedge, \vee, \oplus\} \rightarrow (e \circ e') \in BE.$$

4. if f_i is a symbol for a function with n_i arguments, then we can obtain a Boolean expression $f_i(e_1, \ldots, e_{n_i})$ by substituting the function arguments with Boolean expressions e_j

$$(\forall j \in [1 : n_i] : e_j \in BE) \rightarrow f_i(e_1, \ldots, e_{n_i}) \in BE.$$

5. all Boolean expressions are formed by the above rules.

We call a Boolean expression *pure* if it uses only the predefined connectives.

In order to save brackets one uses the convention that $^-$ binds stronger than $\wedge$ and $\wedge$ binds stronger than $\vee$. Thus $\bar{x_1} \wedge x_2 \vee x_3$ is an abbreviation for

$$\bar{x_1} \wedge x_2 \vee x_3 = ((\bar{x_1}) \wedge x_2) \vee x_3.$$

By $e(x)$ we often denote expressions e depending on variables $x = x[1 : n]$. Variables x_i can take values in $\mathbb{B}$, thus $x = x[1 : n]$ can take values in $\mathbb{B}^n$. By $e(a)$ we denote the result of evaluation of expression $e \in BE$ with a bit string $a \in \mathbb{B}^n$, under the following straightforward set of rules for evaluating expressions:

1. Substitute a_i for x_i

$$x_i(a) = a_i.$$

2. evaluate $(\bar{e})$ by evaluating e and negating the result according to the predefined meaning of negation in Table 3

$$(\bar{e})(a) = \overline{e(a)}.$$

3. evaluate $(e \circ e')$ by evaluating e and e' and then combining the results according to the predefined meaning of $\circ$ in Table 3

$$(e \circ e')(a) = e(a) \circ e'(a).$$

4. expressions of the form $f_i(e_1, \ldots, e_{n_j})$ can only be evaluated if symbol f_i has an interpretation as a function

$$f_i : \mathbb{B}^{n_i} \to \mathbb{B}.$$

In this case evaluate $f_i(e_1, \ldots, e_{n_j})(a)$ by evaluating arguments e_j, substituting the result into f and evaluating f:

$$f_i(e_1, \ldots, e_{n_j})(a) = f_i(e_1(a), \ldots, e_{n_i}(a)).$$

The following small example illustrates that this very formal and detailed set of rules captures our usual way of evaluating expressions.

$$(x_1 \wedge x_2)(0, 1) = x_1(0, 1) \wedge x_2(0, 1)$$
$$= 0 \wedge 1$$
$$= 0.$$

Equations have the form

$$e = e'$$

where e and e' are expressions (involving here variables $x = x[1 : n]$). They come in two flavors:

- identities. An equation $e = e'$ is an identity iff expressions e and e' evaluate to the same value in $\mathbb{B}$ for any substitution of the variables $a = a[1 : n] \in \mathbb{B}^n$:

$$\forall a \in \mathbb{B}^n : e(a) = e'(a),$$

- equations which one wants to solve. A substitution $a = a[1 : n] \in \mathbb{B}^n$ solves equation $e = e'$ if $e(a) = e'(a)$.

We proceed to show where identities and equations we want to solve differ formally: in the *implicit quantification* as explained in Sect. 3.1.1. If not stated otherwise, we usually assume equations to be of the first type, i.e., to be implicitly quantified over all free variables. This is also the case with *definitions* of functions, where the left-hand side of an equation represents an entity being defined. For instance, the following definition of the function

$$f(x_1, x_2) = x_1 \wedge x_2$$

is the same as

$$\forall a,b \in \mathbb{B} : f(a,b) = a \wedge b.$$

We may also write

$$e \equiv e'$$

to stress that a given equation is an identity or to avoid brackets in case this equation is a definition and the right-hand side itself contains an equality sign.

In case we talk about several equations in a single statement (this is often the case when we solve equations), we assume implicit quantification over the whole statement rather than over every single equation. For instance,

$$e_1 = e_2 \leftrightarrow e_3 = 0$$

is the same as

$$\forall a \in \mathbb{B}^n : e_1(a) = e_2(a) \leftrightarrow e_3(a) = 0$$

and means that, for a given substitution a, equations e_1 and e_2 evaluate to the same value iff equation e_3 evaluates to 0.

In Boolean algebra there is a very simple connection between the solution of equations and identities. An identity $e \equiv e'$ holds iff equations $e = 1$ and $e' = 1$ have the same set of solutions.

Lemma 25. *Given Boolean expressions $e(x)$ and $e'(x)$, we have*

$$e \equiv e' \leftrightarrow (e = 1 \leftrightarrow e' = 1).$$

Proof.

$$e = 1 \leftrightarrow e' = 1$$

is implicitly quantified and abbreviates

$$\forall a \in \mathbb{B}^n : e(a) = 1 \leftrightarrow e'(a) = 1.$$

In the statement of the lemma, the direction from left to right is trivial. For the other direction we distinguish cases:

- $e(a) = 1$. Then $e'(a) = 1$ by hypothesis.
- $e(a) = 0$. Then $e'(a) = 1$ would by hypothesis imply the contradiction $e(a) = 1$. Because in Boolean algebra $e'(a) \in \mathbb{B}$ we conclude $e'(a) = 0$.

Thus, we have $e(a) = e'(a)$ for all $a \in \mathbb{B}^n$.

4.3.1 Useful Identities

In the sequel, we provide a list of useful identities of Boolean algebra.

- Commutativity:

$$x_1 \wedge x_2 \equiv x_2 \wedge x_1,$$
$$x_1 \vee x_2 \equiv x_2 \vee x_1,$$
$$x_1 \oplus x_2 \equiv x_2 \oplus x_1.$$

- Associativity:

$$(x_1 \wedge x_2) \wedge x_3 \equiv x_1 \wedge (x_2 \wedge x_3),$$
$$(x_1 \vee x_2) \vee x_3 \equiv x_1 \vee (x_2 \vee x_3),$$
$$(x_1 \oplus x_2) \oplus x_3 \equiv x_1 \oplus (x_2 \oplus x_3).$$

- Distributivity:

$$x_1 \wedge (x_2 \vee x_3) \equiv (x_1 \wedge x_2) \vee (x_1 \wedge x_3),$$
$$x_1 \vee (x_2 \wedge x_3) \equiv (x_1 \vee x_2) \wedge (x_1 \vee x_3).$$

- Identity:

$$x_1 \wedge 1 \equiv x_1,$$
$$x_1 \vee 0 \equiv x_1.$$

- Idempotence:

$$x_1 \wedge x_1 \equiv x_1,$$
$$x_1 \vee x_1 \equiv x_1.$$

- Annihilation:

$$x_1 \wedge 0 \equiv 0,$$
$$x_1 \vee 1 \equiv 1.$$

- Absorption:

$$x_1 \vee (x_1 \wedge x_2) \equiv x_1,$$
$$x_1 \wedge (x_1 \vee x_2) \equiv x_1.$$

- Complement:

$$x_1 \wedge \overline{x_1} \equiv 0,$$
$$x_1 \vee \overline{x_1} \equiv 1.$$

- Double negation:

$$\overline{\overline{x_1}} \equiv x_1.$$

x_1 x_2	$x_1 \wedge x_2$	$\overline{x_1 \wedge x_2}$	$\overline{x_1}$ $\overline{x_2}$ $\overline{x_1} \vee \overline{x_2}$
0 0	0	1	1 1 1
0 1	0	1	1 0 1
1 0	0	1	0 1 1
1 1	1	0	0 0 0

Table 4. Verifying the first of De Morgan's laws

- De Morgan's laws:

$$\overline{x_1 \wedge x_2} \equiv \overline{x_1} \vee \overline{x_2},$$
$$\overline{x_1 \vee x_2} \equiv \overline{x_1} \wedge \overline{x_2}.$$

Each of these identities can be proven in a simple brute force way: if the identity has n variables, then for each of the 2^n possible substitutions of the variables the left- and right-hand sides of the identities are evaluated with the help of Table 3. If for each substitution the left-hand side and the right-hand side evaluate to the same value, then the identity holds. For the first of De Morgan's laws this is illustrated in Table 4.

4.3.2 Solving Equations

We consider expressions e and e_i (where $1 \leq i \leq n$), involving a vector of variables x. We derive three basic lemmas about the solution of Boolean equations. For $a \in \mathbb{B}$ we define

$$e^a = \begin{cases} e, & a = 1, \\ \overline{e}, & a = 0. \end{cases}$$

Inspection of the semantics of $^-$ in Table 3 immediately gives the following.

Lemma 26 (Solving Negation). *Given a Boolean expression $e(x)$, we have*

$$e^0 = 1 \leftrightarrow e = 0.$$

Inspection of the semantics of $\wedge$ in Table 3 gives

$$(e_1 \wedge e_2) = 1 \leftrightarrow e_1 = 1 \wedge e_2 = 1.$$

Induction on n results in the following.

Lemma 27 (Solving Conjunction). *Given Boolean expressions $e_i(x)$, where $1 \leq i \leq n$, we have*

$$\left(\bigwedge_{i=1}^{n} e_i \right) = 1 \leftrightarrow \forall i \in [1:n] : e_i = 1.$$

From the semantics of $\vee$ in Table 3, we have

$$(e_1 \vee e_2) = 1 \leftrightarrow e_1 = 1 \vee e_2 = 1.$$

Induction on n yields the following.

Lemma 28 (Solving Disjunction). *Given Boolean expressions* e_i, *where* $1 \leq i \leq n$, *we have*

$$\left(\bigvee_{i=1}^{n} e_i\right) = 1 \leftrightarrow \exists i \in [1:n] : e_i = 1.$$

4.3.3 Disjunctive Normal Form

Definition 18. *Let* $f : \mathbb{B}^n \to \mathbb{B}$ *be a switching function[3] and let e be a Boolean expression with variables x. We say that e* computes *f iff the identity* $f(x) \equiv e$ *holds.*

Lemma 29. *Every switching function is computed by some Boolean expression:*

$$\forall f : \mathbb{B}^n \to \mathbb{B} \; \exists e : f(x) \equiv e.$$

Moreover, expression e is pure.

Proof. Let $b \in \mathbb{B}$ and let x_i be a variable. We define the *literal*

$$x_i^b = \begin{cases} x_i, & b = 1, \\ \overline{x_i}, & b = 0. \end{cases}$$

Then

$$x_i^1 = 1 \leftrightarrow x_i = 1$$

and using Lemma 26 we have

$$x_i^b = 1 \leftrightarrow x_i = b. \tag{3}$$

Let $a = a[1:n] \in \mathbb{B}^n$ and let $x = x[1:n]$ be a vector of variables. We define the *monomial*

$$m(a) = \bigwedge_{i=1}^{n} x_i^{a_i}.$$

Then

$$\begin{aligned} m(a) = 1 &\leftrightarrow \forall i \in [1:n] : x_i^{a_i} = 1 \quad \text{(Lemma 27)} \\ &\leftrightarrow \forall i \in [1:n] : x_i = a_i \quad \text{(Equation 3)} \\ &\leftrightarrow x = a. \end{aligned}$$

Thus, we have

$$m(a) = 1 \leftrightarrow x = a. \tag{4}$$

We define the *support* $S(f)$ of f as the set of arguments a for which f takes the value $f(a) = 1$:

$$S(f) = \{a \in \mathbb{B}^n \mid f(a)\}.$$

---

[3] The term *switching function* comes from electrical engineering and stands for a Boolean function.

If the support is empty, then $e = 0$ computes f. Otherwise we set

$$e = \bigvee_{a \in S(f)} m(a).$$

Then

$$
\begin{aligned}
e = 1 &\leftrightarrow \exists a \in S(f) : m(a) = 1 \quad \text{(Lemma 28)} \\
&\leftrightarrow \exists a \in S(f) : x = a \quad \text{(Equation 4)} \\
&\leftrightarrow x \in S(f) \\
&\leftrightarrow f(x) = 1.
\end{aligned}
$$

Thus, equations $e = 1$ and $f(x) = 1$ have the same solutions. With Lemma 25 we conclude

$$e \equiv f(x).$$

The expression e constructed in the proof of Lemma 29 is called the *complete disjunctive normal form* of f.

Example: The complete disjunctive normal forms of the sum and carry functions c' defined in Table 2 are

$$
\begin{aligned}
c'(a,b,c) &\equiv \bar{a} \wedge b \wedge c \vee a \wedge \bar{b} \wedge c \vee a \wedge b \wedge \bar{c} \vee a \wedge b \wedge c, \\
s(a,b,c) &\equiv \bar{a} \wedge \bar{b} \wedge c \vee \bar{a} \wedge b \wedge \bar{c} \vee a \wedge \bar{b} \wedge \bar{c} \vee a \wedge b \wedge c.
\end{aligned}
\tag{5}
$$

In algebra, one often omits the multiplication sign in order to simplify notation. Recall that the first binomial formula is usually written as

$$(a+b)^2 = a^2 + 2ab + b^2,$$

where $2ab$ is a shorthand for $2 \cdot a \cdot b$. In the same spirit we often omit the $\wedge$-sign in the conjunction of literals. Thus the above identities can also be written as

$$
\begin{aligned}
c'(a,b,c) &\equiv \bar{a}bc \vee a\bar{b}c \vee ab\bar{c} \vee abc, \\
s(a,b,c) &\equiv \bar{a}\bar{b}c \vee \bar{a}b\bar{c} \vee a\bar{b}\bar{c} \vee abc.
\end{aligned}
$$

Simplified Boolean expressions for the same functions are

$$
\begin{aligned}
c'(a,b,c) &\equiv ab \vee bc \vee ac, \\
s(a,b,c) &\equiv a \oplus b \oplus c.
\end{aligned}
$$

The correctness can be checked in the usual brute force way by trying all eight assignments of values in $\mathbb{B}^3$ to the variables of the expressions, or by applying the identities listed in Sect. 4.3.1.

In the remainder of this book, we return to the usual mathematical notation, using the equality sign for both identities and equations to be solved. Whether we deal with identities or whether we solve equations will (hopefully) be clear from the context. We will also sometimes use the equivalence sign for identities, e.g., to avoid using brackets in case we give a definition with an equality sign in the right-hand side.

4.4 Final Remarks

We recommend to remember the following key technical points:

- about binary numbers
  - the definition of the value of a binary number

$$\langle a \rangle = \sum_{i=0}^{n} a_i \cdot 2^i.$$

  - the decomposition lemma

$$\langle a[n-1:0] \rangle = \langle a[n-1:m] \rangle \cdot 2^m + \langle a[m-1:0] \rangle.$$

    The proof is an exercise, so do not bother to memorize it.
  - the definition of a single step of the addition algorithm

$$\langle c_{i+1}, s_i \rangle = a_i + b_i + c_i.$$

  - the correctness statement of the addition algorithm

$$\langle c_n s \rangle = \langle a \rangle + \langle b \rangle + c_0.$$

    For the proof by induction only remember to decompose the operands first by splitting the leading digits off. Then everything works out.
- about two's complement numbers
  - the definition of the value of a two's complement number

$$[a] = -a_{n-1} \cdot 2^{n-1} + \langle a[n-2:0] \rangle.$$

  - the five basic equations

$$[0a] = \langle a \rangle,$$
$$[a] \equiv \langle a \rangle \bmod 2^n,$$
$$[a] < 0 \leftrightarrow a_{n-1} = 1,$$
$$[a_{n-1}a] = [a],$$
$$[a] = [\bar{a}] + 1.$$

    For the proof only remember

$$x + \bar{x} = 1.$$

    Conclude that everything else is straightforward, because you memorized nothing else.
  - the subtraction algorithm for binary numbers

$$\langle x \rangle - \langle y \rangle \equiv \langle x \rangle + \langle \bar{y} \rangle + 1 \bmod 2^n.$$

    Remember that for $\langle x \rangle \geq \langle y \rangle$ you can solve the congruence in a way you already remember and conclude

$$\langle x \rangle - \langle y \rangle = \langle x \rangle + \langle \bar{y} \rangle + 1.$$

- about Boolean Algebra
  - about identities the fact there there are two distributive laws and De Morgan's laws. Everything else one remembers kind of automatically.
  - how to establish identities by solving equations

$$e \equiv e' \leftrightarrow (e = 1 \leftrightarrow e' = 1).$$

  - for the proof that every switching function f can be computed by a pure Boolean expression e remember the definitions of literals

$$x_i^b = \begin{cases} x_i, & b = 1, \\ \overline{x_i}, & b = 0, \end{cases}$$

monomials

$$m(a) = \bigwedge_{i=1}^{n} x_i^{a_i},$$

the support of S

$$S(f) = \{a : f(a) = 1\},$$

and the definition of e

$$e = \bigvee_{a \in S(f)} m(a).$$

For the correctness proof simply solve successively the obvious equations $Q = 1$ for the expressions Q you defined.
- for parties: why does the school method for addition of decimal numbers work (same proof as for binary numbers, just different basis).

One would expect the results of this chapter to be present in any textbook or lecture on digital design and computer arithmetic, *if* the author wishes his students to learn by understanding and not by gathering experience (we explained the difference in Chap. 2). We encourage readers to inspect a few such books in the library of their university or lecture notes on the web and to check whether the results are there — not only the algorithms and constructions but also explanations for why they always work.

4.5 Exercises

1. Prove that for $n \in \mathbb{N}$, $bin_n(\cdot)$ and $twoc_n(\cdot)$ are the inverse functions of $\langle \cdot \rangle$ and $[\cdot]$ by showing:

$$\begin{aligned} x \in B_n &\rightarrow \langle bin_n(x) \rangle = x, \\ x \in T_n &\rightarrow [twoc_n(x)] = x, \\ a \in \mathbb{B}^n &\rightarrow bin_n(\langle a \rangle) = a, \\ a \in \mathbb{B}^n &\rightarrow twoc_n([a]) = a. \end{aligned}$$

2. Recall from Sect. 3.1.4 that we abbreviate a sequence of i zeros by 0^i. Prove

$$\langle 0^i a \rangle = \langle a \rangle,$$
$$\langle a 0^i \rangle = \langle a \rangle \cdot 2^i,$$
$$n \geq m > 0 \ \rightarrow \ \langle a[m-1:0] \rangle = (\langle a[n-1:0] \rangle \bmod 2^m).$$

3. Prove a decomposition lemma for two's complement numbers: for all $n > m > 0$ and $a \in \mathbb{B}^n$

$$[a[n-1:0]] = [a[n-1:m]] \cdot 2^m + \langle a[m-1:0] \rangle.$$

4. Prove or disprove

$$\langle a \rangle - \langle b \rangle = (\langle a \rangle + \langle \overline{b} \rangle + 1 \bmod 2^n),$$
$$\langle a \rangle - \langle b \rangle = \langle a \rangle + \langle \overline{b} \rangle + 1.$$

5. Prove or disprove

$$x_1 \wedge (x_2 \oplus x_3) \equiv (x_1 \oplus x_2) \wedge (x_1 \oplus x_3),$$
$$x_1 \oplus (x_2 \wedge x_3) \equiv (x_1 \wedge x_2) \oplus (x_1 \wedge x_3).$$

6. Prove by induction on n

$$\overline{\bigvee_{i=1}^{n} x_i} = \bigwedge_{i=1}^{n} \overline{x_i},$$
$$\overline{\bigwedge_{i=1}^{n} x_i} = \bigvee_{i=1}^{n} \overline{x_i}.$$

7. Derive from the identities of Sect. 4.3.1

$$x_1 x_3 \vee x_2 \overline{x_3} \equiv x_1 x_3 \vee x_2 \overline{x_3} \vee x_1 x_2,$$
$$(x_1 \vee x_3) \wedge (x_2 \vee \overline{x_3}) \equiv (x_1 \vee x_3) \wedge (x_2 \vee \overline{x_3}) \wedge (x_1 \vee x_2),$$
$$x_1 \overline{x_2} \vee x_1 x_2 \equiv x_1,$$
$$(x_1 \vee \overline{x_2}) \wedge (x_1 \vee x_2) \equiv x_1.$$

8. We define the connective NAND ($\overline{\wedge}$) by the following table.

x_1	x_2	$x_1 \overline{\wedge} x_2$
0	0	1
0	1	1
1	0	1
1	1	0

Show that every switching function can be computed by a Boolean expression whose only operator is NAND.

9. Show for Boolean expressions e and e'

$$e \equiv e' \leftrightarrow (e = 0 \leftrightarrow e' = 0).$$

10. For $a \in \mathbb{B}^n$ we define the clause

$$c(a) = \bigvee_{i=1}^{n} x_i^{\bar{a}_i}.$$

Recall that the support $S(f)$ was defined as

$$S(f) = \{a \in \mathbb{B}^n \mid f(a)\}.$$

Show — using the previous exercise — that

$$f = \bigwedge_{a \notin S(f)} c(a).$$

The right-hand side of the above equality is called the conjunctive normal form (CNF) of f.

11. Switching functions $f_1, f_2 : \{0, 1\}^3 \rightarrow \{0, 1\}$ are specified by the following table:

x_1	x_2	x_3	f_1	f_2
0	0	0	0	1
0	0	1	0	0
0	1	0	1	0
0	1	1	1	1
1	0	0	1	0
1	0	1	0	0
1	1	0	1	1
1	1	1	0	0

a) Specify the support $S(f_1)$ and $S(f_2)$ of the functions.
b) Write down the disjunctive normal form of the functions.
c) Write down the conjunctive normal form of the functions.

12. The number of occurrences of connectives $\wedge$, $\vee$, and $\neg$ in a pure Boolean expression e is called the cost $L(e)$ of the expression. We define the cost of a switching function f as the cost of the cheapest pure Boolean expression which computes f:

$$L(f) = \min\{L(e) \mid e \in BE \wedge e \text{ is pure} \wedge f \equiv e\}.$$

Show for $f : \mathbb{B}^n \rightarrow \mathbb{B}$

a) $$L(f) \leq n \cdot 2^{n+1}.$$
b) $$f(x_1, \ldots, x_n) = (x_n \wedge f(x_1, \ldots, x_{n-1}, 1)) \vee (\overline{x_n} \wedge f(x_1, \ldots, x_{n-1}, 0)).$$
c) $$L(f) \leq (5/2) \cdot 2^n - 4.$$

Hint: to show (c), apply the formula from (b) recursively and conclude by induction on n.

5

Hardware

In Sect. 5.1 we introduce the classical model of *digital circuits* and show — by induction on the depth of gates — that the computation of signals in this model is well defined. The simple argument hinges on the fact that depth in directed acyclic graphs is defined, which was established in Sect. 3.4.2. We show how to transform Boolean expressions into circuits and obtain circuits for arbitrary switching functions with the help of the disjunctive normal form.

A few basic digital circuits are presented for later use in Sect. 5.2. This is basically the same collection of circuits as presented in [MP00] and [KMP14].

In Sect. 5.3 we introduce a computational model consisting of digital circuits and 1-bit registers as presented in [MP00,MP98,KMP14]. In the very simple Sect. 5.4 we define multiple 1-bit registers which are clocked together as n-bit registers and model the state of such registers R as *single* components $h.R$ of hardware configurations h.

5.1 Gates and Circuits

In a nutshell, we can think of hardware as consisting of three kinds of components which are interconnected by wires: gates, storage elements, and drivers. Drivers are tricky, and their behavior cannot even be properly explained in a purely digital circuit model; for details see [KMP14]. Fortunately we can avoid them in the designs in this book. Gates are: AND-gates, OR-gates, $\oplus$-gates, and inverters. For the purpose of uniform language we sometimes also call inverters NOT-gates. In circuit schematics we use the symbols from Fig. 7(a).

Intuitively, a *circuit* C consists of a finite set H of gates, a sequence of input signals $x[n - 1 : 0]$, a set N of wires that connect them, as well as a sequence of output signals $y[m - 1 : 0]$ chosen from all signals of circuit C (as illustrated in Fig. 7(b)). Special inputs 0 and 1 are always available for use in a circuit. Formally we specify a circuit C by the following components:

- a sequence of *inputs* $C.x[n - 1 : 0]$. We define the corresponding set of inputs of the circuit as

$$In(C) = \{C.x[i] \mid i \in [0 : n - 1]\}.$$

© Springer International Publishing Switzerland 2016
W.J. Paul et al., *System Architecture*, DOI 10.1007/978-3-319-43065-2_5

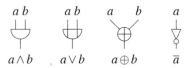

(a) Symbols for gates in circuit schematics

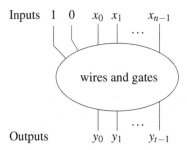

(b) Illustration of inputs and outputs of circuit C

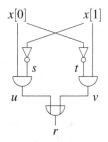

(c) Example of a circuit

Fig. 7. Gates and circuits

- a set $C.H$ of *gates* which is disjoint from the set of inputs

$$C.H \cap In(C) = \emptyset.$$

The *signals* of a circuit then are its inputs and its gates. Moreover the constant signals 0 and 1 are always available. We collect the signals of circuit C in the set

$$Sig(C) = In(C) \cup C.H \cup \{0,1\}.$$

- a labeling function

$$C.\ell : C.H \to \{\wedge, \vee, \oplus, \neg\}$$

specifying for each gate $g \in C.H$ its type $C.\ell(g)$. Thus, a gate g is a $\circ$-gate if $C.\ell(g) = \circ$.
- two functions

$$C.in1, C.in2 : C.H \to Sig(C)$$

specifying for each gate $g \in C.H$ the signals which provide its left input $C.in1(g)$ and its right input $C.in2(g)$. These functions specify the wires interconnecting the gates and inputs. For inverters the second input does not matter.
- a sequence of *outputs* $C.y[m-1:0]$, which are taken from the signals of the circuit:

$$\forall i \in [0 : m-1] : C.y[i] \in Sig(C).$$

As an example we specify the circuit from Fig. 7(c) formally by

$$C.x[1:0] = x[1:0],$$
$$C.H = \{s,t,u,v,r\},$$
$$C.y[0:0] = r,$$

g	s	t	u	v	r
$C.\ell(g)$	$\neg$	$\neg$	$\wedge$	$\wedge$	$\vee$
$C.in1(g)$	x_1	x_0	x_0	x_1	u
$C.in2(g)$			s	t	v

Table 5. Specification of the circuit from Fig. 7(c)

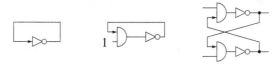

Fig. 8. Examples of cycles in circuits

as well as the function tables in Table 5.

At first glance it is very easy to define how a circuit should work.

Definition 19 (Evaluating Signals of a Circuit). *For a circuit C, we define the values s(a) produced by signals $s \in Sig(C)$ for a given substitution $a = a[n-1:0] \in \mathbb{B}^n$ of the input signals:*

1. if $s = x_i$ is an input:

$$\forall i \in [0:n-1] : x_i(a) = a_i,$$

2. if s is an inverter:

$$s(a) = \overline{in1(s)(a)},$$

3. if s is a $\circ$-gate with $\circ \in \{\wedge, \vee, \oplus\}$:

$$s(a) = in1(s)(a) \circ in2(s)(a).$$

Unfortunately, this is not always a definition. For three counterexamples, see Fig. 8. Due to the cycles, one cannot find an order in which the above definition can be applied. Fortunately, defining and then forbidding cycles solves the problem. With circuits C we associate in an obvious way a graph $G(C)$ by the following rules:

- the nodes of the graph are the signals of the circuit, i.e., inputs, gates, and the constants 0 and 1

$$G(C).V = Sig(C).$$

- we draw an edge (s,t) from signal s to signal t if t is a gate and s is an input for t

$$G(C).E = \{(s,t) \mid t \in C.H \wedge s \in \{C.in1(t), C.in2(t)\}\}.$$

If we apply this definition to the circuit in Fig. 7(c) we obtain the graph from Fig. 4. The sources of such graphs are the inputs as well as the constant signals 0 and 1. We restrict the definition of circuits C by requiring that their graphs $G(C)$ are cycle free. This not only excludes the counterexamples above. We know from Sect. 3.4.2 that in a directed acyclic graph every node has a depth. Evaluating signals s in circuits in order of increasing depth will always work, and we conclude by induction on the depth of gates the following very reassuring lemma.

$a[1:0]$	00	01	10	11
s	1	1	0	0
t	1	0	1	0
u	0	1	0	0
v	0	0	1	0
r	0	1	1	0

Table 6. Values of signals in the circuit from Fig. 7(c)

Lemma 30. *The values $s(a)$ are well defined for all signals s in a circuit C.*

Applying Def. 19 to the circuit of Fig. 7(c) gives the values in Table 6.

Obviously we can treat names g of gates as function symbols with n arguments satisfying with $x = x[n-1:0]$ the identities

1. if s is an inverter:
$$s(x) = \overline{in1(s)(x)},$$

2. if s is a $\circ$-gate with $\circ \in \{\wedge, \vee, \oplus\}$:

$$s(x) = in1(s)(x) \circ in2(s)(x).$$

So we can do Boolean algebra with signal names. We can for instance summarize the last line of Table 6 by

$$\forall a \in \mathbb{B}^2 : r(a) = 1 \leftrightarrow a_1 \oplus a_2 = 1$$

which can be written as the identity

$$r(x) = x_1 \oplus x_2.$$

We can transform Boolean expressions into circuits in a trivial way.

Lemma 31. *Let e be a pure Boolean expression with variables $x[n-1:0]$. Then there is a circuit $C(e)$ with inputs $x[n-1:0]$ and a gate $g \in C.H$ such that*

$$e = g(x)$$

is an identity.

Proof. by induction on the structure of Boolean expressions. If e is a variable x_i or a constant $c \in \{0,1\}$ we take g as the corresponding input x_i or the corresponding constant signal c in the circuit. In the induction step there are two obvious cases

- $e = \overline{e'}$. By induction hypothesis we have circuit $C(e')$ and gate g' such that

$$e' = g'(x).$$

We extend the circuit by an inverter g as shown in Fig. 9(a) and get

$$g(x) = \overline{g'(x)} = \overline{e'} = e.$$

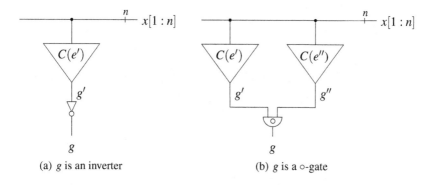

(a) g is an inverter   (b) g is a ∘-gate

Fig. 9. Transforming Boolean expressions into circuits

- $e = e' \circ e''$. By induction hypothesis we have circuits $C(e')$ and $C(e'')$ with gates g' and g'' satisfying

$$e' = g'(x) \quad \text{and} \quad e'' = g''(x).$$

We feed g' and g'' into a ∘-gate as shown in Fig. 9(b) and get

$$g(x) = g'(x) \circ g''(x) = e' \circ e'' = e.$$

Circuits for arbitrary switching functions are obtained with the help of Lemma 29.

Lemma 32. *Let $f : \mathbb{B}^n \to \mathbb{B}$ be a switching function. Then there is a circuit C with inputs $x[n-1:0]$ and a gate g in the circuit computing f, s.t.*

$$f(x) = g(x)$$

is an identity.

Proof. Let e be the complete disjunctive normal form for f using variables $x[n-1:0]$. Then

$$f(x) = e.$$

By Lemma 31, in circuit $C(e)$ there is a gate g with

$$e = g(x).$$

We conclude this section by mentioning two complexity measures for circuits C:

- the number of gates

$$c(C) = \#C.H$$

is an approximation for the cost of the circuit.
- the depth (of the underlying graph $G(C)$)

$$d(C) = d(G(C))$$

is an approximation for the delay of the circuit. It counts the maximal number of gates on a path in the circuit.

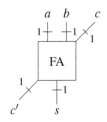

(a) symbol

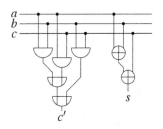

(b) implementation

Fig. 10. Full adder

5.2 Some Basic Circuits

We have shown in Lemma 31 that Boolean expressions can be transformed into circuits in a very intuitive way. In Fig. 10(b) we have transformed the simple formulas from Equation 5 for $c'(a,b,c)$ and $s(a,b,c)$ into a circuit. With inputs (a,b,c) and outputs (c',s) this circuit satisfies

$$\langle c',s \rangle = a+b+c.$$

A circuit satisfying this condition is called a *full adder*. We use the symbol from Fig. 10(a) to represent this circuit in subsequent constructions.

When the b-input of a full adder is known to be zero, the specification simplifies to

$$\langle c',s \rangle = a+c.$$

The resulting circuit is called a *half adder*. Symbol and implementation are shown in Fig. 11.

The circuit in Fig. 12(b) is called a *multiplexer* or for short *mux*. Its inputs and outputs satisfy

$$z = \begin{cases} x, & s=0, \\ y, & s=1. \end{cases}$$

For multiplexers we use the symbol from Fig. 12(a).

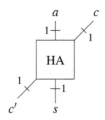

(a) symbol

(b) implementation

Fig. 11. Half adder

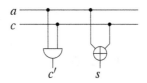

(a) symbol

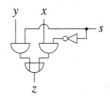

(b) implementation

Fig. 12. Multiplexer

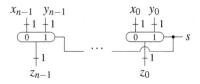

(a) symbol

(b) implementation

Fig. 13. n-bit multiplexer

The n-bit multiplexer or for short n-mux in Fig. 13(b) consists of n multiplexers with a common select signal s. Its inputs and outputs satisfy

$$z[n-1:0] = \begin{cases} x[n-1:0], & s=0, \\ y[n-1:0], & s=1. \end{cases}$$

For n-muxes we use the symbol from Fig. 13(a).

Fig. 14(a) shows the symbol for an n-bit inverter. Its inputs and outputs satisfy

$$y[n-1:0] = \overline{x[n-1:0]}.$$

n-bit inverters are simply realized by n separate inverters as shown in Fig. 14(b).

For $\circ \in \{\wedge, \vee, \oplus\}$, Fig. 15(a) shows symbols for n-bit $\circ$-gates. Their inputs and outputs satisfy

$$z[n-1:0] = x[n-1:0] \circ y[n-1:0],$$
$$u[n-1:0] = v \circ y[n-1:0].$$

n-bit $\circ$-gates are simply realized in the first case by n separate $\circ$-gates as shown in Fig. 15(b). In the second case all left inputs of the gates are connected to the same input v.

An n-bit $\circ$-tree has inputs $a[n-1:0]$ and a single output b satisfying

$$b = \circ_{i=0}^{n-1} a_i.$$

Symbol and recursive construction are shown in Fig. 16.

An (n, A)-OR tree (Fig. 17(a)) simply consists of n A-bit OR-trees in parallel. Formally it has A groups of inputs $a^{(i)} \in \mathbb{B}^n$ for $i \in [A-1:0]$. A recursive construction is shown in Fig. 17(b).

x

n

n

y

(a) symbol

x_{n-1} x_0

1   1

$\cdots$

1   1

y_{n-1} y_0

(b) implementation

Fig. 14. n-bit inverter

The inputs $a[n-1:0]$ and outputs *zero* and *nzero* of an *n-zero tester* shown in Fig. 18 satisfy

$$zero \equiv a = 0^n,$$
$$nzero \equiv a \neq 0^n.$$

The implementation uses

$$nzero(a[n-1:0]) = \bigvee_{i=0}^{n-1} a_i = 1 \leftrightarrow \exists i \in [0:n-1] : a_i = 1,$$
$$zero = \overline{nzero} = 1 \leftrightarrow \forall i \in [0:n-1] : a_i = 0.$$

The inputs $a[n-1:0]$, $b[n-1:0]$ and outputs *eq*, *neq* of an *n*-bit *equality tester* (Fig. 19) satisfy

$$eq \equiv a = b,$$
$$neq \equiv a \neq b.$$

The implementation uses the fact $x \oplus y = 1 \leftrightarrow x \neq y$ for $x, y \in \mathbb{B}$ and

$$neq = nzero(a[n-1:0] \oplus b[n-1:0]) = 1 \leftrightarrow \exists i \in [0:n-1] : a_i \neq b_i,$$
$$eq = \overline{neq} = 1 \leftrightarrow \forall i \in [0:n-1] : a_i = b_i.$$

An *n-decoder* is a circuit with inputs $x[n-1:0]$ and outputs $y[2^n-1:0]$ satisfying for all $i \in [0:2^n-1]$:

$$y_i = 1 \leftrightarrow \langle x \rangle = i.$$

(a) symbol

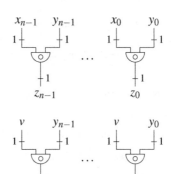

(b) implementation

Fig. 15. Gates for n-bit wide inputs

A recursive construction with $k = \lceil \frac{n}{2} \rceil$ is shown in Fig. 20(b). For the correctness one argues in the induction step

$$
\begin{aligned}
y[i \cdot 2^k + j] = 1 &\leftrightarrow V[i] = 1 \wedge U[j] = 1 \quad \text{(construction)} \\
&\leftrightarrow \langle x[n-1:k] \rangle = i \wedge \langle x[k-1:0] \rangle = j \quad \text{(ind. hypothesis)} \\
&\leftrightarrow \langle x[n-1:k]x[k-1:0] \rangle = i \cdot 2^k + j. \quad \text{(Lemma 18)}
\end{aligned}
$$

5.3 Clocked Circuits

So far we have not treated storage elements. Now we introduce registers, the simplest possible storage elements capable of storing a single bit, and construct a computa-

$$a[n-1:0]$$

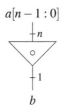

$$b$$

(a) symbol

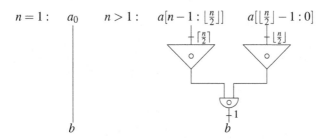

(b) implementation

Fig. 16. n-bit $\circ$-tree of gates for $\circ \in \{\wedge, \vee, \oplus\}$

tional model involving both circuits and these registers. All hardware constructions in this book, including the construction of an entire processor with operating system support, will be done in this model. Of course, we will construct more comfortable storage elements, namely n-bit registers and various flavors of random access memories, on the way.

A *digital clocked circuit* or for short *clocked circuit*, as illustrated in Fig. 21, has four components:

- a special *reset* input,
- special 0 and 1 inputs,
- a sequence $x[n-1:0]$ of 1-bit registers, and
- a circuit with inputs $x[n-1:0]$, *reset*, 0, and 1; and outputs $x[n-1:0]in$ and $x[n-1:0]ce$.

Each register $x[i]$ has

- a data input $x[i]in$,
- a clock enable input $x[i]ce$, and
- a register value $x[i]$, which is also the output signal of the register.

In the digital model we assume that register values as well as all other signals are always in $\mathbb{B}$.

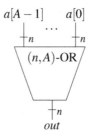

(a) symbol

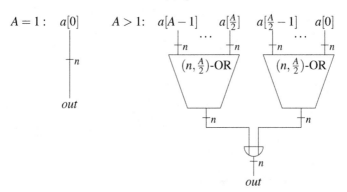

(b) implementation

Fig. 17. Recursive construction of an (n,A)-OR tree

A *hardware configuration* h of a clocked circuit is a snapshot of the current values of the registers:

$$h = x[n-1:0] \in \mathbb{B}^n.$$

A *hardware computation* is a sequence of hardware configurations where the next configuration h' is computed from the current configuration h and the current value of the *reset* signal by a *next hardware configuration* function δ_H:

$$h' = \delta_H(h, reset).$$

In a hardware computation, we count *cycles* (steps of the digital model) using natural numbers $t \in \mathbb{N} \cup \{-1\}$. The hardware configuration in cycle t of a hardware computation is denoted by $h^t = x^t[n-1:0]$ and the value of signal y during cycle t is denoted by y^t.

The values of the *reset* signal are fixed. Reset is on in cycle -1 and off ever after:

$$reset^t = \begin{cases} 1, & t = -1, \\ 0, & t \geq 0. \end{cases}$$

(a) symbol

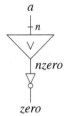

(b) implementation

Fig. 18. *n*-bit zero tester

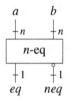

(a) symbol

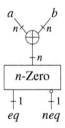

(b) implementation

Fig. 19. *n*-bit equality tester

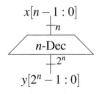

(a) symbol

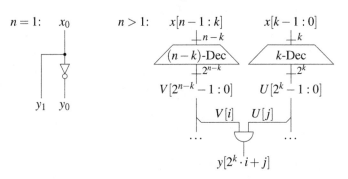

(b) implementation

Fig. 20. n-bit decoder

At power up, register values are binary but unknown.[1] Thus we only know:

$$x^{-1}[n-1:0] \in \mathbb{B}^n.$$

The current value of a circuit signal y in cycle t is defined according to the previously introduced circuit semantics:

$$y^t = \begin{cases} \overline{in1(y)^t}, & y \text{ is an inverter,} \\ in1(y)^t \circ in2(y)^t, & y \text{ is a } \circ\text{-gate.} \end{cases}$$

Let $x[n-1:0]in^t$ and $x[n-1:0]ce^t$ be the register input and clock enable signals computed from the current configuration $x^t[n-1:0]$ and the current value of the reset signal $reset^t$. Then the register value $x^{t+1}[i]$ of the next hardware configuration

$$x^{t+1}[n-1:0] = \delta_H(x^t[n-1:0], reset^t)$$

is defined as

$$x^{t+1}[i] = \begin{cases} x[i]in^t, & x[i]ce^t = 1, \\ x^t[i], & x[i]ce^t = 0, \end{cases}$$

---

[1] In the real world it can happen that a register's value is neither 1 nor 0. This so-called *metastability* however is extremely rare and we do not treat it in this text. For a detailed hardware model including metastability, cf. [KMP14].

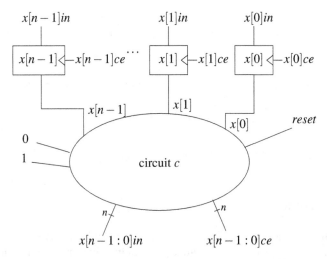

Fig. 21. A digital clocked circuit. Circuit c provides for every register $x[i]$ a pair of output signals: the data input $x[i]in$ and the clock enable $x[i]ce$

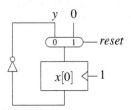

Fig. 22. An example of a clocked circuit with a single register

i.e., when the clock enable signal of register $x[i]$ is active in cycle t, the register value of $x[i]$ in cycle $t+1$ is the value of the data input signal in cycle t; otherwise, the register value does not change.

Example: Consider the digital clocked circuit of Fig. 22. Since the circuit has only one register, we abbreviate $x = x[0]$. At power up, $x^{-1} \in \mathbb{B}$. For cycle -1 we have

$$reset^{-1} = 1,$$
$$xce^{-1} = 1,$$
$$xin^{-1} = 0.$$

Hence, $x^0 = 0$. For cycles $t \geq 0$ we have

$$reset^t = 0,$$
$$xce^t = 1,$$
$$xin^t = y^t.$$

Hence, we get $x^{t+1} = y^t = \overline{x^t}$. An easy induction on t shows that

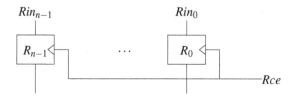

Fig. 23. *n*-bit register

$$\forall t \geq 0 : x^t = (t \bmod 2).$$

5.4 Registers

Although all memory components can be built from 1-bit registers, it is inconvenient to refer to all memory bits in a computer by numbering them with an index i of a clocked circuit input $x[i]$. It is more convenient to deal with hardware configurations h and to gather groups of such bits into certain memory components $h.M$. For M we introduce n-bit registers $h.R$. In Chap. 6 we add to this no fewer than five random access memory (RAM) designs.[2] As before, in a hardware computation with memory components, we have

$$h^{t+1} = \delta_H(h^t, reset^t).$$

An n-bit register R consists simply of n 1-bit registers $R[i]$ with a common clock enable signal Rce as shown in Fig. 23.

Hardware configurations h now have register components

$$h.R \in \mathbb{B}^n,$$

where n is the length of register R. Given input signals $Rin(h^t)$ and $Rce(h^t)$, we obtain from the semantics of the basic clocked-circuit model:

$$h^{t+1}.R = \begin{cases} Rin(h^t), & Rce(h^t) = 1, \\ h^t.R, & Rce(h^t) = 0. \end{cases}$$

Recall, from the initialization rules for 1-bit registers, that after power up register content is binary but unknown:

$$h^{-1}.R \in \mathbb{B}^n.$$

5.5 Final Remarks

We recommend to remember the following few key technical points:

---

[2] In the more advanced text [KMP14] even more memory designs are used.

- the modeling of circuits by directed acyclic graphs.
- the proof that the signals computed are well defined by induction on the depth of the nodes/gates. That the depth exists follows from the bounded path length in finite directed acyclic graphs (you have already memorized the proof of this fact using the pigeonhole principle).
- the extension to the clocked-circuit model. The content of registers $h.R$ is input to a circuit. The outputs of this circuit are the data inputs Rin and clock enable signals Rce of the registers. Signal values in hardware configuration h determine register values in the next hardware configuration h':

$$h'.R = \begin{cases} Rin(h), & Rce(h) = 1, \\ h.R, & \text{otherwise.} \end{cases}$$

- the constructions of basic circuits presented here are completely obvious with one exception: the decoder construction.

A word of caution: hardware is commonly thought 'to be digital'. Of course real gates and registers are physical devices with quite a few extra parameters: propagation delays, set-up times, hold times. Clocked circuits have cycle times. The digital model presented here is an abstraction of physical clocked circuits if the cycle times are large enough and hold times are met. For a long time this was what we call a *folklore theorem*: everybody believes it; nobody has seen the proof. The reader can find a recent fairly straightforward proof of this result in [KMP14]. At the same place the authors also show that drivers and buses *cannot* be modeled by a digital model. They present hardware where bus contention is provably absent in the digital model and present 1/3 of the time in the physical model, resp. the real world. In the real world, unfortunately, bus contention is a nice word for 'short circuit', and circuits which have a 'short circuit 1/3' overheat and destroy themselves.

Because we do not want to introduce a physical circuit model here, we have avoided drivers and buses in this text.

5.6 Exercises

1. Construct circuits computing functions f_1 and f_2.

x_1	x_2	x_3	f_1	f_2
0	0	0	0	1
0	0	1	0	0
0	1	0	1	0
0	1	1	1	1
1	0	0	1	0
1	0	1	0	0
1	1	0	1	1
1	1	1	0	0

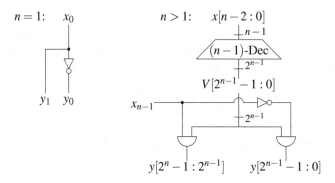

Fig. 24. An alternative decoder construction

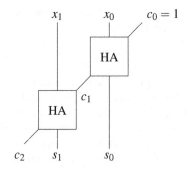

Fig. 25. A 2-bit incrementer

2. Show that the recursive construction in Fig. 24 is a decoder.
3. Prove for the circuit in Fig. 25:

$$\langle c_2 s_1 s_0 \rangle, = \langle x_1, x_0 \rangle + 1,$$
$$\langle s_1 s_0 \rangle, = (\langle x_1, x_0 \rangle + 1 \bmod 4).$$

Hint: proceed as in the correctness proof of the addition algorithm for binary numbers.

4. Recall that we start counting cycles $t \geq 0$ when the reset signal is off

$$reset^t = \begin{cases} 1, & t = -1, \\ 0, & t \geq 0. \end{cases}$$

Prove for the clocked circuit in Fig. 26 for $t \geq 0$:

$$h^t.R = bin_2(t \bmod 4).$$

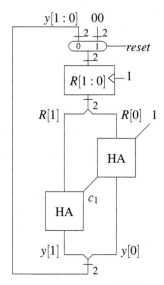

Fig. 26. A modulo 4 counter

6

Five Designs of Random Access Memory (RAM)

Memory components play an important role in the construction of a machine. We start in Sect. 6.1 with a basic construction of (static) random access memory (RAM). Next, we derive four specialized designs: read-only memory (ROM) in Sect. 6.2, combined ROM and RAM in Sect. 6.3, general-purpose register RAM (GPR-RAM) in Sect. 6.4 and special purpose register RAM (SPR-RAM) in Sect. 6.5.

For the correctness proof of a RAM construction, we consider a hardware configuration h which has the abstract state of the RAM $h.S$ as well as the hardware components implementing this RAM. The abstract state of the RAM is coupled with the state of its implementation by means of an *abstraction relation*. Given that both the abstract RAM specification and RAM implementation have the same inputs, we show that their outputs are also always the same.

The material in this section builds clearly on [MP00]. Correctness proofs for the various flavors of RAM are quite similar. Thus, if one lectures on this material, it suffices to present only a few of them in the classroom.

6.1 Basic Random Access Memory

As illustrated in Fig. 27(a), an (n, a)-static RAM S or SRAM is a portion of a clocked circuit with the following inputs and outputs:

- an n-bit data input Sin,
- an a-bit address input Sa,
- a write signal Sw, and
- an n-bit data output $Sout$.

Internally, the static RAM contains 2^a n-bit registers, thus it is modeled as a function

$$h.S : \mathbb{B}^a \to \mathbb{B}^n.$$

The initial content of the SRAM is unknown, thus after the reset for all $x \in \mathbb{B}^a$ we have $h^0.S(x) \in \mathbb{B}^n$. For addresses $x \in \mathbb{B}^a$ we define the next state transition function for SRAM as

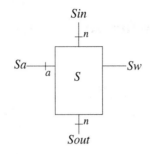

(a) symbol

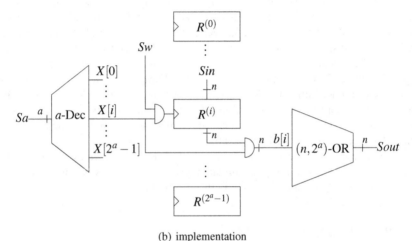

(b) implementation

Fig. 27. (n,a)-RAM

$$h'.S(x) = \begin{cases} Sin(h), & Sa(h) = x \wedge Sw(h) = 1, \\ h.S(x), & \text{otherwise.} \end{cases}$$

Finally, the output of the SRAM is the register content selected by the address input:

$$Sout(h) = h.S(Sa(h)).$$

The implementation of an SRAM is shown in Fig. 27(b). We use 2^a n-bit registers $R^{(i)}$ with $i \in [0 : 2^a - 1]$ and an a-decoder with outputs $X[2^a - 1 : 0]$ satisfying

$$X(i) = 1 \leftrightarrow i = \langle Sa(h) \rangle.$$

The inputs of register $R^{(i)}$ are defined as

$$h.R^{(i)} in = Sin(h),$$
$$h.R^{(i)} ce = Sw(h) \wedge X[i].$$

For the next state computation we get

$$h'.R^{(i)} = \begin{cases} Sin(h), & Sw(h) \wedge i = \langle Sa(h) \rangle, \\ h.R^{(i)}, & \text{otherwise.} \end{cases}$$

The i'th input vector $b[i]$ to the OR-tree is constructed as

$$b[i] = X[i] \wedge h.R^{(i)}$$
$$= \begin{cases} h.R^{(i)}, & i = \langle Sa(h) \rangle, \\ 0^n, & \text{otherwise.} \end{cases}$$

Thus, the output of the SRAM is given by

$$Sout(h) = \bigvee_{i=0}^{2^a-1} b[i]$$
$$= h.R^{(\langle Sa(h) \rangle)}.$$

As a result, when we choose

$$h.S(x) = h.R^{(\langle x \rangle)}$$

as the defining equation of our abstraction relation, the presented construction implements an SRAM.

6.2 Read-Only Memory (ROM)

An (n,a)-ROM is a memory with a drawback and an advantage. The drawback: it can only be read. The advantage: its content is known after power up. It is modeled by a mapping

$$S : \mathbb{B}^a \to \mathbb{B}^n,$$

which does not depend on the hardware configuration h. The construction is obtained by a trivial variation of the basic RAM design from Fig. 27(b): replace each register $R^{(i)}$ by the constant input $S(bin_a(i)) \in \mathbb{B}^n$. Since the ROM cannot be written, there are no data input, write, or clock enable signals; the hardware constructed in this way is a circuit. Symbol and construction are given in Fig. 28.

6.3 Combining RAM and ROM

It is often desirable to implement some small portion of memory by ROM and the remaining large part as RAM. The standard use for this is to store boot code in the ROM. Since, after power up, the memory content of RAM is unknown, computation will not start in a meaningful way unless at least *some* portion of memory contains code that is known after power up. The reset mechanism of the hardware ensures

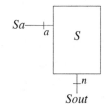

(a) symbol

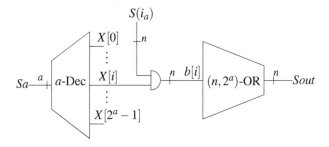

(b) implementation

Fig. 28. (n,a)-ROM

that processors start by executing the program stored in the ROM. This code usually contains a so-called *boot loader* which accesses a large and slow memory device — such as a disk — to load further programs to be executed next, e.g., an operating system.

For $r < a$ we define a combined (n,r,a)-RAM-ROM S as a device that behaves for small addresses $a = 0^{a-r}b$ with $b \in \mathbb{B}^r$ like ROM and on the other addresses like RAM. Just like an ordinary (n,a)-RAM, we model the state of the (n,r,a)-RAM-ROM as

$$h.S : \mathbb{B}^a \to \mathbb{B}^n.$$

Write operations, however, affect only addresses larger than $0^{a-r}1^r$:

$$h'.S(x) = \begin{cases} h^0.S(x), & x[a-1:r] = 0^{a-r}, \\ Sin(h), & x = Sa(h) \wedge Sw(h) \wedge x[a-1:r] \neq 0^{a-r}, \\ h.S(h), & \text{otherwise.} \end{cases}$$

The outputs are defined simply as

$$Sout(h) = h.S(Sa(h)).$$

The symbol for an (n,r,a)-RAM-ROM and a straightforward implementation involving an (n,a)-SRAM, an (n,r)-ROM, and an $(a-r)$-zero tester is shown in Fig. 29.

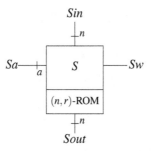

(a) symbol

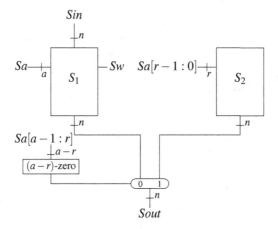

(b) implementation

Fig. 29. (n, r, a)-RAM-ROM

6.4 Three-Port RAM for General-Purpose Registers

An (n, a)-GPR-RAM is a three-port RAM that we use later for general-purpose registers. As shown in Fig. 30(a), it has the following inputs and outputs:

- an n-bit data input Sin,
- three a-bit address inputs Sa, Sb, Sc,
- a write signal Sw, and
- two n-bit data outputs $Souta, Soutb$.

As for ordinary SRAM, the state of the three-port RAM is a mapping

$$h.S : \mathbb{B}^a \to \mathbb{B}^n.$$

However, register 0 has always the content 0^n. We model this by i) requiring an initial content of 0^n

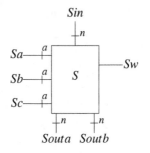

(a) symbol

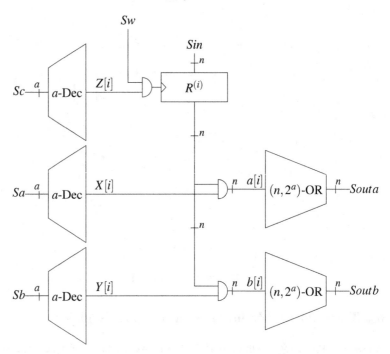

(b) implementation

Fig. 30. (n,a)-GPR-RAM

$$h.S(0^a) = 0^n$$

and ii) by specifying that writes to register 0 have no effect. Writing is performed under control of address input $Sc(h)$:

$$h'.S(x) = \begin{cases} Sin(h), & Sc(h) = x \wedge Sw(h) = 1 \wedge x \neq 0^a, \\ h.S(x), & \text{otherwise,} \end{cases}$$

s.t. a trivial proof by induction gives for all t

$$h^t.S(0^a) = 0^n.$$

Reads are controlled by address inputs $Sa(h)$ and $Sb(h)$:

$$Souta(h) = h.S(Sa(h)),$$
$$Soutb(h) = h.S(Sb(h)).$$

The implementation shown in Fig. 30(b) is a straightforward variation of the design for ordinary SRAM. One uses three different a-decoders with outputs $X[0 : 2^a - 1]$, $Y[0 : 2^a - 1]$, $Z[0 : 2^a - 1]$ satisfying

$$X[i] = 1 \leftrightarrow i = \langle Sa(h) \rangle,$$
$$Y[i] = 1 \leftrightarrow i = \langle Sb(h) \rangle,$$
$$Z[i] = 1 \leftrightarrow i = \langle Sc(h) \rangle.$$

Clock enable signals are derived from the decoded Sc address:

$$R^{(i)}ce = Z[i] \wedge Sw(h).$$

Outputs $Souta$, $Soutb$ are generated by two $(n, 2^a)$-OR trees with inputs $a[i]$, $b[i]$ satisfying

$$a[i] = X[i] \wedge h.R^{(i)},$$
$$b[i] = Y[i] \wedge h.R^{(i)}.$$

Formally, the outputs are

$$Souta(h) = \bigvee a[i],$$
$$Soutb(h) = \bigvee b[i].$$

Register $R^{(0)}$ is replaced by the constant signal 0^n.

6.5 SPR-RAM

An (n, a)-SPR-RAM as shown in Fig. 31(a) is used for the realization of special purpose register files. It behaves both as an (n, a)-RAM and as a set of 2^a n-bit registers. It has the following inputs and outputs:

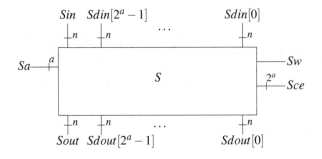

(a) symbol

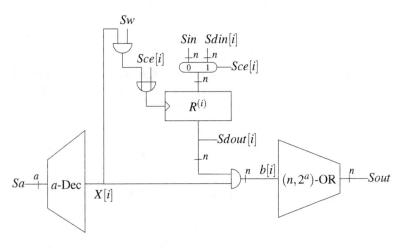

(b) implementation

Fig. 31. (n,a)-SPR-RAM

- an n-bit data input Sin,
- an a-bit address input Sa,
- an n-bit data output $Sout$,
- a write signal Sw,
- for each $i \in [0 : 2^a - 1]$ an individual n-bit data input $Sdin[i]$ for register $R^{(i)}$,
- for each $i \in [0 : 2^a - 1]$ an individual clock enable signal $Sce[i]$ for register $R^{(i)}$, and
- for each $i \in [0 : 2^a - 1]$ an individual n-bit data output $Sdout[i]$ for register $R^{(i)}$.

Register updates to $R^{(i)}$ can be performed either by Sin for regular writes or by $Sdin[i]$ if the special clock enable signals are activated. Special writes take precedence over ordinary writes:

$$h'.S(x) = \begin{cases} Sdin(h)[\langle x \rangle], & Sce(h)[\langle x \rangle] = 1, \\ Sin(h), & Sce(h)[\langle x \rangle] = 0 \wedge x = Sa(h) \wedge Sw(h) = 1, \\ h.S(x), & \text{otherwise.} \end{cases}$$

Ordinary data output is generated as usual, and the individual data outputs are simply the outputs of the internal registers:

$$Sout(h) = h.S(Sa(h)),$$
$$Sdout(h)[i] = h.S(bin_a(i)).$$

For implementation a single address decoder with outputs $X[i]$ and a single OR-tree suffice. Fig. 31(b) shows the construction satisfying

$$R^{(i)}ce = Sce(h)[i] \vee X[i] \wedge Sw(h),$$
$$R^{(i)}in = \begin{cases} Sdin(h)[i], & Sce(h)[i] = 1, \\ Sin(h), & \text{otherwise.} \end{cases}$$

6.6 Final Remarks

We recommend to remember as key technical points

- the specification and construction of basic RAM from Sect. 6.1.
- for multi-port RAM you need an address decoder per port and an OR-tree per output.
- in a RAM which allows multiple writes at the same time — such as the SPR-RAM — these writes can go to the same address. The specification must either forbid these situations or specify what happens. If it is allowed like in our SPR-RAM, the hardware implementation obviously should match the specification (priority to one of the writes).

For the construction of cache coherent shared-memory systems many more multi-port cache designs are used [KMP14], but their construction follows exactly the same pattern.

6.7 Exercises

1. Specify and construct (n,a)-RAM with an extra control input *inv*. Activation of *inv* sets all register contents to 0^n and overrides possible concurrent ordinary writes. Such RAMs are used for instance to store the status of cache lines. Invalid data is signaled by the value 0^n. After power up caches contain no meaningful data, and this is signaled in their status if one activates *inv* together with *reset*.
2. Specify and construct an (n,a)-RAM with two ports A and B which can both be used to read and write. In case of simultaneous writes to the same address the write specified by port A should win.

3. Unless you work closely together with a chip manufacturer you have no control over the gate-level design of RAMs. So you have to construct specialized SRAMs from basic SRAMs. Construct a three-port RAM for general-purpose registers from two basic SRAMS RAM_A and RAM_B plus some extra circuitry. Hint: keep the data in the RAMs identical (except for register 0). For a read from address 0_a force the corresponding output to 0_n.

7

Arithmetic Circuits

In this chapter we collect several basic circuits which are used in the construction of processors with the MIPS instruction set architecture (ISA).

In Sect. 7.1 various flavors of adders and incrementers are presented. In particular, in Sect. 7.1.4 we construct so-called carry-look-ahead adders, which for arguments of length n have cost linear in n and depth logarithmic in n. The slightly advanced construction of these adders is based on the parallel prefix circuits from Sect. 7.1.3. It can be skipped, but we recommend to cover this construction because it is highly useful in practical hardware design.

An arithmetic unit (AU) for binary and two's complement numbers is studied in Sect. 7.2. Understanding the proofs of this section, in our view, is a must for anyone wishing to understand fixed-point arithmetic. With the help of the AU we construct in Sect. 7.3 an arithmetic logic unit (ALU) for the MIPS ISA in a straightforward way. Differences to [MP00] are due to differences in the encoding of ALU operations between the ISA considered here (MIPS) and the ISA considered in [MP00] (DLX).

Section 7.4 contains a rudimentary shifter construction supporting only *logical right shifts*. In textbooks on computer architecture — such as [MP00, KMP14] — the counterpart of this chapter includes general shifter constructions. Shifter constructions are interesting and in lectures on computer architecture we cover them in the classroom. In the short and technical Sect. 7.5 we construct a branch condition evaluation (BCE) unit, which is later used to implement the branch instructions of the MIPS ISA. In lectures this section should be covered by a reading assignment.

7.1 Adder and Incrementer

An *n-adder* is a circuit with inputs $a[n-1:0] \in \mathbb{B}^n$, and $b[n-1:0] \in \mathbb{B}^n$, $c_0 \in \mathbb{B}$ and outputs $c_n \in \mathbb{B}$ and $s[n-1:0] \in \mathbb{B}^n$ satisfying the specification

$$\langle c_n, s[n-1:0] \rangle = \langle a[n-1:0] \rangle + \langle b[n-1:0] \rangle + c_0.$$

An *n-incrementer* is a circuit with inputs $a[n-1:0] \in \mathbb{B}^n$ and $c_0 \in \mathbb{B}$ and outputs $c_n \in \mathbb{B}$ and $s[n-1:0] \in \mathbb{B}^n$ satisfying the specification

© Springer International Publishing Switzerland 2016
W.J. Paul et al., *System Architecture*, DOI 10.1007/978-3-319-43065-2_7

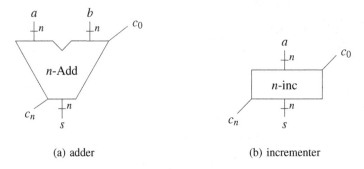

(a) adder   (b) incrementer

Fig. 32. Symbols for n-bit adder and incrementer

$$\langle c_n, s[n-1:0]\rangle = \langle a[n-1:0]\rangle + c_0.$$

Clearly, constructions for the 1-adder and 1-incrementer are given by the *full* and *half adder* resp. from Sect. 5.2. In the following, we present constructions for n-adders and n-incrementers. For n-adders we use the symbol from Fig. 32(a), while for n-incrementers we use the symbol from Fig. 32(b).

7.1.1 Carry Chain Adder and Incrementer

A recursive construction of a very simple carry chain adder is shown in Fig. 33(a). Correctness follows directly from the correctness of the basic addition algorithm for binary numbers (Lemma 20).

Obviously, n-incrementers can be constructed from n-adders by tying the b input to 0^n. As shown in Sect. 5.2, a full adder whose b input is tied to zero can be replaced with a half adder. This yields the construction of carry chain incrementers shown in Fig. 33(b).

In Sect. 5.1 we denoted the cost and delay of circuit S by $c(S)$ resp. $d(S)$. We denote the complexity of n-bit carry chain adders by $c(n)$ and their depth by $d(n)$, and read off from the constructions the following so-called *difference equations*:

$$c(1) = c(FA),$$
$$c(n) = c(n-1) + c(FA),$$
$$d(1) = d(FA),$$
$$d(n) \le d(n-1) + d(FA).$$

A trivial induction shows

$$c(n) = n \cdot c(FA),$$
$$d(n) \le n \cdot d(FA).$$

Replacing full adders by half adders one can make a completely analogous argument for carry chain incrementers. Thus we can summarize the asymptotic complexity of these circuits as follows.

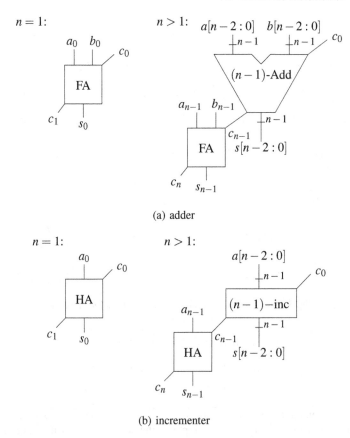

(a) adder

(b) incrementer

Fig. 33. Recursive constructions of carry chain adder and incrementer

Lemma 33. *n-bit carry chain adders and incrementers have cost $O(n)$ and delay $O(n)$.*

Linear cost $O(n)$ is all one can hope for. At first sight, the delay formula also seems optimal, because by definition delay appears to be inherently sequential. One is tempted to argue that one cannot compute the carry c_n at position n before one knows the carry c_{n-1} at position $n-1$. But trying to turn this argument into a proof fails because there are *much* faster adder constructions. The simplest ones are so-called n-bit *conditional-sum adders* or for short n-CSAs.

7.1.2 Conditional-Sum Adders

We define conditional-sum adders inductively. For a 1-CSA we simply take a full adder. For even n the construction of an n-CSA from three $n/2$-CSAs is shown in Fig. 34.

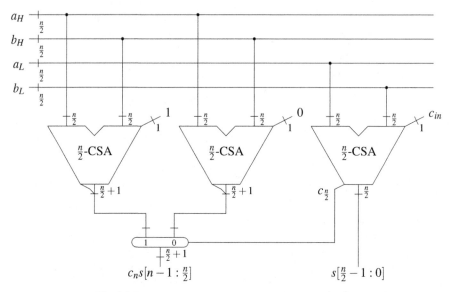

Fig. 34. Recursive construction of a conditional-sum adder

One splits the inputs $a, b \in \mathbb{B}^n$ into upper and lower halves:

$$a_H = a[n-1:\tfrac{n}{2}], \qquad a_L = a[\tfrac{n}{2}-1:0],$$
$$b_H = b[n-1:\tfrac{n}{2}], \qquad b_L = b[\tfrac{n}{2}-1:0].$$

Then, one exploits the fact that the carry of the binary addition of a_L and b_L is either 0 or 1: instead of using carry $c_{\frac{n}{2}}$ as an input to the computation of the upper half of the sum, one simply computes binary representations of both $\langle a_H \rangle + \langle b_H \rangle + 0$ and $\langle a_H \rangle + \langle b_H \rangle + 1$. One of these two will be the correct result, which is chosen with a multiplexer controlled by signal $c_{\frac{n}{2}}$. Correctness of this construction follows from the simple Lemma 21. Now we use now $c(n)$ and $d(n)$ to denote the cost and depth of an n-CSA and obtain from the construction the difference equations:

$$d(1) = d(FA),$$
$$d(n) = d(n/2) + d(MUX),$$
$$c(1) = c(FA),$$
$$c(n) = 3 \cdot c(n/2) + c((n/2+1)\text{-}MUX).$$

In general, difference (in)equations are solved in two steps: i) guessing the solution and ii) showing by induction that the guessed solution satisfies the equations. As is usual in proofs by induction, finding the induction hypothesis is the hard part. Fortunately, there is a simple heuristic that will work for us: we repeatedly apply the inductive definition until we see enough to guess the solution. Let

$$n = 2^k$$

be a power of two so that starting from n we can apply the recursive construction repeatedly until we arrive at $n = 1$. The heuristics gives for the depth of CSA (here we take $d(MUX) = 3$)

$$\begin{aligned} d(n) &= d(n/2) + 3 \\ &= d(n/4) + 3 \cdot 2 \\ &= d(n/8) + 3 \cdot 3. \end{aligned}$$

At this point we might guess for all x:

$$d(n) = d(n/2^x) + 3 \cdot x,$$

which for

$$x = k = \log n$$

gives

$$d(n) = d(FA) + d(MUX) \cdot \log n.$$

That this guess indeed satisfies the difference equations is a trivial exercise that we omit. Unfortunately, conditional-sum adders are quite expensive. Repeated application of the lower bound for their cost gives

$$\begin{aligned} c(n) &\geq 3 \cdot c(n/2) \\ &\geq 3^2 \cdot c(n/4) \\ &\geq 3^x \cdot c(n/2^x) \\ &\geq 3^{\log n} \cdot c(1) \\ &= 3^{\log n} \cdot c(FA). \end{aligned}$$

Again, proving by induction that this guess satisfies the difference equations is easy. We transform the term $3^{\log n}$:

$$\begin{aligned} 3^{\log n} &= (2^{\log 3})^{\log n} \\ &= 2^{(\log 3) \cdot (\log n)} \\ &= (2^{\log n})^{\log 3} \\ &= n^{\log 3}. \end{aligned}$$

We summarize the estimates as follows.

Lemma 34. *n-CSAs have cost at least $C(FA) \cdot n^{\log 3}$ and delay $O(\log n)$.*

As $\log 3 \approx 1.58$ the cost estimate is way above the cost of carry chain adders.

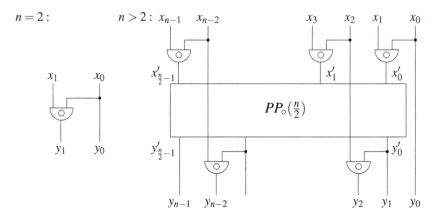

Fig. 35. Recursive construction of an n-bit parallel prefix circuit of the function $\circ$ for an even n

7.1.3 Parallel Prefix Circuits

In life and in science sometimes extremely good things are possible. In life it is possible to be rich and healthy at the same time. In adder construction it is possible to achieve the asymptotic cost of carry chain adders and the asymptotic delay of conditional-sum adders at the same time in so-called n-bit *carry-look-ahead adders* or for short n-CLAs. The fundamental auxiliary circuits that are used in their construction are called n-*parallel prefix circuits*.

An n-parallel prefix circuit for an associative function $\circ : M \times M \to M$ is a circuit with inputs $x[n-1:0] \in M^n$ and outputs $y[n-1:0] \in M^n$ satisfying

$$y_0 = x_0,$$
$$y_{i+1} = x_{i+1} \circ y_i.$$

We construct a slightly generalized circuit model where signals take values in M and computation is done by gates for function $\circ$. Cost and delay in this model are defined by counting $\circ$-gates in the circuit resp. on the longest path. Later we use the construction with $M = \mathbb{B}^2$. For even n a recursive construction of an n-parallel prefix circuit based on $\circ$-gates is shown in Fig. 35. For odd n one can realize an $(n-1)$-bit parallel prefix circuit from Fig. 35 and compute output y_{n-1} using one extra $\circ$-gate:

$$y_{n-1} = x_{n-1} \circ y_{n-2}.$$

For the correctness of the construction we first observe that

$$x_i' = x_{2i+1} \circ x_{2i},$$
$$y_{2i} = x_{2i} \circ y_{i-1}',$$
$$y_{2i+1} = y_i'.$$

We first show that odd outputs of the circuit are computed correctly. For $i = 0$ we have

$$
\begin{aligned}
y_1 &= y_0' && \text{(construction)} \\
&= x_0' && \text{(ind. hypothesis } PP_\circ(n/2)) \\
&= x_1 \circ x_0 && \text{(construction)}.
\end{aligned}
$$

For $i > 0$ we conclude

$$
\begin{aligned}
y_{2i+1} &= y_i' && \text{(construction)} \\
&= x_i' \circ y_{i-1}' && \text{(ind. hypothesis } PP_\circ(n/2)) \\
&= (x_{2i+1} \circ x_{2i}) \circ y_{i-1}' && \text{(construction)} \\
&= x_{2i+1} \circ (x_{2i} \circ y_{i-1}') && \text{(associativity)} \\
&= x_{2i+1} \circ y_{2i} && \text{(construction)}.
\end{aligned}
$$

For even outputs of the circuit we easily conclude

$$
\begin{aligned}
y_0 &= x_0 && \text{(construction)}, \\
i > 0 : \quad y_{2i} &= x_{2i} \circ y_{i-1}' && \text{(construction)} \\
&= x_{2i} \circ y_{2i-1} && \text{(construction)}.
\end{aligned}
$$

We denote by $c(n)$ and $d(n)$ the cost and delay of an n-parallel prefix circuit constructed in this way. From the construction we read off the difference equations:

$$
\begin{aligned}
d(2) &= 1, \\
d(n) &= d(n/2) + 2, \\
c(2) &= 1, \\
c(n) &\leq c(n/2) + n.
\end{aligned}
$$

Applying the heuristics for the depth gives

$$
d(n) = d(n/2^x) + 2 \cdot x.
$$

The recursion must be stopped when

$$
\begin{aligned}
n/2^x &= 2 \\
2^x &= n/2 \\
x &= \log n - 1.
\end{aligned}
$$

This gives the conjecture

$$
d(n) = d(2) + 2 \cdot (\log n - 1) = 2 \cdot \log n - 1,
$$

which can easily be verified by induction.

For the cost we get the estimate

$$c(n) \leq c(n/2) + n$$
$$\leq c(n/4) + n + n/2$$
$$\leq c(n/8) + n + n/2 + n/4$$
$$\leq c(n/x) + n \cdot \sum_{i=0}^{x-1} (1/2)^i$$
$$\leq c(2) + n \cdot \sum_{i=0}^{\log n - 2} (1/2)^i.$$

We guess

$$c(n) \leq 2 \cdot n,$$

which can be verified by an easy induction proof. We summarize the estimates as follows.

Lemma 35. *There are n-parallel prefix circuits with cost $O(n)$ and depth $O(\log n)$.*

7.1.4 Carry-Look-Ahead Adders

For $a[n-1:0] \in \mathbb{B}^n$, $b[n-1:0] \in \mathbb{B}^n$, and indices $i \geq j$ we define

$$p_{i,j}(a,b) \equiv \langle a[i:j] \rangle + \langle b[i:j] \rangle = \langle 1^{i-j+1} \rangle,$$

$$g_{i,j}(a,b) \equiv \begin{cases} \langle a[i:j] \rangle + \langle b[i:j] \rangle \geq \langle 10^{i-j+1} \rangle, & j > 0, \\ \langle a[i:j] \rangle + \langle b[i:j] \rangle + c_0 \geq \langle 10^{i-j+1} \rangle, & j = 0. \end{cases}$$

We abbreviate $p_{i,j}(a,b)$ and $g_{i,j}(a,b)$ by $p_{i,j}$ and $g_{i,j}$ resp. If predicate $p_{i,j}$ holds, we say that carry c_j is *propagated* from positions $[j:i]$ of the operands to position $i+1$, s.t.

$$c_{i+1} = c_j.$$

If predicate $g_{i,j}$ holds, we say that positions $[j:i]$ of the operands *generate* a carry independent of c_j. For $j = i$ we observe:

$$p_{i,i} = a_i \oplus b_i,$$

$$g_{i,i} = \begin{cases} a_i \wedge b_i, & i > 0, \\ a_0 \wedge b_0 \vee a_0 \wedge c_0 \vee b_0 \wedge c_0, & i = 0. \end{cases}$$

Now consider indices k splitting interval $[j:i]$ of indices into two adjacent intervals $[j:k]$ and $[k+1:i]$, and assume that we already know functions $g_{i,k+1}$, $p_{i,k+1}$, $g_{k,j}$, and $p_{k,j}$. Then the signals for the joint interval $[j:i]$ can be computed as

$$p_{i,j} = p_{i,k+1} \wedge p_{k,j},$$

$$g_{i,j} = g_{i,k+1} \vee g_{k,j} \wedge p_{i,k+1}.$$

This computation can be performed by the circuit in Fig. 37, which takes inputs $(g_2, p_2) \in M$ and $(g_1, p_1) \in M$ and produces as output $(g, p) \in M$ (here $M = \mathbb{B}^2$):

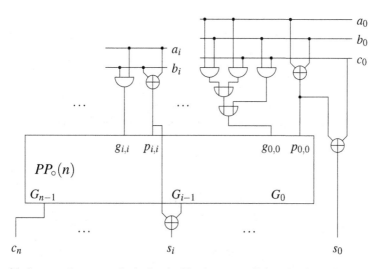

Fig. 36. Constructing a carry-look-ahead adder from a parallel prefix circuit for $\circ$-gates

$$(g,p) = (g_2 \vee g_1 \wedge p_2, p_2 \wedge p_1)$$
$$= (g_2, p_2) \circ (g_1, p_1).$$

An easy calculation shows that the function $\circ$ defined in this way is associative (for details see, e.g., [KP95]). Hence we can substitute in Fig. 35 the $\circ$-gates by the circuit of Fig. 37. One $\circ$-gate now produces a cost of three and a delay of two ordinary gates. For the resulting parallel prefix circuit n-GP we conclude

$$d(n\text{-}GP) \leq 4 \cdot \log n - 2,$$
$$c(n\text{-}GP) \leq 6 \cdot n.$$

The point of the construction is that output i of circuit n-GP computes

$$(G_i, P_i) = (g_{i,i}, p_{i,i}) \circ \ldots \circ (g_{0,0}, p_{0,0}) = (g_{i,0}, p_{i,0}) = (c_{i+1}, p_{i,0}),$$

thus the circuit from Fig. 36 is an n-adder. We summarize as follows.

Lemma 36. *n-CLAs have cost $O(n)$ and delay $O(\log n)$.*

7.2 Arithmetic Unit

The symbol of an n-arithmetic unit or for short n-AU is shown in Fig. 38. It is a circuit with the following inputs:

- operands $a = a[n-1:0] \in \mathbb{B}^n$ and $b = b[n-1:0] \in \mathbb{B}^n$,

Fig. 37. Circuit replacing ∘-gate in the parallel prefix circuit

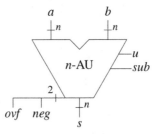

Fig. 38. Symbol of an n-arithmetic unit

- control signal $u \in \mathbb{B}$ indicating whether operands have to be treated as unsigned integers,[1]
- control signal $sub \in \mathbb{B}$ indicating whether input b has to be subtracted from input a,

and the following outputs:

- result $s[n-1:0] \in \mathbb{B}^n$,
- negative bit $neg \in \mathbb{B}$, and
- overflow bit $ovf \in \mathbb{B}$.

We define the *exact result* $S \in \mathbb{Z}$ of an arithmetic unit as

$$S = \begin{cases} S_0, & u = 0, \\ S_1, & u = 1, \end{cases}$$

where

$$S_0 = \begin{cases} [a] + [b], & sub = 0, \\ [a] - [b], & sub = 1, \end{cases} \qquad S_1 = \begin{cases} \langle a \rangle + \langle b \rangle, & sub = 0, \\ \langle a \rangle - \langle b \rangle, & sub = 1. \end{cases}$$

For the result of the AU, we pick the representative of the exact result in T_n resp. B_n and represent it in the corresponding format

$$s = \begin{cases} twoc_n(S \text{ tmod } 2^n), & u = 0, \\ bin_n(S \bmod 2^n), & u = 1, \end{cases}$$

---

[1] When $u = 0$, we treat operands as signed integers represented in two's complement number format, otherwise we treat them as unsigned integers represented in binary.

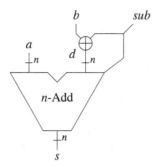

Fig. 39. Data paths of an n-arithmetic unit

i.e., we require

$$[s] = (S \text{ tmod } 2^n) \text{ if } u = 0,$$

$$\langle s \rangle = (S \bmod 2^n) \text{ if } u = 1.$$

Negative and overflow signals are defined with respect to the exact result. The overflow bit is computed only for the case of two's complement numbers; for binary numbers it is always 0 since the architecture we introduce later does not consider unsigned overflows:

$$neg \leftrightarrow S < 0,$$

$$ovf \leftrightarrow \begin{cases} S \notin T_n, & u = 0, \\ 0, & u = 1. \end{cases}$$

Data Paths

The main data paths of an n-AU are shown in Fig. 39. The following lemma asserts that the sum bits are computed correctly for binary numbers ($u = 1$).

Lemma 37. *The sum bits $s[n-1:0]$ in Fig. 39 satisfy*

$$\langle s \rangle = S_1 \bmod 2^n.$$

Proof. From the construction of the circuit, we have

$$d = b \oplus sub$$

$$= \begin{cases} b, & sub = 0, \\ \overline{b}, & sub = 1. \end{cases}$$

From the specification of an n-adder, Lemma 19, and the subtraction algorithm for binary numbers (Lemma 24), we conclude

$$\langle s \rangle = \left(\left(\begin{cases} \langle a \rangle + \langle b \rangle, & sub = 0 \\ \langle a \rangle + \langle \overline{b} \rangle + 1, & sub = 1 \end{cases} \right) \bmod 2^n \right)$$

$$= \left(\left(\begin{cases} \langle a \rangle + \langle b \rangle, & sub = 0 \\ \langle a \rangle - \langle b \rangle, & sub = 1 \end{cases} \right) \bmod 2^n \right)$$

$$= S_1 \bmod 2^n.$$

The following lemma asserts that the result s can be computed independently of u in exactly the same way for both signed and unsigned numbers.

Lemma 38.

$$\langle s \rangle = S_1 \bmod 2^n \ \to \ [s] = S_0 \text{ tmod } 2^n.$$

Proof. By Lemmas 23 and 9 we have:

$$[s] \equiv \langle s \rangle \bmod 2^n$$
$$\equiv S_1 \bmod 2^n$$
$$\equiv S_0 \bmod 2^n.$$

From $[s] \in T_n$ and Lemma 12 we conclude

$$[s] = (S_0 \text{ tmod } 2^n).$$

We introduce special symbols $+_n$ and $-_n$ to denote operations of binary addition and subtraction:

$$a +_n b = bin_n(\langle a \rangle + \langle b \rangle \bmod 2^n),$$
$$a -_n b = bin_n(\langle a \rangle - \langle b \rangle \bmod 2^n).$$

By the last lemma it suffices to specify the sum bits of an arithmetic unit by requiring

$$s = \begin{cases} a +_n b, & sub = 0, \\ a -_n b, & sub = 1. \end{cases}$$

Negative Bit

We start with the case $u = 0$, i.e., with two's complement numbers. We have

$$S = [a] \pm [b]$$
$$= [a] + [d] + sub$$
$$\leq (2^{n-1} - 1) + (2^{n-1} - 1) + 1$$
$$= 2^n - 1,$$
$$S \geq -2^{n-1} - 2^{n-1}$$
$$= -2^n.$$

Thus

$$S \in T_{n+1}.$$

According to Lemma 23 we use sign extension to extend operands to $n+1$ bits:

$$[a] = [a_{n-1}a],$$
$$[d] = [d_{n-1}d].$$

By the basic addition algorithm, the extra sum bit s_n in this case would be

$$s_n = a_{n-1} \oplus d_{n-1} \oplus c_n,$$

s.t. the exact result

$$S = [s[n : 0]].$$

Again by Lemma 23 this is negative if and only if the sign bit s_n is 1:

$$S < 0 \leftrightarrow s_n = 1.$$

For the case $u = 1$, i.e., for binary numbers, a negative result can only occur in the case of subtraction, i.e., if $sub = 1$. In this case we argue along the lines of the correctness proof for the subtraction algorithm:

$$\begin{aligned}
S &= \langle a \rangle - \langle b \rangle \\
&= \langle a \rangle - [0b] \\
&= \langle a \rangle + [1\bar{b}] + 1 \\
&= \langle a \rangle + \langle \bar{b} \rangle + 1 - 2^n \\
&= \langle c_n s[n-1 : 0] \rangle - 2^n \\
&= 2^n \cdot (c_n - 1) + \underbrace{\langle s[n-1 : 0] \rangle}_{\in B_n}.
\end{aligned}$$

Clearly, if $c_n = 1$, we have $S = \langle s \rangle$ and $S \geq 0$. Otherwise, if $c_n = 0$, we have

$$\begin{aligned}
S &= -2^n + \langle s[n-1 : 0] \rangle \\
&\leq -2^n + 2^n - 1 \\
&= -1.
\end{aligned}$$

For binary numbers we conclude that the result is negative iff we subtract and the carry bit c_n is 0:

$$S < 0 \leftrightarrow c_n = 0.$$

We summarize as follows.

Lemma 39.

$$\begin{aligned}
neg &= \bar{u} \wedge (a_{n-1} \oplus d_{n-1} \oplus c_n) \vee \\
&\quad u \wedge sub \wedge \bar{c_n}.
\end{aligned}$$

Overflow Bit

We compute the overflow bit only for the case of two's complement numbers, i.e., when $u = 0$. We have

$$
\begin{aligned}
S &= [a] + [d] + sub \\
&= -2^{n-1}(a_{n-1} + d_{n-1}) + \langle a[n-2:0] \rangle + \langle d[n-2:0] \rangle + sub \\
&= -2^{n-1}(a_{n-1} + d_{n-1} + c_{n-1}) + 2^{n-1}c_{n-1} + \langle c_{n-1}s[n-2:0] \rangle \\
&= -2^{n-1}\langle c_n s_{n-1} \rangle + 2^{n-1}(c_{n-1} + c_{n-1}) + \langle s[n-2:0] \rangle \\
&= -2^n c_n - 2^{n-1}s_{n-1} + 2^n c_{n-1} + \langle s[n-2:0] \rangle \\
&= -2^n(c_n - c_{n-1}) + [s[n-1:0]].
\end{aligned}
$$

Obviously, if $c_n = c_{n-1}$, we have $S = [s]$ and $S \in T_n$. Otherwise, if $c_n = 1$ and $c_{n-1} = 0$, we have

$$
\begin{aligned}
S &= -2^n + [s] \\
&\le -2^n + 2^{n-1} - 1 \\
&= -2^{n-1} - 1 \\
&< -2^{n-1},
\end{aligned}
$$

and if $c_n = 0$ and $c_{n-1} = 1$, we have

$$
\begin{aligned}
S &= 2^n + [s] \\
&\ge 2^n - 2^{n-1} \\
&> 2^{n-1} - 1.
\end{aligned}
$$

For two's complement numbers we conclude that the result overflows iff the carry bits c_n and c_{n-1} differ:

$$
S \notin T_n \leftrightarrow c_{n-1} \ne c_n,
$$

and summarize as follows.

Lemma 40.

$$
ovf = \bar{u} \wedge (c_n \oplus c_{n-1}).
$$

7.3 Arithmetic Logic Unit (ALU)

Fig. 40 shows the symbol for the n-ALU constructed here. Width n should be even. The circuit has the following inputs:

- operand inputs $a = a[n-1:0] \in \mathbb{B}^n$ and $b = b[n-1:0] \in \mathbb{B}^n$,
- control inputs $af[3:0] \in \mathbb{B}^4$ and $i \in \mathbb{B}$ specifying the operation that the ALU performs with the operands,

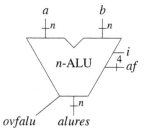

Fig. 40. Symbol of an n-arithmetic logic unit

$af[3:0]$	i	$alures[n-1:0]$	$ovfalu$
0000	*	$a +_n b$	$[a] + [b] \notin T_n$
0001	*	$a +_n b$	0
0010	*	$a -_n b$	$[a] - [b] \notin T_n$
0011	*	$a -_n b$	0
0100	*	$a \wedge b$	0
0101	*	$a \vee b$	0
0110	*	$a \oplus b$	0
0111	0	$\overline{a \vee b}$	0
0111	1	$b[n/2-1:0]0^{n/2}$	0
1010	*	$0^{n-1}([a] < [b] \,?\, 1:0)$	0
1011	*	$0^{n-1}(\langle a \rangle < \langle b \rangle \,?\, 1:0)$	0

Table 7. Specification of ALU operations

and the following outputs:

- result $alures[n-1:0] \in \mathbb{B}^n$,
- overflow bit $ovfalu \in \mathbb{B}$.

The results that must be generated are specified in Table 7. There are three groups of operations:

- arithmetic operations,
- logical operations.

  At first sight, the result $b[n/2:0]0^{n/2}$ might appear odd. This ALU function is later used to compute the upper half of an n-bit constant using the immediate fields of an instruction,

- test-and-set instructions.

  These compute an n-bit result $0^{n-1}z$, where only the last bit is of interest. These instructions can be computed by performing a subtraction in the AU and then testing the negative bit.

Fig. 41 shows the fairly obvious data paths of an n-ALU. The missing signals are easily constructed. We subtract if $af[1] = 1$. For test-and-set operations with $af[3] = 1$, the signal z is the negative bit neg that we compute for unsigned numbers if $af[0] = 1$

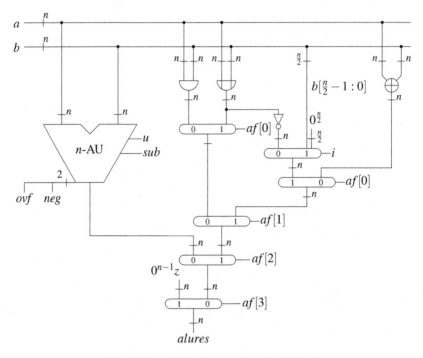

Fig. 41. Data paths of an n-arithmetic logic unit

and for signed numbers otherwise. The overflow bit can only differ from zero if we perform an arithmetic operation. Thus, we have

$$u = af[0],$$
$$sub = af[1],$$
$$z = neg,$$
$$ovfalu = ovf \wedge \neg af[3] \wedge \neg af[2].$$

7.4 Shifter

For $a \in \mathbb{B}^n$ and $i \in [0 : n-1]$ we define the *logical right shift* $srl(a,i)$ of a by i positions by

$$srl(a,i) = 0^i a[n-1 : i].$$

Thus operand a is shifted by i positions to the right. Unused bits are filled in by zeros. For fixed $i \in [0 : n-1]$ the circuit in Fig. 42 satisfies

$$a' = \begin{cases} srl(a,i), & s = 1, \\ a, & s = 0. \end{cases}$$

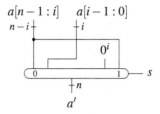

Fig. 42. Construction of an (n,i)-logical right shifter

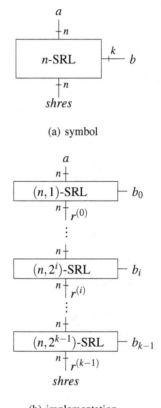

(a) symbol

(b) implementation

Fig. 43. Construction of an n-logical right shifter using a stack of (n,i)-SRLs

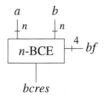

Fig. 44. Symbol of an n-branch condition evaluation unit

We call it an (n, i)-logical right shifter or for short (n, i)-SRL.

Let $n = 2^k$ be a power of two. An n-logical right shifter or for short n-SRL (see Fig. 43(a)) is a circuit with inputs

- $a[n-1:0] \in \mathbb{B}^n$,
- $b[k-1:0] \in \mathbb{B}^k$,

and output

- $shres[n-1:0] \in \mathbb{B}^n$

satisfying

$$shres(a, b) = srl(a, \langle b \rangle).$$

Fig. 43(b) shows the construction of an n-logical right shifter (n-SRL) using a stack of k (n, i)-SRLs. An obvious induction shows

$$r^{(i)} = srl(a, \langle b[i:0] \rangle).$$

7.5 Branch Condition Evaluation Unit

An n-BCE (see Fig. 44) is a circuit with inputs

- $a[n-1:0] \in \mathbb{B}^n$ and $b[n-1:0] \in \mathbb{B}^n$,
- $bf[3:0] \in \mathbb{B}^4$ selecting the condition to be tested,

and output

- $bcres \in \mathbb{B}$

specified by Table 8.
The auxiliary circuit in Fig. 45 computes obvious auxiliary signals satisfying

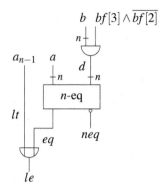

Fig. 45. Computation of auxiliary signals in an n-branch condition evaluation unit

$$d \equiv b \wedge (bf[3] \wedge \overline{bf[2]})$$

$$\equiv \begin{cases} b, & bf[3:2] = 10, \\ 0_n, & \text{otherwise,} \end{cases}$$

$$eq \equiv a = d$$

$$\equiv \begin{cases} a = b, & bf[3:2] = 10, \\ [a] = 0, & \text{otherwise,} \end{cases}$$

$$neq \equiv \overline{eq},$$

$$lt \equiv [a] < 0,$$

$$le \equiv lt \vee \begin{cases} a = b, & bf[3:2] = 10, \\ [a] = 0, & \text{otherwise.} \end{cases}$$

Then the result $bcres$ can be computed as

$bf[3:0]$	$bcres$
0010	$[a] < 0$
0011	$[a] \geq 0$
100*	$a = b$
101*	$a \neq b$
110*	$[a] \leq 0$
111*	$[a] > 0$

Table 8. Specification of branch condition evaluation

$$bcres \equiv bf[3:1] = 001 \wedge (\overline{bf[0]} \wedge lt \vee bf[0] \wedge \overline{lt})$$
$$\vee\ bf[3:2] = 10 \wedge (\overline{bf[1]} \wedge eq \vee bf[1] \wedge \overline{eq})$$
$$\vee\ bf[3:2] = 11 \wedge (\overline{bf[1]} \wedge le \vee bf[1] \wedge \overline{le})$$

$$\equiv \overline{bf[3]} \wedge \overline{bf[2]} \wedge bf[1] \wedge (bf[0] \oplus lt)$$
$$\vee\ bf[3] \wedge \overline{bf[2]} \wedge (bf[1] \oplus eq)$$
$$\vee\ bf[3] \wedge bf[2] \wedge (bf[1] \oplus le).$$

7.6 Final Remarks

As key technical points we recommend to remember

- about adders
    - nothing about carry chain adders. They are a trivial implementation of the basic addition algorithm. If you have seen it once and forget it, you have never understood it.
    - the recursive construction of conditional-sum adders.
    - the recursive construction of parallel prefix circuits.
    - the definition of generate and propagate signals and how to define an associative operation ∘ for their computation, and that one gets the sum bits trivially once one has the generate bits.
    - how to read off difference equations from recursive circuit constructions.
    - how to solve a difference equation by expanding it a few times, making a guess about its solution and proving by induction that your guess satisfies the difference equation.
- about arithmetic units
    - the construction with an n-bit-wide $\oplus$ before the second operand of an adder. That this works for binary numbers follows directly from the subtraction algorithm for binary numbers, which you have already remembered.
    - that the same n-AU works for two's complement numbers follows from

$$[a] \equiv \langle a \rangle \bmod 2^n$$

and the solution of congruences with system of representatives

$$T_n = [-2^{n-1} : 2^{n-1} - 1].$$

    - the argument about the computation of the overflow signal for two's complement numbers starts with the obvious split of operands into the leading bit and the others (which are interpreted as a binary number). Add and subtract $2^{n-1} \cdot c_n$ and then identify $[s[n-1:0]]$ in the result. Then discuss the four combinations of values for carries $c[n, n-1]$.
- about ALUs very little. Only that an ALU is realized with an arithmetic unit, vector operations of logical connectives, and a tree of muxes. The exact format has to match the intended instruction set.

- about shifters that they are constructed using a stack of shifters whose shift distances are powers of two.
- about the branch condition evaluation unit nothing except its existence. It is adapted to the intended instruction set.
- for parties with computer science students: why does the same hardware add and subtract both binary and two's complement numbers correctly?

Floating-point units can be treated with the same precision. A formalization of the IEEE floating-point standard and the construction of a fully IEEE compatible floating-point unit in 120 pages of lecture notes can be found in [MP00]. Teaching that material takes about half a semester. After very minor corrections it was formally verified [BJ01].

7.7 Exercises

1. Prove that an n-CSA has depth

$$d(n) = d(FA) + 3 \cdot \log n.$$

2. Prove Lemma 35.
3. Show that function $\circ : \mathbb{B}^2 \times \mathbb{B}^2 \to \mathbb{B}^2$ which computes generate and propagate signals (g, p) from pairs of such signals in the construction of carry-look-ahead adders is indeed associative.
4. Define an overflow signal for unsigned numbers by

$$ovfu \equiv u = 1 \wedge S \notin B_n.$$

   Compute $ovfu$.
   Hint: based on the text your solution should be *very* short.
5. For the computation of the overflow signal one uses carry signals c_n and c_{n-1}. If the AU construction is based on a conditional-sum adder, signal c_{n-1} is not directly available. Show how to compute this signal with one additional $\oplus$ gate.
   Hint: you are allowed to use internal signals of the construction.
6. An n-'find first one' circuit has input $a \in \mathbb{B}^n$ and output $y \in B^n$ satisfying

$$y_i = 1 \leftrightarrow (a_i = 1 \wedge \forall j < i : a_j = 0).$$

   a) Construct an n-'find first one' circuit.
   b) Construct such a circuit with cost $O(n)$ and delay $O(\log(n))$.
      Hint: compute a parallel prefix OR first.
7. Write down and solve difference equations for cost and depth for the construction of
   a) the decoder in Fig. 20(b).
   b) the 'slow-decoder' in Fig. 24.

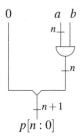

Fig. 46. Constructing an $(n,1)$-multiplier $M_{n,1}$ just amounts to a vector operation

8. An n-compound adder is a circuit with inputs $a, b \in \mathbb{B}^n$ and outputs $S^0, S^1 \in \mathbb{B}^{n+1}$ satisfying for $i \in \{0,1\}$

$$S^i = \langle a \rangle + \langle b \rangle + i,$$

   i.e., a compound adder computes the sum and the incremented sum of the input operands in binary number format.
   a) Give a recursive construction of n-compound adders in the spirit of conditional-sum adders.
      Hint: in the recursion step you need only two $n/2$-compound adders.
   b) Read off difference equations for cost and delay.
   c) Solve the difference equations. Compare with the cost of conditional-sum adders.

9. Show for $a \in \mathbb{B}^n$ and $b \in \mathbb{B}^m$:

$$\langle a \rangle \cdot \langle b \rangle \le 2^{n+m} - 1.$$

10. We define

$$\cdot_n : \mathbb{B}^n \times \mathbb{B}^n \to \mathbb{B}^{2n}$$

   by

$$a \cdot_n b = bin_{2n}(\langle a \rangle \cdot \langle b \rangle \bmod 2^{2n}).$$

   Prove
   a)

$$\langle a \cdot_n b \rangle = (\langle a \rangle \cdot \langle b \rangle \bmod 2^{2n}).$$

   b)

$$[a \cdot_n b] = ([a] \cdot [b] \text{ tmod } 2^{2n}).$$

11. An (n,m)-multiplier is a circuit with inputs $a \in \mathbb{B}^n$ and $b \in \mathbb{B}^m$ and output $p \in \mathbb{B}^{n+m}$ satisfying

$$\langle a \rangle \cdot \langle b \rangle = \langle p \rangle.$$

   An n-multiplier is an (n,n)-multiplier. Show that
   a) the circuit $M_{n,1}$ defined in Fig. 46 is an $(n,1)$-multiplier.
   b) the circuits $M_{n,m}$ defined in Fig. 47 are (n,m)-multipliers.
   c) the cost of n-multipliers constructed in this way is $O(n^2)$.

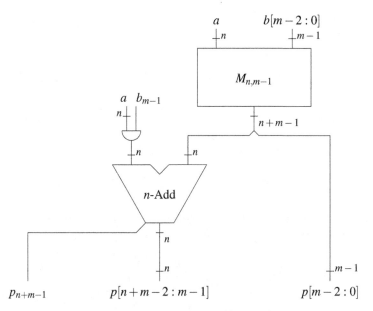

Fig. 47. Constructing an (n,m)-multiplier $M_{n,m}$ from an $(n,m-1)$-multiplier $M_{n,m-1}$, essentially by the school method

Now assume that the adders used are carry chain adders. Show that the construction has depth $O(n+m)$.

Hint: denote by A^m the adder in recursion step m and by $FA^{m,i}$ the i'th full adder from the right in this adder. Show for any signal S in this adder

$$d(S) \leq 1+(n+m)\cdot d(FA).$$

12. Multipliers with logarithmic delay.

a) An n-carry-save adder is a circuit with inputs $a,b,c \in \mathbb{B}^n$ and outputs $s,t \in \mathbb{B}^{n+1}$ satisfying

$$\langle a \rangle + \langle b \rangle + \langle c \rangle = \langle s \rangle + \langle t \rangle.$$

Show that the circuit in Fig. 48 is an n-carry-save adder.

b) An (n,k)-addition tree $AT(n,k)$ is a circuit with inputs $p^i \in \mathbb{B}^n$ for $i \in [0 : k-1]$ and outputs $s,t \in \mathbb{B}^n$ satisfying

$$\langle s \rangle + \langle t \rangle = (\sum_{i=0}^{k-1} \langle p^i \rangle \bmod 2^n).$$

For powers of two $k \geq 4$ a recursive construction of (n,k)-addition trees $T(n,k)$ is given in Figs. 49 and 50. Prove

$$d(T(n,k)) = O(\log k).$$

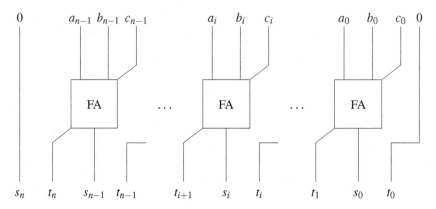

Fig. 48. n full adders in parallel form an n-carry-save adder. Omit the leading output bits s_n and t_n to obtain an $(n,3)$-addition tree $T(n,3)$

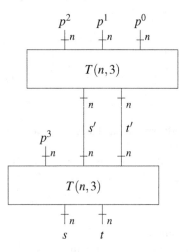

Fig. 49. Constructing an $(n,4)$-addition tree from two $(n,3)$-addition trees

c) Connect the outputs s and t of a tree $T(2n,n)$ with the inputs of a $2n$-carry-look-ahead adder as shown in Fig. 51. Feed the inputs $i \in [n-1:0]$ of the addition tree with

$$p^i = 0^{n-i-1} (a \wedge b_i) 0^i.$$

Show that the resulting circuit is an m-multiplier and estimate its depth.

The use of carry-save adders in addition trees is due to Wallace [Wal64]. The use of $(n,4)$-addition trees in the induction step was introduced by Becker [Bec87]. It does not strictly minimize the depth, but it makes constructions much more regular and easy to analyze.

13. Construction of $(n+1)$-multipliers from n-multipliers. Show: there is a constant α such that every n-multiplier M_n can be turned into an $(n+1)$-multiplier of cost

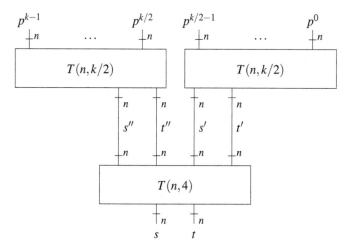

Fig. 50. The obvious recursive construction of (n,k)-addition trees

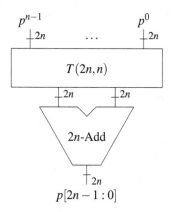

Fig. 51. Using an addition tree to construct a multiplier

$$c(M_{n+1}) \leq c(M_n) + \alpha \cdot n.$$

Hint: first decompose $a, b \in \mathbb{B}^{n+1}$ as

$$a = a_n a[n-1:0],$$
$$b = b_n b[n-1:0],$$

then compute $\langle a \rangle \cdot \langle b \rangle$.

14. (Karatsuba multipliers, [KO63]). Assume n is a power of two. For (n,n)-multipliers $K(n)$ we decompose operands $a, b \in \mathbb{B}^n$ into lower half and upper half as for conditional-sum adders.

$$a = a_H \circ a_L,$$
$$b = b_H \circ b_L.$$

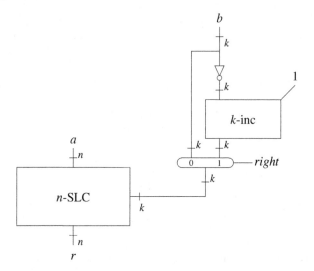

Fig. 52. Transforming right shifts into equivalent left shifts in the construction of an n-cyclic left/right shifter

a) Show

$$\langle a \rangle \cdot \langle b \rangle = \langle a_H \rangle \cdot \langle b_H \rangle \cdot 2^n + \langle a_L \rangle \cdot \langle b_L \rangle +$$
$$(\langle a_H \rangle + \langle a_L \rangle) \cdot (\langle b_H \rangle + \langle b_L \rangle) \cdot 2^{n/2} -$$
$$(\langle a_H \rangle \cdot \langle b_H \rangle + \langle a_L \rangle \cdot \langle b_L \rangle) \cdot 2^{n/2}.$$

b) Construct an n-multiplier from two $n/2$-multipliers, one $(n/2+1)$-multiplier and extra adders and subtractors of total cost $\beta \cdot n$.

c) Replacing the $(n/2+1)$-multiplier by an $n/2$-multiplier with $\alpha \cdot n$ gates, you can now conclude for the cost of the multipliers M_n constructed in this way

$$c(M_n) \leq 3 \cdot c(M_{n/2}) + \gamma \cdot n.$$

Using $c(M_1) = 1$ show
$$c(M_n) = O(n^{\log 3}).$$

15. With Roman numbers addition and subtraction was OK, division by 2 was OK too, but multiplication was very hard. Unfortunately, at the market one needs to multiply the price of items by the number of items sold. Installing large multiplication tables at the market would have been impractical. Instead, the ancient Romans installed large tables of square numbers. Why did this solve the problem?

Hint: the ancient Romans *did* know the Binomial Formula.

16. For $a \in \mathbb{B}^n$ and $i < n$ we define the cyclic left shift $slc(a,i)$ and the cyclic right shift $src(a,i)$:

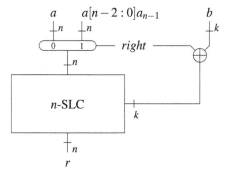

Fig. 53. By preshifting the a-operand in case of a right shift one saves the delay of an incrementer

$$slc(a,i) = a[n-i-1:0] \circ a[n-1:n-i],$$
$$src(a,i) = a[i-1:0] \circ a[n-1:i].$$

a) Show
$$src(a,i) = slc(a, (n-i) \bmod n).$$

b) For fixed i construct a cyclic left shifter (n,i)-SLC with input $a \in \mathbb{B}^n$ and output $a' \in \mathbb{B}^n$ satisfying

$$a' = \begin{cases} slc(a,i), & i=1, \\ a, & i=0. \end{cases}$$

c) For $n = 2^k$ construct an n-cyclic left shifter n-SLC with inputs $a \in \mathbb{B}^n$ and $b \in \mathbb{B}^k$, and output $r \in \mathbb{B}^n$ satisfying

$$r = slc(a, \langle b \rangle).$$

d) An n-cyclic left/right shifter is a circuit with inputs $a \in \mathbb{B}^n$, $b \in \mathbb{B}^k$, control signal $right \in \mathbb{B}$, and output r satisfying

$$r = \begin{cases} slc(a, \langle b \rangle), & right = 0, \\ src(a, \langle b \rangle), & right = 1. \end{cases}$$

Prove that the construction in Fig. 52 is an n-cyclic left/right shifter and determine its depth.

e) Preshifting: prove that the construction in Fig. 53 is an n-cyclic left/right shifter. Compare its depth to the depth of the construction in Fig. 52.

8

A Basic Sequential MIPS Machine

We define the basic *sequential* MIPS instruction set architecture (ISA)[1] for user programmers and present a gate-level construction of a processor that implements it. The first section of this chapter (Sect. 8.1) contains a very compact summary of the instruction set architecture and the assembly language in the form of tables, which define the ISA *if* one knows how to interpret them. In Sect. 8.2 we provide a succinct and completely precise interpretation of the tables with a small exception: treatment of the three coprocessor instructions and the system call instruction is postponed until Chap. 14 on operating system support. From the mathematical description of the ISA we derive in Sect. 8.3 the hardware of a sequential — i.e., non-pipelined — MIPS processor and argue that this processor construction is correct.

Because the simple processor construction presented here follows the ISA specification very closely, most of the correctness proof is reduced to very simple bookkeeping. Pure bookkeeping ends where the construction deviates from the ISA or is not implicitly taken from it: i) the implementation of predicates (Lemma 43) ii) the use of 30-bit adders for computation of aligned addresses (Lemmas 50 and 51) iii) accesses to memory, where the specification memory is byte addressable and the hardware memory is word addressable (Lemmas 41 and 56). In the classroom we treat the lemmas concerned in some detail and for the pure bookkeeping we refer to the lecture notes.

For brevity we treat in this text only one shift operation (srl) and no loads or stores of bytes and half words. A full implementation of all loads, stores, and shifts is not as straightforward as one would expect. The interested reader can find processor constructions for the MIPS instruction set with these instructions in [KMP14]. Chapter 9 contains some example programs written in MIPS assembly language.

---

[1] Pipelined MIPS processors use delayed branch.

© Springer International Publishing Switzerland 2016
W.J. Paul et al., *System Architecture*, DOI 10.1007/978-3-319-43065-2_8

8.1 Tables

In the table for I-type instructions we use the following shorthand

$$m = c.m_4(ea(c))$$
$$= c.m_4(c.gpr(rs(c)) +_{32} sxtimm(c)).$$

8.1.1 I-type

opc	rt	Mnemonic	Assembler-Syntax	Effect
Data Transfer				
100 011		lw	lw *rt rs imm*	rt = m
101 011		sw	sw *rt rs imm*	m = rt
Arithmetic, Logical Operation, Test-and-Set				
001 000		addi	addi *rt rs imm*	rt = rs + sxt(imm)
001 001		addiu	addiu *rt rs imm*	rt = rs + sxt(imm)
001 010		slti	slti *rt rs imm*	rt = (rs < sxt(imm) ? 1_{32} : 0_{32})
001 011		sltiu	sltiu *rt rs imm*	rt = (rs < sxt(imm) ? 1_{32} : 0_{32})
001 100		andi	andi *rt rs imm*	rt = rs ∧ zxt(imm)
001 101		ori	ori *rt rs imm*	rt = rs ∨ zxt(imm)
001 110		xori	xori *rt rs imm*	rt = rs ⊕ zxt(imm)
001 111		lui	lui *rt imm*	rt = $imm0^{16}$
Branch				
000 001	00000	bltz	bltz *rs imm*	pc = pc + (rs < 0 ? $imm00$: 4_{32})
000 001	00001	bgez	bgez *rs imm*	pc = pc + (rs ≥ 0 ? $imm00$: 4_{32})
000 100		beq	beq *rs rt imm*	pc = pc + (rs = rt ? $imm00$: 4_{32})
000 101		bne	bne *rs rt imm*	pc = pc + (rs ≠ rt ? $imm00$: 4_{32})
000 110	00000	blez	blez *rs imm*	pc = pc + (rs ≤ 0 ? $imm00$: 4_{32})
000 111	00000	bgtz	bgtz *rs imm*	pc = pc + (rs > 0 ? $imm00$: 4_{32})

8.1.2 R-type

opcode	fun	rs	Mnemonic	Assembler-Syntax	Effect
Shift Operation					
000000	000 010		srl	srl *rd rt sa*	rd = srl(rt, sa)
Arithmetic, Logical Operation					
000000	100 000		add	add *rd rs rt*	rd = rs + rt
000000	100 001		addu	addu *rd rs rt*	rd = rs + rt
000000	100 010		sub	sub *rd rs rt*	rd = rs − rt
000000	100 011		subu	subu *rd rs rt*	rd = rs − rt
000000	100 100		and	and *rd rs rt*	rd = rs ∧ rt
000000	100 101		or	or *rd rs rt*	rd = rs ∨ rt
000000	100 110		xor	xor *rd rs rt*	rd = rs ⊕ rt
000000	100 111		nor	nor *rd rs rt*	rd = $\overline{rs \lor rt}$
Test-and-Set Operation					
000000	101 010		slt	slt *rd rs rt*	rd = (rs < rt ? 1_{32} : 0_{32})
000000	101 011		sltu	sltu *rd rs rt*	rd = (rs < rt ? 1_{32} : 0_{32})
Jumps, System Call					
000000	001 000		jr	jr *rs*	pc = rs
000000	001 001		jalr	jalr *rd rs*	rd = pc + 4_{32} pc = rs
000000	001 100		sysc	sysc	System Call
Coprocessor Instructions					
010000	011 000	10000	eret	eret	Exception Return
010000		00100	movg2s	movg2s *rd rt*	spr[rd] := gpr[rt]
010000		00000	movs2g	movs2g *rd rt*	gpr[rt] := spr[rd]

8.1.3 J-type

opc		Mnemonic	Assembler-Syntax	Effect
Jumps				
000 010		j	j *iindex*	pc = bin_{32}(pc+4_{32})[31:28]iindex00
000 011		jal	jal *iindex*	R31 = pc + 4_{32}, pc = bin_{32}(pc+4_{32})[31:28]iindex00

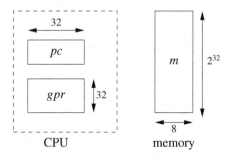

Fig. 54. Visible data structures of MIPS ISA

8.2 MIPS ISA

8.2.1 Configuration and Instruction Fields

A basic *MIPS configuration* c has only three user visible data structures (Fig. 54):

- $c.pc \in \mathbb{B}^{32}$: the program counter (PC),
- $c.gpr : \mathbb{B}^5 \to \mathbb{B}^{32}$: the general-purpose register file consisting of 32 registers, each 32 bits wide. For register addresses $x \in \mathbb{B}^5$ the content of general-purpose register x in configuration c is denoted by $c.gpr(x) \in \mathbb{B}^{32}$. General-purpose register 0_5 is always 0_{32}:

$$c.gpr(0^5) = 0^{32}.$$

- $c.m : \mathbb{B}^{32} \to \mathbb{B}^8$: the processor memory. It is byte addressable; addresses have 32 bits. Thus, for memory addresses $a \in \mathbb{B}^{32}$ the content of memory location a in configuration c is denoted by $c.m(a) \in \mathbb{B}^8$.

Program counter and general-purpose registers belong to the central processing unit (CPU).

Let K be the set of all basic MIPS configurations. A mathematical definition of the ISA will be given by a function

$$\delta : K \to K,$$

where

$$c' = \delta(c, reset)$$

is the configuration reached from configuration c, if the next instruction is executed. An ISA computation is a sequence (c^i) of ISA configurations with $i \in \mathbb{N}$ satisfying

$$c^0.pc = 0^{32},$$
$$c^{i+1} = \delta(c^i, 0),$$

i.e., initially the program counter points to address 0^{32} and in each step one instruction is executed. In the remainder of this section we specify the ISA simply by specifying function δ, i.e., by specifying $c' = \delta(c, 0)$ for all configurations c.

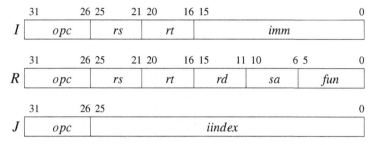

Fig. 55. Types and fields of MIPS instructions

Recall, that for numbers $y \in \mathbb{B}^n$ we abbreviate the binary representation of y with n bits as

$$y_n = bin_n(y),$$

e.g., $1_8 = 00000001$ and $3_8 = 00000011$. For memories $m : \mathbb{B}^{32} \to \mathbb{B}^8$, addresses $a \in \mathbb{B}^{32}$, and numbers $d \geq 1$ of bytes we denote the content of d consecutive memory bytes starting at address a by

$$m_1(a) = m(a),$$
$$m_{d+1}(a) = m(a +_{32} d_{32}) \circ m_d(a).$$

The current instruction $I(c)$ to be executed in configuration c is defined by the four bytes in memory addressed by the current program counter:

$$I(c) = c.m_4(c.pc).$$

Because all instructions are four bytes long, one requires that instructions are *aligned* on four byte boundaries, or equivalently that

$$c.pc[1:0] = 00.$$

In case this condition is violated a so-called misalignment interrupt is raised. The six high-order bits of the instruction are called the opcode of the instruction:

$$opc(c) = opc(c)[5:0] = I(c)[31:26].$$

There are three instruction types: R-, J-, and I-type. The instruction type is determined by the following predicates:

$$rtype(c) \equiv opc(c) = 0*0^4,$$
$$jtype(c) \equiv opc(c) = 0^41*,$$
$$itype(c) \equiv \overline{rtype(c) \vee jtype(c)}.$$

Depending on the instruction type, the bits of the instruction are subdivided as shown in Fig. 55. Addresses of registers in the general-purpose register file are specified in the following fields of the instruction:

$$rs(c) = I(c)[25:21],$$
$$rt(c) = I(c)[20:16],$$
$$rd(c) = I(c)[15:11].$$

Field

$$sa(c) = I(c)[10:6]$$

specifies the shift distance (shift amount) in the one shift operation (*srl*) that we implement here. The field is later also used as an address for a special purpose register file. For R-type instructions, ALU-functions to be applied to the register operands can be specified in the function field:

$$fun(c) = I(c)[5:0].$$

Two kinds of immediate constants can be specified: the immediate constant *imm* in I-type instructions, and an instruction index *iindex* in J-type (like jump) operations:

$$imm(c) = I(c)[15:0],$$
$$iindex(c) = I(c)[25:0].$$

Immediate constant *imm* has 16 bits. In order to apply ALU functions to it, the constant can be extended with 16 high-order bits in two ways: zero extension and sign extension:

$$zxtimm(c) = 0^{16}imm(c),$$
$$sxtimm(c) = imm(c)[15]^{16}imm(c)$$
$$= I(c)[15]^{16}imm(c).$$

The value of the constant interpreted as a two's complement number does not change after sign extension:

$$[sxtimm(c)] = [imm(c)].$$

8.2.2 Instruction Decoding

For every mnemonic *mn* of a MIPS instruction from the tables above we define a predicate *mn(c)* which is true, if $I(c)$ is an *mn* instruction. For instance,

$$lw(c) \equiv opc(c) = 100011,$$
$$bltz(c) \equiv opc(c) = 0^51 \wedge rt(c) = 0^5,$$
$$add(c) \equiv rtype(c) \wedge fun(c) = 10^5.$$

The remaining predicates directly associated with the mnemonics of the assembly language are derived in the same way from the tables. We split the basic instruction set into different groups and define for each group a predicate that holds if an instruction from that group is to be executed:

- ALU operations of I-type are recognized by the leading three opcode bits, i.e., $I(c)[31:29]$; ALU operations of R-type by the two leading bits of the function code, i.e., $I(c)[5:4]$:

$$alur(c) \equiv rtype(c) \wedge fun(c)[5:4] = 10$$
$$\equiv rtype(c) \wedge I(c)[5:4] = 10,$$
$$alui(c) \equiv itype(c) \wedge opc(c)[5:3] = 001$$
$$\equiv itype(c) \wedge I(c)[31:29] = 001,$$
$$alu(c) \equiv alur(c) \vee alui(c).$$

- loads and stores are of I-type and are recognized by the three leading bits of the opcode:

$$l(c) \equiv opc(c)[5:3] = 100$$
$$\equiv I(c)[31:29] = 100,$$
$$s(c) \equiv opc(c)[5:3] = 101$$
$$\equiv I[31:29] = 101,$$
$$ls(c) \equiv l(c) \vee s(c)$$
$$\equiv opc(I)[5:4] = 10$$
$$\equiv I(c)[31:30] = 10.$$

- branches are of I-type and are recognized by the three leading bits of the opcode:

$$b(c) \equiv itype(c) \wedge opc(c)[5:3] = 000,$$
$$\equiv itype(c) \wedge I(c)[31:29] = 000.$$

We define jumps in a brute force way:

$$jump(c) \equiv jr(c) \vee jalr(c) \vee j(c) \vee jal(c),$$
$$jb(c) \equiv jump(c) \vee b(c).$$

8.2.3 ALU Operations

We can now go through the ALU operations in the tables one by one and give them precise interpretations. We do this for two examples.

$add(c)$

The table specifies the effect as $rd = rs + rt$. This is to be interpreted as the corresponding register contents: on the right-hand side of the equation for c, i.e., before execution of the instruction; on the left-hand side for c'. Writes to register 0_5 have no effect. Other registers are not changed.

$$c'.gpr(x) = \begin{cases} c.gpr(rs(c)) +_{32} c.gpr(rt(c)), & x = rd(c) \wedge x \neq 0_5, \\ c.gpr(x), & \text{otherwise.} \end{cases}$$

The memory content does not change:

$$c'.m = c.m.$$

The program counter is advanced by four bytes to point at the next instruction:

$$c'.pc = c.pc +_{32} 4_{32}.$$

addi(c)

The second operand is now the sign-extended immediate constant:

$$c'.gpr(x) = \begin{cases} c.gpr(rs(c)) +_{32} sxtimm(c), & x = rt(c) \wedge x \neq 0_5, \\ c.gpr(x), & \text{otherwise}, \end{cases}$$

$$c'.m = c.m,$$

$$c'.pc = c.pc +_{32} 4_{32}.$$

It is clear how to derive precise specifications for the remaining ALU operations, but we take a shortcut exploiting the fact that we have already constructed an ALU that was specified in Table 7. This table defines functions $alures(a,b,af,i)$ and $ovf(a,b,af,i)$. As we do not treat interrupts (yet) we use only the first of these functions here. We observe that in all ALU operations a function of the ALU is performed. The left operand is always

$$lop(c) = c.gpr(rs(c)).$$

For R-type operations, the right operand is the register specified by the *rt* field of the R-type instruction. For I-type instructions it is the sign-extended immediate operand, if $opc(c)[2] = I(c)[28] = 0$, or zero-extended immediate operand, if $opc(c)[2] = 1$.[2] Thus, we define the immediate fill bit $ifill(c)$, extended immediate constant $xtimm(c)$, and right operand $rop(c)$ in the following way:

---

[2] Note that for instructions *addiu* and *sltiu* this is counterintuitive. The letter u in the instruction name suggests unsigned arithmetic, and arithmetic with binary numbers is indeed performed by these operations. But the immediate operand for these instructions is sign extended and not zero extended, thus we have

$$\langle sxtimm(c) \rangle \neq \langle imm(c) \rangle \quad \text{if} \quad imm(c)[15] = 1.$$

The instruction set manual [MIP01] acknowledges this in the documentation of the *addiu* instruction and calls the term 'unsigned' in the instruction name a misnomer.

$$ifill(c) \equiv \begin{cases} 0, & opc(c)[2] = 1, \\ imm(c)[15], & \text{otherwise} \end{cases}$$

$$\equiv I(c)[15] \wedge \overline{I(c)[28]},$$

$$xtimm(c) \equiv \begin{cases} sxtimm(c), & opc(c)[2] = 0, \\ zxtimm(c), & opc(c)[2] = 1 \end{cases}$$

$$\equiv ifill(c)^{16} imm(c)$$

$$rop(c) \equiv \begin{cases} c.gpr(rt(c)), & rtype(c), \\ xtimm(c), & \text{otherwise.} \end{cases}$$

Comparing Table 7 with the tables for I-type and R-type instructions we see that bits $af[2:0]$ of the ALU control can be taken from the low-order fields of the opcode for I-type instructions and from the low-order bits of the function field for R-type instructions:

$$af(c)[2:0] \equiv \begin{cases} fun(c)[2:0], & rtype(c), \\ opc(c)[2:0], & \text{otherwise} \end{cases}$$

$$\equiv \begin{cases} I(c)[2:0], & rtype(c), \\ I(c)[28:26], & \text{otherwise.} \end{cases}$$

For bit $af[3]$ things are more complicated. For R-type instructions it can be taken from the function code. For I-type instructions it must only be forced to 1 for the two test-and-set operations, which can be recognized by $opc(c)[2:1] = 01$:

$$af(c)[3] \equiv \begin{cases} fun(c)[3], & rtype(c), \\ \overline{opc(c)[2]} \wedge opc(c)[1], & \text{otherwise} \end{cases}$$

$$\equiv \begin{cases} I(c)[3], & rtype(c), \\ \overline{I(c)[28]} \wedge I(c)[27], & \text{otherwise.} \end{cases}$$

The i-input of the ALU distinguishes for $af[3:0] = 1111$ between the lui-instruction of I-type for $i = 0$ and the nor-instruction of R-type for $i = 1$. Thus we set it to $itype(c)$.

The result of the ALU computed with these inputs is denoted by

$$ares(c) \equiv alures(lop(c), rop(c), af(c), itype(c)).$$

Depending on instruction type the destination register $rdes$ is specified by the rd field or the rt field:

$$rdes(c) \equiv \begin{cases} rd(c), & rtype(c), \\ rt(c), & \text{otherwise.} \end{cases}$$

A summary of all ALU operations is then

$$alu(c) \rightarrow \quad \begin{aligned} c'.gpr(x) &= \begin{cases} ares(c), & x = rdes(c), \\ c.gpr(x), & \text{otherwise}, \end{cases} \\ c'.m &= c.m, \\ c'.pc &= c.pc +_{32} 4_{32}. \end{aligned}$$

8.2.4 Shift

We implement only a single shift operation. The left operand $slop(c)$ to be shifted is taken from the register specified by the rt field

$$slop(c) = c.gpr(rt(c)).$$

The shift distance is taken from the register specified by the sa instruction field and the result of the shifter from Sect. 7.4 with these operands is

$$sres(c) = shres(slop(c), sa(c)).$$

The destination register is specified by the rd field. We summarize

$$srl(c) \rightarrow \quad \begin{aligned} c'.gpr(x) &= \begin{cases} sres(c), & x = rd(c) \wedge x \neq 0_5, \\ c.gpr(x), & \text{otherwise}, \end{cases} \\ c'.m &= c.m, \\ c'.pc &= c.pc +_{32} 4_{32}. \end{aligned}$$

8.2.5 Branch and Jump

A branch condition evaluation unit was specified in Table 8. It computes a function $bcres(a, b, bf)$. We use this function with the following parameters:

$$\begin{aligned} blop(c) &= c.gpr(rs(c)), \\ brop(c) &= c.gpr(rt(c)), \\ bf(c) &= opc(c)[2:0] \circ rt(c)[0] \\ &= I(c)[28:26]I[16]. \end{aligned}$$

and define the result of a branch condition evaluation as

$$bres(c) = bcres(blop(c), brop(c), bf(c)).$$

The next program counter $c'.pc$ is usually computed as $c.pc +_{32} 4_{32}$. This order is only changed in jump instructions or in branch instructions where the branch is taken, i.e., the branch condition evaluates to 1. We define

$$jbtaken(c) \equiv jump(c) \vee b(c) \wedge bres(c).$$

In case of a jump or a branch taken, there are three possible jump targets.

Branch Instructions

involve a *relative* branch. The PC is incremented by a branch distance:

$$b(c) \wedge bcres(c) \rightarrow \qquad \begin{aligned} bdist(c) &= imm(c)[15]^{14}imm(c)00, \\ btarget(c) &= c.pc +_{32} bdist(c). \end{aligned}$$

Note that the branch distance is a kind of sign-extended immediate constant, but due to the alignment requirement the low-order bits of the jump distance must be 00. Thus, one uses the 16 bits of the immediate constant for bits $[17:2]$ of the jump distance. Sign extension is used for the remaining bits. Note also that address arithmetic is modulo 2^n. We have

$$\begin{aligned} \langle c.pc \rangle + \langle bdist(c) \rangle &\equiv [c.pc] + [bdist(c)] \bmod 2^n \\ &= [c.pc] + [imm(c)00]. \end{aligned}$$

Thus, backward jumps are realized with negative $[imm(c)]$.

R-type Jumps

for instructions *jr* and *jalr*. The branch target is specified by the *rs* field of the instruction:

$$jr(c) \vee jalr(c) \rightarrow \qquad btarget(c) = c.gpr(rs(c)).$$

J-type Jumps

for instructions *j* and *jal*. The branch target is computed in a rather peculiar way: *first* the PC is incremented by 4. *Then* bits $[27:0]$ are replaced by the *iindex* field of the instruction:

$$j(c) \vee jal(c) \rightarrow \qquad btarget(c) = (c.pc +_{32} 4_{32})[31:28]iindex(c)00.$$

Now we can define the next PC computation for *all* instructions as

$$btarget(c) = \begin{cases} c.pc +_{32} imm(c)[15]^{14}imm(c)00, & b(c), \\ c.gpr(rs(c)), & jr(c) \vee jalr(c), \\ (c.pc +_{32} 4_{32})[31:28]iindex(c)00, & \text{otherwise}, \end{cases}$$

$$c'.pc = \begin{cases} btarget(c), & jbtaken(c), \\ c.pc +_{32} 4_{32}, & \text{otherwise}. \end{cases}$$

Jump and Link

The two jump instructions *jal* and *jalr* are used to implement calls of procedures. Besides setting the PC to the branch target they prepare the so-called *link address* by saving the incremented PC

$$linkad(c) = c.pc +_{32} 4_{32}$$

in a register. For the R-type instruction *jalr* this register is specified by the *rd* field. J-type instruction *jal* does not have an *rd* field, and the incremented PC is stored in register 31 $(= \langle 1^5 \rangle)$.

Branch and jump instructions do not change the memory. For the update of registers in branch and jump instructions we therefore have

$$jb(c) \rightarrow \quad c'.gpr(x) = \begin{cases} linkad(c), & jalr(c) \wedge x = rd(c) \wedge x \neq 0_5 \vee \\ & jal(c) \wedge x = 1^5, \\ c.gpr(x), & \text{otherwise,} \end{cases}$$
$$c'.m = c.m.$$

8.2.6 Loads and Stores

The load and store operations of the basic MIPS instruction set each access four bytes of memory starting at a so-called *effective address ea(c)*. Addressing is always relative to a register specified by the *rs* field. The offset is specified by the immediate field:

$$ea(c) = c.gpr(rs(c)) +_{32} sxtimm(c).$$

Note that the immediate constant was defined as

$$imm(c) = I(c)[15:0],$$

independent of the instruction type. Thus the effective address $ea(c)$ is always defined, even if an illegal instruction was fetched. In the computation of the effective address the immediate constant is sign extended, thus negative offsets can be realized in the same way as negative branch distances. Effective addresses are required to be *aligned*. If we interpret them as binary numbers they have to be divisible by 4:

$$4 \mid \langle ea(c) \rangle$$

or, equivalently,

$$ea(c)[1:0] = 00.$$

If this condition is violated a misalignment interrupt *mal* is raised. The treatment of interrupts is postponed to Chap. 14. Observe that the effective address computation defined above together with the fact, that register 0_5 is always zero provides two so-called *addressing modes*

- *relative to a register* content if $rs \neq 0_5$,
- *direct* as specified by the immediate constant if $rs = 0_5$.

We will finally use the direct addressing mode for the first time as late as in Sect. 15.5.4 of this text (where we still could avoid it). The time where we cannot avoid it is even later, namely in Sect. 15.5.7.

Stores

A store instruction takes the register specified by the *rt* field and stores it in four consecutive bytes of memory starting at address $ea(c)$, i.e., it overwrites $m_4(ea(c))$. Other memory bytes and register values are not changed. The PC is incremented by 4 (but we have already defined that).

$$s(c) \rightarrow \quad \begin{aligned} c'.m(x) &= \begin{cases} byte(i, c.gpr(rt(c))), & x = ea(c) +_{32} i_{32} \wedge i < 4, \\ c.m(x), & \text{otherwise,} \end{cases} \\ c'.gpr &= c.gpr. \end{aligned}$$

A Word of Caution

in case you plan to enter this into a CAV system: the first case of the 'definition' of $c'.m(x)$ is very well understandable for humans, but actually it is a shorthand for the following. If

$$\exists i \in [0:3] : x = ea(c) +_{32} i_{32},$$

then update $c.m(x)$ with the, hopefully, unique i satisfying this condition. In this case we can compute this i by solving the equation

$$x = ea(c) +_{32} i_{32}$$

resp.

$$\langle x \rangle = (\langle ea(c) \rangle + i \bmod 2^{32}).$$

From alignment we conclude

$$\langle ea(c) \rangle + i \leq 2^{32} - 1,$$

hence

$$(\langle ea(c) \rangle + i \bmod 2^{32}) = \langle ea(c) \rangle + i.$$

And we have to solve

$$\langle x \rangle = \langle ea(c) \rangle + i$$

as

$$i = \langle x \rangle - \langle ea(c) \rangle.$$

This turns the above definition into

$$c'.m(x) = \begin{cases} byte(\langle x \rangle - \langle ea(c) \rangle, c.gpr(rt(c))), & \langle x \rangle - \langle ea(c) \rangle \in [0:3], \\ c.m(x), & \text{otherwise,} \end{cases}$$

which is not so readable for humans.

Loads

Loads like stores access four bytes of memory starting at address $ea(c)$. The result is stored in the destination register, which is specified by the rt field of the instruction. The load result $lres(c) \in \mathbb{B}^{32}$ is computed as

$$lres(c) = c.m_4(ea(c)).$$

The general-purpose register specified by the rt field is updated. Other registers and the memory are left unchanged:

$$l(c) \rightarrow \quad c'.gpr(x) = \begin{cases} lres(c), & x = rt(c) \wedge x \neq 0_5, \\ c.gpr(x), & \text{otherwise,} \end{cases}$$
$$c'.m = c.m.$$

8.2.7 ISA Summary

We collect all previous definitions of destination registers for the general-purpose register file into

$$cad(c) = \begin{cases} 1^5, & jal(c), \\ rd(c), & rtype(c), \\ rt(c), & \text{otherwise.} \end{cases}$$

For technical reasons we first define an intermediate result C as

$$C(c) = \begin{cases} linkad(c), & jal(c) \vee jalr(c), \\ sres(c), & srl(c), \\ ares(c), & \text{otherwise.} \end{cases}$$

Using this intermediate result we easily specify the data input to the general-purpose register file:

$$gprin(c) = \begin{cases} lres(c), & l(c), \\ C(c), & \text{otherwise.} \end{cases}$$

Finally, we collect in a general-purpose register write signal all situations, in which some general-purpose register is updated:

$$gprw(c) \equiv alu(c) \vee srl(c) \vee l(c) \vee jal(c) \vee jalr(c).$$

Now we can summarize the MIPS ISA in three rules concerning the updates of PC, general-purpose registers, and memory:

$$c'.pc = \begin{cases} btarget(c), & jbtaken(c), \\ c.pc +_{32} 4_{32}, & \text{otherwise}, \end{cases}$$

$$c'.gpr(x) = \begin{cases} gprin(c), & x = cad(c) \wedge x \neq 0_5 \wedge gprw(c), \\ c.gpr(x), & \text{otherwise}, \end{cases}$$

$$c'.m(x) = \begin{cases} byte(i, c.gpr(rt(c))), & x = ea(c) +_{32} i_{32} \wedge i < 4 \wedge s(c), \\ c.m(x), & \text{otherwise}. \end{cases}$$

8.3 A Sequential Processor Design

From the ISA spec we derive a hardware implementation of the basic MIPS processor. It will execute every MIPS instruction in two hardware cycles: in a *fetch* cycle, the hardware machine fetches an instruction from memory; in an *execute* cycle, the machine executes the current instruction.

In order to prove correctness of the hardware implementation with respect to the MIPS ISA specification, we prove simulation as follows: given a hardware computation

$$h^0, h^1, \ldots, h^{2t}, h^{2t+1}, h^{2t+2}, \ldots$$

we show that there is a corresponding ISA computation

$$c^0, c^1, \ldots, c^t, c^{t+1}, \ldots$$

s.t. a simulation relation

$$c^i \sim h^{2i}$$

holds for all $i \in \mathbb{N}$.

We proceed by giving the hardware construction and simulation relation, arguing that the given construction for each individual component is indeed correct at the time that component is introduced.

8.3.1 Hardware Configuration

A hardware configuration h of the MIPS implementation contains among others the following components:

- program counter register $h.pc \in \mathbb{B}^{32}$,
- general-purpose register RAM $h.gpr : \mathbb{B}^5 \to \mathbb{B}^{32}$,
- *word addressable* $(32, r, 30)$-RAM-ROM $h.m : \mathbb{B}^{30} \to \mathbb{B}^{32}$.

Here a memory *word* denotes four consecutive bytes, i.e., a bit string $x \in \mathbb{B}^{32}$. More components will be specified shortly.

Recall that the ISA specification states that the hardware implementation only needs to work if all memory accesses of the ISA computation (c^i) are *aligned*, i.e., $\forall i \geq 0$:

$$c^i.pc[1:0] = 00,$$
$$ls(c^i) \rightarrow ea(c^i)[1:0] = 00.$$

Now consider that the memory we use for our hardware machine — RAM-ROM — is word addressable, in contrast to the byte addressable memory of our ISA specification. Having a word addressable hardware memory has the advantage that we can serve any properly aligned memory access up to word size in a single cycle. The reason why the hardware memory contains a ROM portion was already explained in Sect. 6.3: after power up/reset we need some known portion of memory where the boot loader resides.

We define the simulation relation $c \sim h$ which states that hardware configuration h encodes ISA configuration c by

$$c \sim h \equiv h.pc = c.pc \wedge h.gpr = c.gpr \wedge c.m \sim_M h.m,$$

where the simulation relation $\sim_M$ for the memory is given as the following simple embedding:

$$c.m \sim_M h.m \equiv \forall a \in \mathbb{B}^{30} : c.m_4(a00) = h.m(a),$$

i.e., for word addresses a, $h.m(a)$ contains the four bytes of ISA memory starting at address $a00$.

In order to prove correctness of the hardware construction, we perform induction on the number of steps of the ISA specification machine: we need to show for initial configurations c^0 and h^0 that $c^0 \sim h^0$ holds (we assume that in cycle -1 the reset signal is active). In the induction step, we show that, given c^i and h^{2i} with $c^i \sim h^{2i}$, the simulation relation is maintained when we perform two steps of the hardware machine and a single step of the ISA specification machine, resulting in $c^{i+1} \sim h^{2(i+1)}$.

Since we use a RAM-ROM to implement the memory of the machine, there is another software condition that we need to obey in order to be able to maintain the memory simulation relation:

$$\forall i \geq 0 : s(c^i) \rightarrow ea(c^i)[31:(r+2)] \neq 0^{32-(r+2)}.$$

That is, every write access to the memory must go to an address that does not belong to the ROM, since writes to the ROM do not have any effect (recall that ROM stands for read-only memory).

We prove correctness of the hardware construction only for those executions of the ISA that obey our software conditions.

Definition 20 (Hardware Correctness Software Conditions). *The software conditions on ISA computation* (c^i) *are:*

1. all memory accesses are aligned properly:

$$\forall i \geq 0 : c^i.pc[1:0] = 00 \, \wedge \, (ls(c^i) \rightarrow ea(c^i)[1:0] = 00),$$

2. there are no writes to the ROM portion of memory:

$$\forall i \geq 0 : s(c^i) \rightarrow ea(c^i)[31:(r+2)] \neq 0^{32-(r+2)}.$$

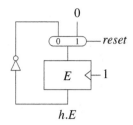

Fig. 56. Computation of the execute signal E

8.3.2 Fetch and Execute Cycles

Since a single MIPS instruction may perform up to two memory accesses and our main memory can only serve a single request per cycle, we construct the hardware machine in such a way that it performs *fetch* cycles and *execute* cycles in an alternating fashion. This is done by introducing register E ($E = 1$ stands for *execute*) and the surrounding circuitry given in Fig. 56. Recall that this was the introductory example when we introduced the model of clocked circuits. Formally, we have to include

$$h.E \in \mathbb{B}$$

as a component of the hardware configuration. In Sect. 5.3 we already showed

$$h^t.E = t \bmod 2.$$

Even cycles t (with $h^t.E = 0$) will be called *fetch cycles* and odd cycles (with $h^t.E = 1$) will be called *execute cycles*.

The portion of the hardware relevant for reset and instruction fetch is shown in Fig. 57. Portions with the three dots will be filled in later. The *nextpc* circuit computing the next program counter will be discussed in Sect. 8.3.8.

8.3.3 Reset

Recall from the clocked-circuit model that the reset signal is active in cycle $t = -1$ and inactive afterwards. The PC is clocked during cycle -1 and thus initially, i.e., after reset, we have

$$h^0.pc = 0^{32} = c^0.pc.$$

Hence the first part of the simulation relation holds for $i = 0$. We take no precautions to prevent writes to $h.gpr$ or $h.m$ during cycle -1 and define

$$c^0.gpr = h^0.gpr,$$
$$c^0.m_4(a00) = h^0.m(a).$$

We can conclude

$$c^0 \sim h^0.$$

From now on, let $i \geq 0$ and assume $c^i \sim h^{2i}$. We construct the hardware in such a way that we can conclude $c^{i+1} \sim h^{2(i+1)}$.

8.3.4 Instruction Fetch

At the end of fetch cycles, instructions fetched from memory are clocked into a register I of the hardware, which is called the instruction register. Formally it has to be included as a component of the hardware configuration

$$h.I \in \mathbb{B}^{32}.$$

Assume that the simulation $c \sim h$ holds for some ISA configuration c. For the hardware in Fig. 57 we show the following: if in configuration h reset is off ($reset(h) = 0$) while fetch is performed ($h.E = 0$), then in the next hardware configuration h' the simulation is maintained and the ISA instruction $I(c)$ has been clocked into the instruction register.

Lemma 41 (Correctness of Fetch Cycles).

$$c \sim h \wedge \neg reset(h) \wedge \neg h.E \rightarrow c \sim h' \wedge h'.I = I(c).$$

This of course implies the induction step of the processor correctness proof

$$c^i \sim h^{2i+1} \wedge h^{2i+1}.I = I(c^i).$$

Input for the instruction register comes from the data output $mout$ of memory m

$$Iin(h) = mout(h).$$

Because $h.E = 0$, the hardware memory is addressed with bits

$$h.pc[31:2].$$

From $c.m \sim_M h.m$, we have

$$h.m(h.pc[31:2]) = c.m_4(h.pc[31:2]00),$$

and from $c \sim h$, we know

$$h.pc[31:2] = c.pc[31:2].$$

We conclude that the hardware instruction $Iin(h)$ fetched by the circuitry in Fig. 57 is

$$
\begin{aligned}
Iin(h) &= mout(h.m) \\
&= h.m(h.pc[31:2]) \\
&= c.m_4(c.pc[31:2]00) \\
&= c.m_4(c.pc) \quad \text{(alignment)} \\
&= I(c).
\end{aligned}
$$

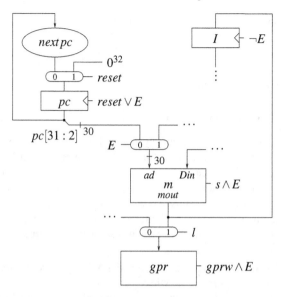

Fig. 57. Portion of the MIPS hardware relevant for reset and instruction fetch

Note that, by construction, the instruction fetched from memory is stored in the instruction register $h.I$ at the end of the cycle, since for the clock enable signal Ice of the instruction register we have

$$Ice(h) = \neg h.E = 1,$$

thus

$$h'.I = I(c).$$

The main memory and the general-purpose register file are written under control of hardware signals

$$mw(h) = h.E \wedge s(h),$$
$$gw(h) = h.E \wedge gprw(h).$$

Hardware signals $s(h)$ and $gprw(h)$ will be defined later as the obvious counterparts of the ISA predicates with the same names. Without knowing their exact definition we can already conclude that hardware memory and hardware GPR can only be written in execute cycles, so we have

$$
\begin{aligned}
h'.m &= h.m \\
&\sim_M c.m \quad \text{(induction hypothesis)}, \\
h'.gpr &= h.gpr \\
&= c.gpr \quad \text{(induction hypothesis)}.
\end{aligned}
$$

Finally, in fetch cycles the PC is not updated and we get

$$h'.pc = h.pc$$
$$= c.pc \quad \text{(induction hypothesis)}.$$

This completes the proof of Lemma 41.

8.3.5 Proof Goals for the Execute Stage

In the analysis of the subsequent execute cycle $2i+1$ we assume Lemma 41 and need to establish $c^{i+1} \sim h^{2i+2}$. For numerous functions $f(c)$ which are already defined as functions of MIPS ISA configurations c we have to consider their obvious counterparts $f_h(h)$, which are functions of hardware configurations h. In order to avoid cluttered notation caused by the subscript h we often overload the notation and use for both functions the same name f. Confusion will not arise, because we will use such functions f only with arguments c or h, where MIPS configurations are denoted by c and hardware configurations by h.

For the remainder of this chapter we will consider MIPS ISA configuration c which simulates hardware configuration h, where reset is off $(reset(h) = 0)$, an execute cycle is performed $(h.E = 1)$, and the instruction register $h.I$ contains the ISA instruction $I(c)$:

$$c \sim h \wedge \neg reset(h) \wedge h.E \wedge h.I = I(c). \tag{6}$$

Our general goal will be to establish the equality of ISA functions $f(c)$ and their counterparts $f(h)$ in the hardware, i.e.

$$f(h) = f(c).$$

Hardware will be constructed such that we can show: if Condition 6 holds, then after clocking the next hardware configuration h' codes the next ISA configuration c'.

Lemma 42 (Correctness of Execute Cycles).

$$c \sim h \wedge \neg reset(h) \wedge h.E \wedge h.I = I(c) \rightarrow c' \sim h'.$$

We know that the hypothesis of this lemma is fulfilled for

$$c = c^i,$$
$$h = h^{2i+1},$$

hence for the induction step of the processor correctness proof we conclude

$$c^{i+1} \sim h^{2i+2}$$
$$\sim h^{2(i+1)}.$$

8.3.6 Instruction Decoder

The instruction decoder shown in Fig. 58 computes the hardware version of functions $f(c)$ that only depend on the current instruction $I(c)$, i.e., which can be written as

$$f(c) = f'(I(c)).$$

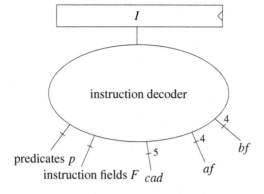

Fig. 58. Instruction decoder

For example,

$$rtype(c) \equiv opc(c) = 0*0^4,$$
$$\equiv I(c)[31:26] = 0*0^4,$$
$$rtype'(I[31:0]) \equiv I[31:26] = 0*0^4,$$

or

$$rd(c) = I(c)[15:11],$$
$$rd'(I[31:0]) = I[15:11].$$

Predicates

Let p be an ISA predicate. By Lemma 32 there is a circuit C with inputs $I[31:0]$ of the hardware instruction register and a gate g in the circuit such that

$$g(I) = p'(I).$$

Connecting inputs $I[31:0]$ to the instruction register $h.I$ of the hardware gives

$$g(h) = g(h.I) = p'(h.I).$$

Condition (6) gives

$$g(h) = p'(h.I)$$
$$= p'(I(c))$$
$$= p(c).$$

Renaming hardware signal g to p gives the following.

Lemma 43. *For all predicates p:*

$$p(h) = p(c).$$

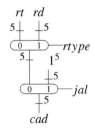

Fig. 59. *C*-address computation

Instruction Fields

All instruction fields F have the form

$$F(c) = I(c)[m:n].$$

Compute the hardware version as

$$F(h) = h.I[m:n].$$

Condition 6 gives

$$\begin{aligned} F(h) &= h.I[m:n] \\ &= I(c)[m:n] \\ &= F(c). \end{aligned}$$

Thus we have the following.

Lemma 44. *For all function fields F:*

$$F(h) = F(c).$$

C-address

The output $cad(h)$ in Fig. 59 computes the C-address for the general-purpose register file. Using Lemmas 43 and 44 we have

$$\begin{aligned} cad(h) &= \begin{cases} 1^5, & jal(h), \\ rd(h), & rtype(h), \\ rt(h), & \text{otherwise} \end{cases} \\ &= \begin{cases} 1^5, & jal(c), \\ rd(c), & rtype(c), \\ rt(c), & \text{otherwise} \end{cases} \\ &= cad(c). \end{aligned}$$

Thus we have the following.

Lemma 45.

$$cad(h) = cad(c).$$

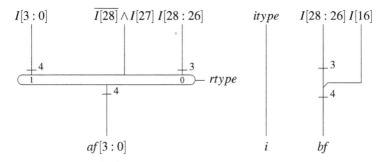

Fig. 60. Computation of function fields for ALU and BCE

Extended Immediate Constant

The fill bit $ifill(c)$ is a predicate and $imm(c)$ is a field of the instruction. Thus, we can compute the extended immediate constant in hardware as

$$xtimm(h) = ifill(h)^{16}imm(h)$$
$$= ifill(c)^{16}imm(c) \quad \text{(Lemmas 43 and 44)}$$
$$= xtimm(c).$$

Hence we have the following.

Lemma 46.
$$xtimm(h) = xtimm(c).$$

Function Fields for ALU and BCE

Fig. 60 shows the computation of the function fields af, i, and bf for the ALU and the branch condition evaluation unit. By Lemmas 43 and 44 outputs $af(h)[2:0]$ satisfy

$$af(h)[3:0] = \begin{cases} I(h)[3:0], & rtype(h), \\ \overline{I(h)[28]} \wedge I(h)[27] \circ I(h)[28:26], & \text{otherwise} \end{cases}$$
$$= \begin{cases} I(c)[3:0], & rtype(c), \\ \overline{I(c)[28]} \wedge I(c)[27] \circ I(c)[28:26], & \text{otherwise} \end{cases}$$
$$= af(c)[3:0].$$

Because $i(c) = itype(c)$ is a predicate, directly from Lemma 43 we get

$$i(h) = i(c).$$

For function fields of the branch condition evaluation unit one shows in the same way:

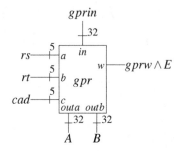

Fig. 61. General-purpose register file

$$bf(h) = I(h)[28:26]I(h)[16]$$
$$= I(c)[28:26]I(c)[16]$$
$$= bf(c).$$

We summarize this as follows.

Lemma 47.

$$af(h) = af(c),$$
$$i(h) = i(c),$$
$$bf(h) = bf(c).$$

That finishes the bookkeeping of what the instruction decoder does.

8.3.7 Reading from General-Purpose Registers

The general-purpose register file $h.gpr$ of the hardware implementation as shown in Fig. 61 is a three-port GPR-RAM with two read ports and one write port. The a and b addresses of the file are connected to $rs(h)$ and $rt(h)$. For the data outputs $gprouta$ and $gproutb$ we introduce the shorthands A and B resp.

Then, we have

$$A(h) = gprouta(h)$$
$$= h.gpr(rs(h)) \quad \text{(construction)}$$
$$= c.gpr(rs(h)) \quad \text{(Equation 6)}$$
$$= c.gpr(rs(c)) \quad \text{(Lemma 44)},$$

and, in an analogous way,

$$B(h) = c.gpr(rt(c)).$$

Thus, we have the following.

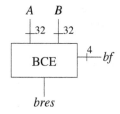

Fig. 62. The branch condition evaluation unit and its operands

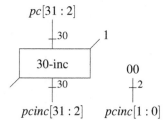

Fig. 63. Incrementing an aligned PC with a 30-incrementer

Lemma 48 (Correctness of Values Read From the GPR).

$$A(h) = c.gpr(rs(c)),$$
$$B(h) = c.gpr(rt(c)).$$

8.3.8 Next PC Environment

Branch Condition Evaluation Unit

The BCE unit is wired as shown in Fig. 62. By Lemmas 48 and 43 as well as the correctness of the BCE implementation from Sect. 7.5 we have

$$\begin{aligned}
bres(h) &= bceres(A(h), B(h), bf(h)) \quad \text{(construction)} \\
&= bceres(c.gpr(rs(c)), c.gpr(rt(c)), bf(c)) \quad \text{(Lemmas 48, 47)} \\
&= bres(c) \quad \text{(ISA definition)}.
\end{aligned}$$

Thus, we have the following.

Lemma 49 (Branch Result Correctness).

$$bres(h) = bres(c).$$

Incremented PC

The computation of an incremented PC as needed for the next PC environment as well as for the link instructions is shown in Fig. 63. Because the PC can be assumed

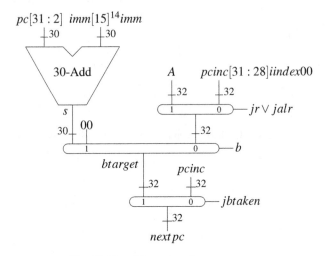

Fig. 64. Next PC computation

to be aligned[3] the use of a 30-incrementer suffices. Using the correctness of the incrementer from Sect. 7.1 we get

$$
\begin{aligned}
pcinc(h) &= (h.pc[31:2] +_{30} 1_{30})00 \\
&= (c.pc[31:2] +_{30} 1_{30})00 \quad \text{(Equation 6)} \\
&= c.pc[31:2]00 +_{32} 1_{30}00 \quad \text{(Lemma 21)} \\
&= c.pc +_{32} 4_{32} \quad \text{(alignment)}.
\end{aligned}
$$

Thus, we have the following.

Lemma 50 (Incremented Program Counter Correctness).

$$
pcinc(h) = c.pc +_{32} 4_{32}.
$$

Next PC Computation

The circuit computing the next PC input, which was left open in Fig. 57 when we treated instruction fetch, is shown in Fig. 64.

Predicates $p \in \{jr, jalr, jump, b\}$ are computed in the instruction decoder. Thus, by Lemma 43 we have

$$
p(h) = p(c).
$$

We compute *jbtaken* in the obvious way and conclude with Lemma 49:

$$
\begin{aligned}
jbtaken(h) &= jump(h) \vee b(h) \wedge bres(h) \\
&= jump(c) \vee b(c) \wedge bres(c) \\
&= jbtaken(c).
\end{aligned}
$$

---

[3] Otherwise a misalignment interrupt would be signaled.

We have

$$A(h) = c.gpr(rs(c)) \quad \text{(by Lemma 48)},$$
$$pcinc(h) = c.pc +_{32} 4_{32} \quad \text{(by Lemma 50)},$$
$$imm(h)[15]^{14}imm(h)00 = imm(c)[15]^{14}imm(c)00 \quad \text{(by Lemma 44)}$$
$$= bdist(c).$$

For the computation of the 30-bit adder we argue as in Lemma 50:

$$\begin{aligned}
s(h)00 &= (h.pc[31:2] +_{30} imm(h)[15]^{14}imm(h))00 \\
&= (c.pc[31:2] +_{30} imm(c)[15]^{14}imm(c))00 \quad \text{(Lemma 44)} \\
&= c.pc[31:2]00 +_{32} imm(c)[15]^{14}imm(c)00 \quad \text{(Lemma 21)} \\
&= c.pc +_{32} bdist(c) \quad \text{(alignment)}.
\end{aligned}$$

We conclude

$$btarget(h) = \begin{cases} c.pc +_{32} bdist(c), & b(c), \\ c.gpr(rs(c)), & jr(c) \vee jalr(c), \\ (c.pc +_{32} 4_{32})[31:28]iindex(c)00, & \text{otherwise} \end{cases}$$
$$= btarget(c).$$

Exploiting $reset(h) = 0$ as well as the semantics of register updates we conclude

$$\begin{aligned}
h'.pc &= nextpc(h) \\
&= \begin{cases} btarget(c), & jbtaken(c), \\ c.pc +_{32} 4_{32}, & \text{otherwise} \end{cases} \\
&= c'.pc.
\end{aligned}$$

Thus, we have shown the following.

Lemma 51 (Next Program Counter Correctness).

$$h'.pc = c'.pc.$$

This shows the first part of $c' \sim h'$.

8.3.9 ALU Environment

The ALU environment is shown in Fig. 65. For the ALU's left operand we have

$$\begin{aligned}
lop(h) &= A(h) \quad \text{(construction)} \\
&= c.gpr(rs(c)) \quad \text{(Lemma 48)} \\
&= lop(c).
\end{aligned}$$

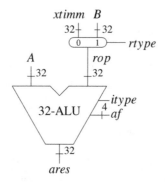

Fig. 65. ALU environment

The argument for the right operand follows via Lemmas 48, 46, and 43

$$rop(h) = \begin{cases} B(h), & rtype(h), \\ xtimm(h), & \text{otherwise} \end{cases}$$

$$= \begin{cases} c.gpr(rt(c)), & rtype(c), \\ xtimm(c), & \text{otherwise} \end{cases}$$

$$= rop(c).$$

For the result *ares* of the ALU we get

$$ares(h) = alures(lop(h), rop(h), itype(h), af(h)) \quad \text{(Sect. 7.3)}$$
$$= alures(lop(c), rop(c), itype(c), af(c)) \quad \text{(Lemmas 48, 47)}$$
$$= ares(c).$$

We summarize this as follows.

Lemma 52 (ALU Result Correctness).

$$ares(h) = ares(c).$$

Note that in contrast to previous lemmas the proof of this lemma is not just bookkeeping; it involves not so trivial correctness of the ALU implementation from Sect. 7.3.

8.3.10 Shifter Environment

The shifter environment is shown in Fig. 66. For the shifter's left operand we have

$$slop(h) = B(h) \quad \text{(construction)}$$
$$= c.gpr(rt(c)) \quad \text{(Lemma 48)}$$
$$= slop(c).$$

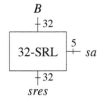

Fig. 66. Shifter environment

By Lemma 44 the shift distance is

$$sa(c) = sa(h).$$

The result of the shifter with these operands is

$$
\begin{aligned}
sres(h) &= shres(slop(h), sa(h)) \quad \text{(Sect. 7.4)} \\
&= shres(slop(c), sa(c)) \\
&= sres(c).
\end{aligned}
$$

We summarize as follows.

Lemma 53 (Shift Result Correctness).

$$sres(c) = sres(h).$$

8.3.11 Jump and Link

The value *linkad* that is saved in jump-and-link instructions is identical to the incremented PC *pcinc* from the next PC environment:

$$
\begin{aligned}
linkad(h) &= pcinc(h) \\
&= h.pc +_{32} 4_{32} \\
&= c.pc +_{32} 4_{32} \\
&= linkad(c).
\end{aligned}
\tag{7}
$$

8.3.12 Collecting Results

Fig. 67 shows two multiplexers collecting results *linkad*, *sres*, and *ares* into an intermediate result C. Using Lemmas 52, 53 and 43 as well as Equation 7 we conclude

$$
\begin{aligned}
C(h) &= \begin{cases}
linkad(h), & jal(h) \lor jalr(h), \\
sres(h), & srl(h), \\
ares(h), & \text{otherwise}
\end{cases} \\
&= C(c).
\end{aligned}
$$

Thus, we have the following.

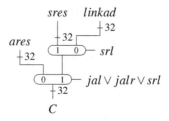

Fig. 67. Collecting results into signal C

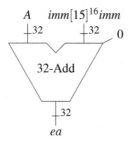

Fig. 68. Effective address computation

Lemma 54 (C Result Correctness).

$$C(h) = C(c).$$

8.3.13 Effective Address

The effective address computation is shown in Fig. 68. We have

$$
\begin{aligned}
ea(h) &= A(h) +_{32} imm(h)[15]^{16}imm(h) \quad \text{(Sect. 7.1)}\\
&= c.gpr(rs(c)) +_{32} sxtimm(c) \quad \text{(Lemmas 48 and 44)}\\
&= ea(c).
\end{aligned}
$$

Thus, we have the following.

Lemma 55 (Effective Address Correctness).

$$ea(h) = ea(c).$$

8.3.14 Memory Environment

We implement only word accesses. Fig. 69 shows the inputs to the ROM-RAM used as memory. Here, one has to show the following.

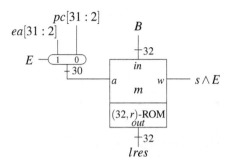

Fig. 69. Memory environment

Lemma 56 (Memory Implementation Correctness).

$$lres(h) = lres(c)$$

and

$$h'.m \sim_M c'.m.$$

Proof. The proof of the first statement is completely analogous to the analysis of instruction fetch using $ea(h) = ea(c)$ and concluding $ma(h) = ea(h)[31 : 2]$ from $h.E = 1$. For the second statement we have to consider an obvious case split:

- $s(h) = 0 \rightarrow mw(h) = 0$, i.e., no store is performed. From Lemma 43 we get $s(c) = 0$. Using hypothesis $c \sim_M h$ we conclude for all word addresses a

$$
\begin{aligned}
h'.m(a) &= h.m(a) \\
&= c.m_4(a00) \\
&= c'.m_4(a00).
\end{aligned}
$$

- $s(h) = 1 \rightarrow mw(h) = 1$, i.e., a store with address $ea(h)[31 : 2]$ is performed. From Lemma 43 we get $s(c) = 1$. For word addresses $a \neq ea(h)[31 : 2]$ one argues as above

$$h'.m(a) = c'.m_4(a00).$$

For $a = ea(h)[31 : 2]$ we use Lemma 48

$$min(h) = c.gpr(rt(c))$$

and conclude

$$
\begin{aligned}
h'.m(a) &= c.gpr(rt(c)) \\
&= c'.m_4(ea(c)) \quad \text{(software condition about } ea \text{ of stores)} \\
&= c'.m_4(a00) \quad \text{(alignment).}
\end{aligned}
$$

Note that in this proof we argued that *both* the ISA memory and the hardware memory are updated at the locations concerned. This only works if the effective address does not lie in the ROM.

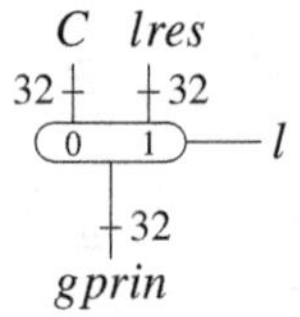

$$C \quad lres$$

Fig. 70. Computing the data input of the *gpr*

8.3.15 Writing to the General-Purpose Register File

Fig. 70 shows the last multiplexer connecting the data input of the general-purpose register file with intermediate result C and the result $lres$ coming from the memory. Using Lemmas 54 and 43 we conclude

$$gprin(h) = \begin{cases} lres(h), & l(h), \\ C(h), & \text{otherwise} \end{cases}$$
$$= \begin{cases} lres(c), & l(c), \\ C(c), & \text{otherwise} \end{cases}$$
$$= gprin(c).$$

Using $h.E = 1$ and Lemma 43 we get

$$gprw(h) = gprw(c).$$

With hypothesis $c \sim h$ we conclude for the general-purpose register file:

$$h'.gpr(x) = \begin{cases} gprin(h), & gprw(h) \wedge x = cad(h) \wedge x \neq 0_5, \\ h.gpr(x), & \text{otherwise} \end{cases}$$
$$= \begin{cases} gprin(c), & gprw(c) \wedge x = cad(c) \wedge x \neq 0_5, \\ c.gpr(x), & \text{otherwise} \end{cases}$$
$$= c'.gpr(x).$$

This concludes the proof of Lemma 42 and also the induction step of the correctness proof of the entire (simple) processor.

8.4 Final Remarks

Do not bother to memorize the entire tables of Sect. 8.1. In our view, the key technical points to remember are

- about the ISA design:

- the tables have a structure with groups of instructions. Groups of similar instructions like *alu* and *alui* have a similar encoding, so they can be supported by the same functional units.
- the coding of operations of functional units is similar to the coding of the instructions the unit supports. This makes the computation of the function signals like *af* for the units easier.
- immediate operands can be extended in different ways.
- the formal specification was the key to the hardware design. Just remember some order in which it can be obtained from the tables, such as
  - instruction.
  - instruction fields, predicates decoding single instructions and groups of instructions, function signals, *gpr* addresses, jump or branch taken, branch target, next PC, extension of immediate operands, left and right operand, results of functional units, effective address, load result and effect of stores, update of the *gpr*.
  - in the end there is one equation describing the update of each component of an ISA configuration.
- about the processor design:
  - it hinges on software conditions, here alignment and the absence of writes to the ROM.
  - the translation from specification to hardware is most of the time completely straightforward. This is reflected in the simulation relation

$$X \neq m \rightarrow h.X = c.X.$$

The only exception is the hardware memory. It is wider than ISA memory, so aligned (!) word accesses can be performed in one cycle. In the simulation relation this is reflected as

$$h.m(a) = d.m_4(a00).$$

  - alignment occasionally allows use of shorter adders in address computations
  - for every computation, memory content $c^0.m$ and register contents $c^0.gpr$ of the initial ISA configuration c^0 are constructed from the initial hardware configuration h^0. In general these contents are different every time you turn on the computer. This is why the ROM must be there.

Mathematical correctness proofs for much more complex single-core processors have been produced and formally verified [MP00, Krö01, SH02, JR05]. The situation is very different for multi-core processors. A correctness proof for the hardware of a multi-core processor with pipelined processor cores but no operating system support has only very recently been produced [KMP14].

8.5 Exercises

1. Change the MIPS processor design such that it executes one instruction per cycle.

Hint: you need to replace the hardware memory with a two-port RAM with i) a read port for instruction fetch and ii) a read/write port for load/store.

2. Extend the design of the MIPS processor with a hardware multiplier for binary numbers. Make it store the result of a multiplication in two 32-bit-wide registers *HI* and *LO*. The content of these registers is transferred to the general-purpose registers by instructions *mfhi* and *mflo*.

opcode	fun	Mnemonic	Assembler-Syntax	Effect
000000	011001	multu	multu *rs rt*	HI,LO = rs · rt
000000	010000	mfhi	mfhi *rd*	rd = HI
000000	010010	mflo	mflo *rd*	rd = LO

a) Specify the exact semantics of the three new instructions.
b) What new components are present in the configurations of ISA and hardware?
c) Specify the effect of the new instructions.
d) Design the hardware extensions. Don't forget clock enable signals for *HI*, *LO* and the write signal for *gpr*.

3. Add also support for two's complement multiplication.

opcode	fun	Mnemonic	Assembler-Syntax	Effect
000000	01100	mult	mult *rs rt*	HI,LO = rs · rt

a) Convert two's complement numbers to sign and magnitude format, perform unsigned integer multiplication of the magnitudes, and convert back.
b) Zero extend binary numbers and sign extend two's complement numbers. Attention: a binary multiplier will not do this for you. You need a multiplier for two operands of length $2n$ which computes modulo 2^{2n}.

For both problems specify the hardware and prove that it works. You can do this in two ways.

4. The DLX ISA [PH90] is (more or less) given by Tables 9 – 11. Proceed in the following order to design a DLX processor.
a) Extract from Table 10 a specification of an ALU for DLX. It should have four control inputs $af[3:0]$ and should handle arithmetic, logical and test-and-set operations.
b) Design this ALU.
Hint: it is crucial to understand the coding of the test conditions. First compute signals

$$L \equiv [a] < [b],$$
$$E \equiv a = b,$$
$$G \equiv [a] > [b],$$

then compute the result bit of the test-and-set operation as

$$z = L \wedge af[2] \vee E \wedge af[1] \vee G \wedge af[0].$$

c) Adapt the specification. Write down only the portions of the specification which differ from the corresponding part for MIPS. Hint: be careful about the extensions of the immediate constants. Also observe that jump distances are given in bytes and that the jump distance for J-type instructions is computed in a more straightforward way.

d) Finally, specify the changes in the hardware.

opc	Mnemonic	Assembler-Syntax	Effect
Data Transfer			
100011	lw	lw *rt rs imm*	rt = m
101011	sw	sw *rt rs imm*	m = rt
Arithmetic, Logical			
001000	addi	addi *rt rs imm*	rt = rs + sxt(imm)
001001	addiu	addiu*rt rs imm*	rt = rs + zxt(imm)
001010	subi	subi *rt rs imm*	rt = rs − sxt(imm)
001011	subiu	subiu*rt rs imm*	rt = rs − zxt(imm)
001100	andi	andi *rt rs imm*	rt = rs ∧ zxt(imm)
001101	ori	ori *rt rs imm*	rt = rs ∨ zxt(imm)
001110	xori	xori *rt rs imm*	rt = rs ⊕ zxt(imm)
001111	lui	lui *rt imm*	$rt = imm0^{16}$
Test-and-Set			
001000	clri	clri *rt*	$rt = 0_{32}$
001001	sgri	sgri *rt rs*	$rt = (rs > sxt(imm) \ ? \ 1_{32} : 0_{32})$
001010	seqi	seqi *rt rs*	$rt = (rs = sxt(imm) \ ? \ 1_{32} : 0_{32})$
001011	sgei	sgei *rt rs*	$rt = (rs \geq sxt(imm) \ ? \ 1_{32} : 0_{32})$
001100	slsi	slsi *rt rs*	$rt = (rs < sxt(imm) \ ? \ 1_{32} : 0_{32})$
001101	snei	snei *rt rs*	$rt = (rs \neq sxt(imm) \ ? \ 1_{32} : 0_{32})$
001101	slei	slei *rt rs*	$rt = (rs < sxt(imm) \ ? \ 1_{32} : 0_{32})$
001111	seti	seti *rt*	$rt = 1_{32}$
Control Operations			
000010	beqz	beqz *rs*	$pc = pc + (rs1 = 0? \ sxt(imm): 4_{32})$
000101	bnez	bnez *rs*	$pc = pc + (rs1 \neq 0? \ sxt(imm): 4_{32})$
010110	jr	jr *rs*	pc = rs
010111	jalr	jalr *rs*	$R31 = pc + 4_{32}, \ pc = rs$

Table 9. I-type instructions of DLX

In Table 9 we used a shorthand:

$$m = c.m_4(ea(c)),$$
$$ea(c) = c.gpr(rs(c)) +_{32} sxtimm(c).$$

opcode	fun	Mnemonic	Assembler-Syntax	Effect
Shift Operation				
000000	000011	sra	sra *rd rt sa*	rd = sra(rt,sa)
Arithmetic, Logical				
000000	100000	add	add *rd rs rt*	rd = rs + rt
000000	100001	addu	addu *rd rs rt*	rd = rs + rt
000000	100010	sub	sub *rd rs rt*	rd = rs − rt
000000	100011	subu	subu *rd rs rt*	rd = rs − rt
000000	100100	and	and *rd rs rt*	rd = rs $\wedge$ rt
000000	100101	or	or *rd rs rt*	rd = rs $\vee$ rt
000000	100110	xor	xor *rd rs rt*	rd = rs $\oplus$ rt
000000	100111	lu	lu *rd rs*	rd = rs[15:0]0^{16}
Test-and-Set				
000000	101000	clr	clr *rd*	rd $=0_{32}$
000000	101001	sgr	sgr *rd rs rt*	rd $=(rs > rt \; ? \; 1_{32}:0_{32})$
000000	101010	seq	seq *rd rs rt*	rd $=(rs = rt \; ? \; 1_{32}:0_{32})$
000000	101011	sge	sge *rd rs rt*	rd $=(rs \geq rt \; ? \; 1_{32}:0_{32})$
000000	101100	sls	sls *rd rs rt*	rd $=(rs < rt \; ? \; 1_{32}:0_{32})$
000000	101100	sne	sne *rd rs rt*	rd $=(rs \neq rt \; ? \; 1_{32}:0_{32})$
000000	101101	sle	sle *rd rs rt*	rd $=(rs < rt \; ? \; 1_{32}:0_{32})$
000000	101111	set	set *rd*	rd $=1_{32}$

Table 10. R-type instructions of DLX

opc	Mnemonic	Assembler-Syntax	Effect
Jumps			
000 010	j	j *iindex*	pc = pc + sxt(iindex)
000 011	jal	jal *iindex*	R31 = pc + 4_{32}, pc = pc + sxt(iindex)

Table 11. J-type instructions of DLX

9

Some Assembler Programs

This text is not meant to replace classes on programming. In Saarbrücken, where we teach, a class focusing on programming in MIPS assembly and C is presently taught in parallel with our system architecture class.[1] The main purpose of this small chapter is to present programs for multiplication and division of binary numbers and of two's complement numbers. These programs are later used in the compiler that we construct, as the code generated for the corresponding operations in the source language. We go over the material of this chapter in the classroom very quickly. We do however present the correctness proof for the school method of division from Sect. 9.3.1.

Before we present example programs, we have to say a few words about how we write assembler programs.

Assembler Syntax

An assembler program is a sequence of assembler instructions. The syntax of assembler instructions is given by the instruction-set-architecture tables in Sect. 8.1.[2] The assembler translates instructions into machine words according to the MIPS instruction layout (see Fig. 55) in the straightforward way.

Representation of Constants in Assembly Language

We represent register numbers and immediate constants in decimal notation. Conversion to two's complement representation is done by the assembler. Additionally, our assembler syntax allows us to specify immediate constants $i[15:0]$ and instruction indices $ii[25:0]$ given in binary representation by prepending 0b:

$$0bi[15:0] \qquad 0bii[25:0].$$

---

[1] In a hardware design lab the students can also have the experience that transferring the processor design of Chap. 8 literally to an FPGA board results in ... running hardware.

[2] Note that most assemblers reference register i by $\$i$. We omit the dollar sign in subsequent assembly programs for simplicity.

© Springer International Publishing Switzerland 2016
W.J. Paul et al., *System Architecture*, DOI 10.1007/978-3-319-43065-2_9

Other Conventions

We put comments behind a double slash '//'. Sometimes we number lines. These numbers are comments too; they just serve to compute jump distances.

9.1 Simple MIPS Programs

We present a few simple MIPS programs to be used in implementing assembler programs for software multiplication and division.

Storing a 32-bit Constant in a Register

Let $S[31:0] \in \mathbb{B}^{32}$. The following program loads this constant in register k using two instructions. First, it stores the upper half and then the zero-extended lower half is ORed with the register content:

```
lui k 0bS[31:16]
ori k k 0bS[15:0]
```

We abbreviate this program with

$$\text{gpr}(k) = S.$$

Computing Sign and Absolute Value of an Integer

Assume a two's complement number is stored in GPR register i. We want to store the sign and absolute value of this number in registers j and k. That is, after executing the program starting from configuration c, we reach a configuration c' with

$$c'.gpr(j_5) = \begin{cases} 1, & [c.gpr(i_5)] < 0, \\ 0, & \text{otherwise} \end{cases} \quad \text{and} \quad \langle c'.gpr(k_5) \rangle = |[c.gpr(i_5)]|.$$

For the sign we simply compare register i with zero:

```
0: slti j i 0
```

To compute the absolute value, we invert and increment in case the number is negative. We invert by XORing with the mask 1^{32} which is obtained by a NOR of the mask 0^{32} with itself. In case the number is positive, we simply assign it to register k:

```
1: blez j 5      // if j=0 continue after line 5
2: nor  k k k    // generate mask
3: xor  k k i    // negate
4: addi k k 1
5: blez 0 2      // gpr(0) is always zero
6: addi k i 0
```

We abbreviate this program by

$$\text{sign-and-abs}(j,k,i).$$

Initializing Memory Cells with Zeros

Suppose we have in configuration c in GPR register i a word-aligned address a and in register j the two's complement representation of a number of words $\ell > 0$:

$$c.gpr(i_5) = a \wedge a[1:0] = 00 \wedge [c.gpr(j_5)] = \ell.$$

We want to initialize ℓ memory words with zeros, starting at address a. This is achieved by the following program:

```
1: sw    0 i 0
2: addi i i 4    // gpr(i) = gpr(i) + 4
3: addi j j -1   // gpr(j) = gpr(j) - 1
4: bne   0 i -3  // if gpr(j)!=0 go to line 1
```

We abbreviate this program by

$$\texttt{zero(i,j)}.$$

9.2 Software Multiplication

We present an assembler program that performs multiplication of the binary numbers given by registers i and j. It computes the product mod 2^{32} of these numbers in register k. That is, given $a, b \in \mathbb{B}^{32}$ and an initial configuration c (which is about to execute the assembler program) with

$$c.gpr(i_5) = a,$$
$$c.gpr(j_5) = b,$$

we want to show that, in the final configuration c' (after executing the program), we have

$$\langle c'.gpr(k_5) \rangle = (\langle a \rangle \cdot \langle b \rangle \bmod 2^{32}).$$

We use four general-purpose registers as auxiliary registers, namely registers 24, 25, 26, and 27. In comments, we abbreviate them by

$$x = gpr(24_5),$$
$$y = gpr(25_5),$$
$$z = gpr(26_5),$$
$$u = gpr(27_5).$$

We initialize x with $0^{31}1$, z with a, and u with 0^{32}. Then we AND b and x together and store the result in y:

```
1: addi 24 0 1    // x = 1
2: addi 26 i 0    // z = a
3: addi 27 0 0    // u = 0
```

We add z to u if $b[0] = 1$:

```
4: and   25 j 24   // y = b AND x
5: beq   25 0 2    // skip next instruction if b[0]=0
6: add   27 27 26  // u = u + z
```

Afterwards, we have

$$x = 0^{31} 1,$$
$$z = a,$$
$$\langle u \rangle = \langle a \rangle \cdot b[0].$$

The next lines of the program turn it into a loop. After the loop has been processed n times the following will hold:

$$x = (0^{31} 10^n)[31 : 0],$$
$$\langle z \rangle = (\langle a \rangle \cdot 2^n \bmod 2^{32}),$$
$$\langle u \rangle = (\langle a \rangle \cdot \langle b[n-1 : 0] \rangle) \bmod 2^{32}).$$

As we have seen, this is true initially (for $n = 0$), and we program the loop in such a way that this invariant is maintained:

```
7: add   24 24 24  // double x mod 2^32
8: add   26 26 26  // double z mod 2^32
9: bne   24 0 -5   // go back to line 4
                   // if 32 iterations are not over
10: addi k 27 0    // gpr(k) = u
```

For every iteration $n \in [1 : 32]$ we have

$$y \neq 0 \leftrightarrow b[n-1] = 1.$$

We add z to u if $y \neq 0$, double x and z, and if $n < 32$ we go five instructions back to the start of the loop. After 32 iterations we have the desired product in u. In the end, we copy the result to register k. We abbreviate the above program with

$$\text{mul}(k, i, j).$$

Note that the program can be used unchanged for integer multiplication tmod 2^{32} as well. The argument is known from the correctness of arithmetic units.[3]

---

[3] By Lemmas 23 and 9 we have $[a] \cdot [b] \equiv \langle a \rangle \cdot \langle b \rangle \bmod 2^{32}$. Moreover $[u] \equiv \langle u \rangle \bmod 2^{32} \equiv \langle a \rangle \cdot \langle b \rangle \bmod 2^{32}$. Since $[u] \in T_{32}$ we use Lemma 12 to obtain $[u] = [a] \cdot [b]$ tmod 2^{32}.

9.3 Software Division

We first discuss the basic binary division algorithm following the integer division algorithm learned in school. Then, we proceed by implementing the algorithm for non-negative integer division in assembler. Finally, we briefly discuss division of negative integers.

9.3.1 School Method for Non-negative Integer Division

For natural numbers a and b with $b > 0$, the integer division of a by b can be defined as

$$a/b = \max\{C \in \mathbb{N} \mid b \cdot C \le a\}.$$

Lemma 57. *Let* $a, b, d \in \mathbb{N}$ *and* $d \cdot b \le a$. *Then*

$$(a - d \cdot b)/b = a/b - d.$$

Proof. We have

$$
\begin{aligned}
x = (a - d \cdot b)/b &\leftrightarrow x \cdot b \le a - d \cdot b \wedge (x+1) \cdot b > a - d \cdot b \\
&\leftrightarrow (x+d) \cdot b \le a \wedge (x+d+1) \cdot b > a \\
&\leftrightarrow x + d = a/b \\
&\leftrightarrow x = a/b - d.
\end{aligned}
$$

For the division algorithm, if $a < b$, then $a/b = 0$ and we are done. Otherwise, we adapt the school algorithm for long division of decimal numbers to binary numbers. A single step of the binary school division algorithm is justified as follows.

Lemma 58. *For* $a > b$ *let*

$$t(a,b) = \max\{k \in \mathbb{N} \mid 2^k \cdot b \le a\}.$$

Then, we have

$$a/b = 2^{t(a,b)} + (a - 2^{t(a,b)} \cdot b)/b.$$

Proof. This follows from Lemma 57 with

$$d = 2^{t(a,b)}.$$

For $a > b$ we obtain the school algorithm for iterative binary division in the obvious way:

- we start by setting the current remainder of the division to a:

$$A(0) = a.$$

- as long as $A(i) \geq b$ (the remainder after the i'th step is still greater or equal to b), we define

$$t(i) = t(A(i),b),$$
$$A(i+1) = A(i) - 2^{t(i)} \cdot b.$$

From

$$A(i) < 2^{t(i)+1} \cdot b$$

we get

$$A(i+1) = A(i) - 2^{t(i)} \cdot b$$
$$< 2^{t(i)} \cdot b.$$

Hence

$$t(i+1) < t(i),$$
$$t(i) \leq t(0) - i.$$

With

$$A(0) \leq 2^{t(o)} \cdot b$$

we conclude

$$A(i) \leq 2^{t(0)-i} \cdot b$$

and the algorithm terminates after step u with

$$u = \max\{i \mid A(i) \geq b\}.$$

An easy induction on i with Lemma 58 gives

$$a/b = \sum_{j=0}^{i} 2^{t(i)} + A(i+1)/b,$$

thus we have the following.

Lemma 59.

$$a/b = \sum_{i=0}^{u} 2^{t(i)}.$$

9.3.2 'Small' Unsigned Integer Division

The following algorithm takes two binary numbers in registers i and j and computes the result of the integer division in register k. That is, let $p, q \in \mathbb{B}^{32}$ and let c be the initial configuration before execution of the program and c' be the final configuration after execution of the program. Then, the program should obey

$$c.gpr(i_5) = p,$$
$$c.gpr(j_5) = q,$$
$$\langle c'.gpr(k_5)\rangle = \langle p\rangle/\langle q\rangle.$$

Obviously, we assume $q \neq 0^{32}$. In order to avoid a tedious case distinction in the case $t(0) = 31$, and to save registers, we also assume that the dividend $\langle p\rangle$ is not too large by requiring that the leading bit of p is zero:

$$q \neq 0^{32} \wedge p[31] = 0.$$

This implies

$$t(\langle p\rangle, \langle q\rangle) < 31.$$

We abbreviate

$$a = \langle p\rangle \quad \text{and} \quad b = \langle q\rangle.$$

We use registers 23 through 27 as auxiliary registers and we abbreviate

$$A = gpr(23_5),$$
$$B = gpr(24_5),$$
$$X = gpr(25_5),$$
$$C = gpr(26_5),$$
$$U = gpr(27_5).$$

We initialize the result register k with zero. With an initial test we handle the case $a < b$ and simply jump to the end of the routine.

```
1: add  k 0 0      // gpr(k) = 0
2: sltu 23 i j     // A = (a < b)
3: bgtz 23 17      // quit, i.e., go to line 20 if a<b
```

We initialize

```
4: add  23 i 0     // A = p
5: add  24 j 0     // B = q
6: addi 25 0 1     // X = 0...01
7: add  26 0 0     // C = 0
```

and have after execution of this code

$$\langle A\rangle = a,$$
$$\langle B\rangle = 2^0 \cdot b,$$
$$X[j] = 1 \leftrightarrow j = 0,$$
$$\langle C\rangle = 0.$$

Next we successively double $\langle X\rangle$ and $\langle B\rangle$ until $\langle B\rangle > \langle A\rangle$. This will succeed after some number x of iterations because p has a leading zero while $q[31:0]$ is not equal to zero.

```
 8: add   25 25 25     // X = X + X
 9: add   24 24 24     // B = B + B
10: sltu 27 23 24      // U = (A < B)
11: blez 27 -3         // repeat if A>=B
```

This loop ends after x iterations for some x and we have

$$\langle A \rangle = a,$$
$$\langle B \rangle = 2^x \cdot b,$$
$$X[j] = 1 \leftrightarrow j = x,$$
$$2^x \cdot b > a,$$
$$x = t(0) + 1.$$

We write the main loop such that after i iterations the following is satisfied:

$$\langle A \rangle = A(i),$$
$$\langle B \rangle = \begin{cases} 2^{t(0)+1} \cdot b, & i = 0, \\ 2^{t(i-1)} \cdot b, & i \geq 1, \end{cases}$$
$$X[j] = 1 \leftrightarrow j = \begin{cases} t(0) + 1, & i = 0, \\ t(i-1), & i \geq 1, \end{cases}$$
$$\langle C \rangle = \sum_{j=0}^{i-1} 2^{t(i)}.$$

This obviously holds for $i = 0$ and is maintained in iteration $i + 1$ by the following code:

```
12: srl   25 25 1      // X = srl(X,1)
13: srl   24 24 1      // B = srl(B,1)
14: sltu 27 23 24      // U = (A < B)
15: bgtz 27 -3         // repeat until A>=B
16: or    26 26 25     // C = C OR X
17: sub   23 23 24     // A = A - B
18: sltu 27 23 24      // U = (A < B)
19: blez 27 -7         // repeat outer loop if A>=B
20: add   k 26 0       // gpr(k) = C
```

In the inner loop (lines 12-15) B and X are shifted to the right until

$$\langle B \rangle < \langle A \rangle = A(i).$$

At this point, we have

$$X[j] = 1 \leftrightarrow j = t(i).$$

We OR X into C which gives a new value

$$\langle C \rangle = \sum_{j=0}^{i-1} 2^{t(j)} + 2^{t(i)}$$

$$= \sum_{j=0}^{i} 2^{t(j)}.$$

Finally, we subtract $\langle B \rangle$ from $\langle A \rangle$, obtaining a new value A which satisfies

$$\langle A \rangle = A(i) - 2^{t(i)} \cdot b$$

$$= A(i+1).$$

The last line copies the result into the register k.

We abbreviate the above program for 'short' division by

$$\text{sdiv}(k,i,j).$$

9.3.3 Unsigned Integer Division

We drop the hypothesis $p[31] = 0$ and extend the program $sdiv$ of Sect. 9.3.2 to handle this case. Some care has to be taken to avoid overflow of the integer arithmetic provided by the MIPS ISA. We use two additional auxiliary registers

$$T = gpr(21_5),$$
$$V = gpr(22_5).$$

We leave lines 1 to 7 unchanged and initialize T with 10^{31}:

```
8-9: gpr(21) = 10^31   // T = 10...0
```

In the loop where we successively double $\langle X \rangle$ and $\langle B \rangle$ we have to add an exit condition for the case $B[31] = 1$. The corresponding predicate is computed in register V.

```
10: and   22 24 21    // V = B[31]0...0
11: bgtz  22 5        // exit if B[31]=1
12: add   25 25 25    // X = X + X
13: add   24 24 24    // B = B + B
14: sltu  27 23 24    // U = (A < B)
15: blez  27 -5       // repeat if A>=B
```

This loop ends after x iterations for some x and we have

$$\langle A \rangle = a,$$
$$\langle B \rangle = 2^x \cdot b,$$
$$X[j] = 1 \leftrightarrow j = x,$$
$$2^x \cdot b > a \wedge x = t(0) + 1 \ \vee \ 2^x \cdot b \le a \wedge x = t(0).$$

In order to test for $2^x \cdot b \le a$ we need to compute $\langle B \rangle \le \langle A \rangle$, but our MIPS ISA does not provide such an R-type instruction. Fortunately, using

$$2^x \cdot b \le a \leftrightarrow 2^x \cdot b - 1 < a$$

and $b \ne 0$ the test can be performed with the *sltu* instruction:

```
14: addi 24 24 -1   // B = B - 1, no underflow as B>0
15: sltu 22 24 23   // V = (B < A)
16: addi 24 24 1    // B = B + 1
17: blez 22 3       // skip next 2 instructions if B>A
```

We handle the new case by setting $\langle C \rangle$ to $2^x = 2^{t(0)}$ and computing $A(1) = a - 2^{t(0)} \cdot b$. In case b is very big from the beginning, i.e., if $a/b = 1$, we are done and jump to the last instruction:

```
18: or   26 26 25   // C = C OR X
19: sub  23 23 24   // A = A - B
20: sltu 27 23 24   // U = (A < B)
21: blez 27 9       // continue only if A>=B
```

After this, we append in lines 22 to 29 the main loop which was previously in lines 12 to 19. Now we have two cases.

- If $V = 0$, lines 18-21 are skipped and the old analysis applies.
- If $V = 1$, execution of lines 18-21 amounts to one less iteration of the body of the main loop. This simply shifts the indices by 1 and after i iterations of the new main loop we have

$$\langle A \rangle = A(i+1),$$
$$\langle B \rangle = 2^{t(i)} \cdot b,$$
$$X[j] = 1 \leftrightarrow j = t(i),$$
$$\langle C \rangle = \sum_{j=0}^{i} 2^{t(j)}.$$

As before, we end with

```
30: add  k 26 0     // gpr(k) = C
```

We abbreviate the above program for the division of unsigned integers by

```
divu(k,i,j).
```

9.3.4 Integer Division

We implement integer division for use in a compiler of a C-like programming language. The C11 standard [ISO11] specifies that the result of integer division done by C programs is rounded towards zero, i.e., for integers $a, b \in \mathbb{Z}$ with $b \neq 0$, the result of their division is

$$divt(a,b) = \begin{cases} a/b, & a > 0 \wedge b > 0, \\ (-1)^{sign(a) \oplus sign(b)} \cdot (|a|/|b|), & \text{otherwise}, \end{cases}$$

where

$$sign(z) = \begin{cases} 0, & z \geq 0, \\ 1, & z < 0. \end{cases}$$

We implement this behavior in a fairly brute force way:

1. We treat the special case which produces an overflow in the division of two's complement numbers, i.e.

$$a/b = -2^{31}/(-1) = 2^{31} \equiv -2^{31} \text{ tmod } 2^{32},$$

   explicitly by

```
1-2: gpr(21)  = 10^31  // gpr(21) = 10...0
  3: bne   21 i 5      // go to line 8 if a!=-2^31
  4: nor   21 0 0      // gpr(21) = 11...1
  5: bne   21 j 3      // go to line 8 if b!=-1
  6: add   k i 0       // gpr(k) = gpr(i)
  7: beq   0 0 51      // skip remaining instructions
                       // of the program
```

2. Using program *sign-and-abs* from Sect. 9.1, we convert operands in registers i and j into sign and absolute value. We overwrite operands with their absolute values and store the signs in registers 22 and 21. Then, we compute the sign bit of the result in register 20.

```
 8-15: sign-and-abs(21,i,i)
16-22: sign-and-abs(22,j,j)
   23: xor   20 21 22
```

3. We compute the absolute value of the result in register k using program *divu* for unsigned integers from Sect. 9.3.3.

```
24-53: divu(k,i,j)
```

We now have

$$\langle gpr(k_5) \rangle = |a|/|b| \quad \text{and} \quad |a|/|b| \in [0 : 2^{31} - 1],$$

hence

$$gpr(k_5)[31] = 0 \quad \text{and} \quad [gpr(k_5)] = \langle gpr(k_5) \rangle = |a|/|b|.$$

4. If the sign bit of the result is 1, we invert and increment register k:

```
54: blez 20 4
55: nor  22 0 0      // gpr(22)  = 1^32
56: xor  k k 22      // gpr(k)   = !gpr(k)
57: addi k k 1       // gpr(k)   = gpr(k) + 1
```

Finally, we have the correct result in register k:

$$[c'.gpr(k_5)] = divt([c.gpr(i_5)], [c.gpr(j_5)]).$$

We abbreviate this program by

```
divt(i,j,k).
```

9.4 Final Remarks

In general, never try to memorize particular programs for later use. Memorize the algorithm instead. We recommend to remember the following key technical points:

- the definition of integer division

$$a/b = \max\{C \in \mathbb{N} \mid b \cdot C \le a\}.$$

- the definition of the distance by which the b operand is shifted to the left

$$t(a,b) = \max\{k \in \mathbb{N} \mid 2^k \cdot b \le a\}.$$

- the statement of Lemma 58 formulating the correctness of a single step of the school method of division

$$a/b = 2^{t(a,b)} + (a - 2^{t(a,b)} \cdot b)/b.$$

9.5 Exercises

1. Write an assembly program which
   a) tests whether bit 11 of register 1 is one.
   b) computes the smallest i such that bit i of register 2 is one.

c) shifts bits $[19:0]$ of register 3 by 12 positions to the left and fills the remaining bits with zeros.

d) shifts the content of register 4 by one position to the right without using the *srl* instruction.

Hint: you can easily determine the bits of the register.

2. Write an assembly program which cyclically shifts register 4 by $\langle gpr(5_5) \rangle$ positions to the left and stores the result in register 5.

Attention: the number in register 5 can be larger than 31.

3. Write an assembly program which writes the absolute value of $[gpr(6_5)]$ to register 1.

4. From time to time one sees students and others (!) write the following simple program for the division of binary numbers a and b with the intention of running it:

```
1: addi  3 0 0      // gpr(3) = 0
2: blez  0 3        // go to line 5
3: sub   1 1 2      // gpr(1) = gpr(1) - gpr(2)
4: addi  3 3 1      // gpr(3) = gpr(3) + 1
5: slt   4 1 2      // gpr(4) = ( gpr(1)<gpr(2) )
6: blez  4 -3       // if gpr(1)>=gpr(2) go to line 3
```

Initially $a = \langle gpr(1_5) \rangle$ and $b = \langle gpr(2_5) \rangle$, s.t., $a \geq b$.

a) Prove that the program performs integer division:

$$\langle gpr(3_5) \rangle = a/b.$$

Hint: you have to write an invariant about the configuration after i iterations of the loop.

b) Count the number of instructions executed if $a = 2^{32} - 1$ and $b = 2$.

c) Assuming that our processor hardware is clocked at 1 GHz, determine the time it takes to run the program with these inputs. Recall that $G = 10^9$, $1 \text{ Hz} = 1/s$, and that our processor hardware needs two cycles per instruction.

d) Count instructions and determine the running time for the program from Sect. 9.3.3.

e) Repeat for a 64-bit machine and $a = 2^{64} - 1$.

Try to remember this example. Analyzing the run time of algorithms is *not* abstract nonsense. Try to notice immediately, when you begin programming something like this. This is particularly important if you are new in an organization, because: *there is no second chance to make a first impression.*

10

Context-Free Grammars

In Chap. 11 we will define the syntax and semantics of a C dialect called $C0$, and in Chap. 12 we will construct a non-optimizing compiler translating programs from $C0$ to MIPS ISA. Both chapters will rely heavily on the derivation trees of $C0$ programs for a certain context-free grammar: both the semantics of programs and the code generation of compiled programs are defined by induction over these derivation trees. Therefore we present in Sect. 10.1 an introduction to context-free grammars with a strong emphasis on derivation trees. The classical graph theoretical model of rooted trees (Sect. 3.4.3) does not specify the order of the sons of nodes. In contrast, this order is essential in derivation trees. Moreover, when one treats algorithms operating on trees — such as a compiler — it is desirable to have a concise notation for specifying the i'th son of node v in the tree. In Sect. 10.1.3 we present such a model: *tree regions*.

In general context-free grammars permit us to derive the same word w with different derivation trees, but if one wishes to define the semantics of expressions e or of programs p with the help of derivation trees, one prefers to make use of *unambiguous* grammars G, in which derivation trees are unique, such that the definition of semantics for e or p can hinge on *the* derivation tree for e or p in grammar G. In particular we study in Sect. 10.2.2 a well-known grammar G_A for arithmetic expressions with the usual priorities between operators and incomplete bracket structure, and prove that it is unambiguous. This result is rarely presented in textbooks, although it is of fundamental importance for our daily life whenever we interpret expressions like $3 \cdot x + 5$ as $(3 \cdot x) + 5$ and not as $3 \cdot (x + 5)$. It shows that with proper knowledge of the priority rules no confusion can arise by leaving out brackets. Proper knowledge of these rules is usually *not* taught in high school, by the way. The proof presented here is taken from [LMW89].

In the classroom we introduce tree regions but do not prove lemmas about them. We do present the proof that grammar G_A is unambiguous, one may say for philosophical reasons.

© Springer International Publishing Switzerland 2016
W.J. Paul et al., *System Architecture*, DOI 10.1007/978-3-319-43065-2_10

10.1 Introduction to Context-Free Grammars

10.1.1 Syntax of Context-Free Grammars

A context-free grammar G consists of the following components:

- a finite set of symbols $G.T$, called the alphabet of *terminal* symbols,
- a finite set of symbols $G.N$, called the alphabet of *nonterminal* symbols,
- a *start symbol* $G.S \in G.N$, and
- a finite set $G.P \subset G.N \times (G.N \cup G.T)^*$ of *productions*.

Symbols are either terminal or nonterminal, never both:

$$G.T \cap G.N = \emptyset.$$

If (n, w) is a production of grammar G, i.e., $(n, w) \in G.P$, we say that the string w is *directly derived* in G from the nonterminal symbol n. As a shorthand, we write

$$n \rightarrow_G w.$$

If there are several strings $w^1, \ldots, w^s$ that are directly derived in G from n, we write

$$n \rightarrow_G w^1 \mid \ldots \mid w^s.$$

Note that with this notation we risk ambiguity if the symbol '$\mid$' used to separate the possible right-hand sides lies in $G.N \cup G.T$. Productions with an empty right-hand side are called epsilon rules. They are written in the form

$$n \rightarrow_G \varepsilon,$$

but we will avoid them. If it is clear which grammar is meant we abbreviate

$$T = G.T,$$
$$N = G.N,$$
$$S = G.S,$$
$$P = G.P,$$
$$\rightarrow = \rightarrow_G.$$

Before we present an example, let us briefly discuss notation. In mathematics, there is a constant shortage of symbols that can be easily remembered: one uses capital and small letters of the Latin and Greek alphabets. This is not enough. One borrows in set theory from the Hebrew alphabet. One introduces funny letters like $\mathbb{B}$ and $\mathbb{N}$. The shortage remains. Modification of symbols with accents and tildes is a more radical step: using a symbol naked, with an accent, or with a tilde immediately triples the number of available symbols. In computer science it is not uncommon to take the following brute force measure to generate arbitrarily many symbols: take a non-empty alphabet A that does not contain the symbols '$\langle$' and '$\rangle$'. There are arbitrarily

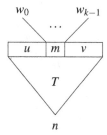

Fig. 71. Extending derivation tree T by application of production $m \to w_0 \dots w_{k-1}$

many strings $w \in A^*$. For each such w treat $\langle w \rangle$ as a new symbol. Given a string of such symbols, the pointed brackets permit one to decide where symbols begin and end.

In the following sections, we explain derivation trees using the following example grammar:

$$
\begin{aligned}
T &= \{0, 1, X\}, \\
N &= \{V, B, \langle BS \rangle\}, \\
S &= V, \\
B &\to 0 \mid 1, \\
\langle BS \rangle &\to B \mid B \langle BS \rangle, \\
V &\to X \langle BS \rangle \mid 0 \mid 1.
\end{aligned}
$$

10.1.2 Quick and Dirty Introduction to Derivation Trees

Let $G = (T, N, S, P)$ be a context-free grammar. Intuitively and with the help of drawings, it is easy to explain what a derivation tree T for grammar G is. For nonterminals $n \in N$ and strings $w \in (N \cup T)^*$, we generate the derivation trees T with root n and border word w by the following rules:

1. a single nonterminal n is a derivation tree with root n and border word n.
2. if T is a derivation tree with root n and border word umv and $m \to w \in P$ is a production with $w = w_0 \dots w_{k-1}$, then the tree illustrated in Fig. 71 is a derivation tree with root n and border word uwv.[1]
3. all derivation trees can be generated by finitely many applications of the above two rules.

In the example grammar above, we can construct a derivation tree with root V and border word $X01$ by six applications of the above rules as illustrated in Fig. 72.

Derivation trees can be composed under certain conditions.

---

[1] For epsilon rules one has to set $w = w_0 = \varepsilon$.

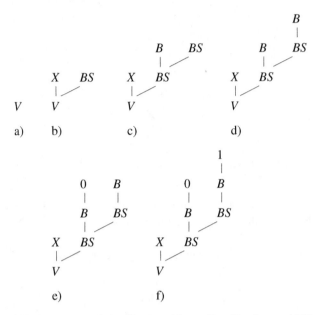

Fig. 72. Generating a derivation tree with root V and border word $X01$

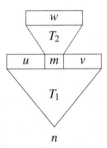

Fig. 73. Composing derivation tree T_1 with root n and border word umv with derivation tree T_2 with root m and border word w

Lemma 60. *If T_1 is a derivation tree with root n and border word umv and T_2 is a derivation tree with root m and border word w, then the tree T from Fig. 73 is a derivation tree with root n and border word uwv.*

Proof. Generate tree T by first applying the rules for generating T_1 and then using the rules for generating T_2.

We can define that a derivation tree T_2 is a subtree of derivation tree T if T can be generated by the composition of T_2 with some derivation tree T_1, as in Lemma 60.

This concept of derivation trees goes a long way. However, problems with the given informal definition arise when we try to argue about the behavior of algorithms which

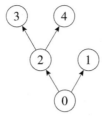

Fig. 74. Drawing the rooted tree from Fig. 3 in a different way

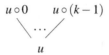

Fig. 75. Standardized names of nodes specify the order of sons

traverse and process derivation trees. When we try to argue what happens when the algorithm visits a certain node, the informal definition provides us with no language to do so — simply because nodes do not have names. In order to later argue about compilers (which implement an algorithm that traverses and processes the derivation tree of a program in order to generate machine code), we give in the next subsections a clean definition of derivation trees and relate it to the basic graph theory from Sect. 3.4.

10.1.3 Tree Regions

The graph theoretic definition of rooted trees from Sect. 3.4.3 does not specify the order of sons. It does not distinguish between the graphs drawn in Fig. 3 and in Fig. 74. In what follows however we cannot ignore that order. Therefore we will formalize trees such that

- for each node u with outdegree k the sons of u are numbered from 0 to $k-1$,
- nodes have standardized names $u \in \mathbb{N}^*$. The root has name $r = \varepsilon$ and the sons of a node u with outdegree k are called $u \circ 0, \ldots, u \circ k - 1$. When we draw trees, node $u \circ i$ is drawn to the left of $u \circ (i+1)$ (see Fig. 75).

Actually this naming is motivated by graph algorithms working on graphs starting at the root. If one labels the edges of a node with k sons from 0 to $k-1$, then the name v of a node is simply the sequence of edge labels encountered on the path from the root to v. For the graphs in Figs. 3 and 74 this is illustrated in Figs. 76 a) and b).

For a formal definition of tree regions we define from sets A of sequences of natural numbers

$$A \subset \mathbb{N}^*$$

graphs $g(A)$ in the following way:

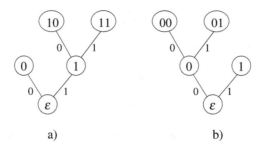

a)   b)

Fig. 76. Names of nodes correspond to edge labels on the paths from the root to the nodes

- the set of nodes of $g(A)$ is A

$$g(A).V = A,$$

- for $u, v \in A$ there is an edge from u to v if $v = u \circ i$ for an $i \in \mathbb{N}$

$$(u,v) \in g(A).E \leftrightarrow \exists i \in \mathbb{N} : v = u \circ i.$$

For nodes $v \in A$ we can characterize their indegree and outdegree in the graph $g(A)$ as follows.

Lemma 61.

$$outdeg(v, g(A)) = \#\{i \in \mathbb{N} \mid v \circ i \in A\},$$

$$indeg(v, g(A)) = \begin{cases} 1, & \exists v' \in A, i \in \mathbb{N} : v = v' \circ i, \\ 0, & otherwise. \end{cases}$$

Observe that with this definition $indeg(\varepsilon, g(A)) = 0$. We again define a node v to be a source of A if $indeg(v, g(A)) = 0$, i.e., if there is no node $u \in A$ and no $i \in \mathbb{N}$ such that $v = u \circ i$. We define u to be a sink of A if $outdeg(u, g(A)) = 0$, i.e., if for no $i \in \mathbb{N}$ we have $u \circ i \in A$.

Now we define tree regions as the subsets $A \subset \mathbb{N}^*$ such that the graphs $g(A)$ are rooted trees with root ε. Formally

1. $A = \{\varepsilon\}$ is a tree region.
2. if A is a tree region, v is a sink of A, and $k \in \mathbb{N}$ then

$$A' = A \cup \{v \circ 0, \ldots, v \circ (k-1)\}$$

    is a tree region.
3. all tree regions can be generated by finitely many applications of these rules.

A trivial argument shows the following.

Lemma 62. *If A is a tree region, then $g(A)$ is a rooted tree.*

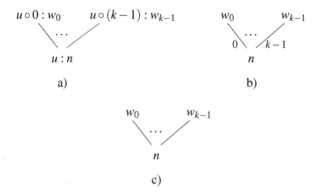

Fig. 77. In derivation trees, nodes and their sons are labeled by productions of the grammar. Here $n \rightarrow w_0 \ldots w_{k-1}$ must hold

Formally the tree region of Fig. 76 a) is specified by

$$A = \{\varepsilon, 0, 1, 10, 11\},$$

while the tree region of Fig. 76 b) is specified by

$$A = \{\varepsilon, 0, 1, 00, 01\}.$$

The father *father*(u) of a node $u = u[1:n]$ in a tree region is defined as

$$father(u[1:n]) = u[1:n-1].$$

Any node $v = u[1:s]$ with $s \leq n$, i.e., satisfying *prefix*(v, u) is then an ancestor of u.

10.1.4 Clean Definition of Derivation Trees

A derivation tree T for grammar G has the following components:

- a tree region $T.A$,
- a labeling $T.\ell : T.A \rightarrow G.N \cup G.T \cup \{\varepsilon\}$ of the elements of $T.A$ by symbols of the grammar satisfying: if $outdeg(u, g(T.A)) = k$, and $T.\ell(u) = n$, and for all $i \in [0 : k-1]$ we have $T.\ell(u \circ i) = w_i$, then

$$n \rightarrow_G w_0 \ldots w_{k-1},$$

i.e., the sequence of the labels of the sons of u is directly derived in G from the label of u (see Fig. 77 a).

There are three obvious ways to draw derivation trees T:

- for each node $u \in T.A$ in the drawing, we write down $u : T.\ell(u)$, i.e., the name and its label as in Fig. 77 a).

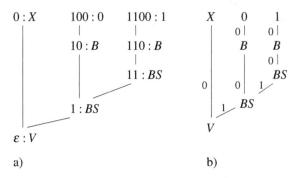

Fig. 78. A derivation tree for the grammar of the introductory example. We have written BS instead of $\langle BS \rangle$

- we can omit the names of the nodes as shown in Fig. 77 b) and only draw edge labels.
- we can omit the edge labels as shown in Fig. 77 c). This is justified as edges starting at the same node are labeled from left to right starting with 0. As long as the entire tree is drawn the names of the nodes can easily be reconstructed from the drawing. Also we know that node labels in derivation trees are single symbols. If such a symbol happens to be an artificially generated symbol of the form $\langle w \rangle$ we will take the freedom to omit the pointed brackets.

A derivation tree for the grammar of the introductory example is shown in Fig. 78 a) with node names and in Fig. 78 b) with edge labels.
Let T be a derivation tree and $u \in T.A$ be a node in T. The border word $bw(u, T)$ is obtained by concatenating the labels of all leaves reachable from u from left to right. Formally

$$bw(u, T) = \begin{cases} bw(u \circ 0, T) \circ \ldots \circ bw(u \circ (k-1), T), & k = outdeg(u, g(T.A)) > 0, \\ T.\ell(u), & \text{otherwise.} \end{cases}$$

If it is clear which tree is meant we drop the T. In the derivation tree T of Fig. 78 we have for instance

$$\begin{aligned}
bw(\varepsilon) &= bw(0) \circ bw(1) \\
&= \ell(0) \circ bw(10) \circ bw(11) \\
&= X \circ bw(100) \circ bw(110) \\
&= X \circ \ell(100) \circ bw(1100) \\
&= X0 \circ \ell(1100) \\
&= X01.
\end{aligned}$$

With the new precise definition of derivation trees we redefine the relation $n \to_G w$: we say that a word w is derivable from n in G and write $n \to_G^* w$ if there is a derivation tree T for G with border word w whose root is labeled with n, i.e.,

Fig. 79. Composition of trees T_1 and T_2 at leaf u of T_1

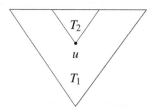

Fig. 80. Decomposing a derivation tree T into T_1 and T_2 at node u

$$n \to_G^* w \leftrightarrow \exists T : T.\ell(\varepsilon) = n \wedge bw(\varepsilon, T) = w.$$

10.1.5 Composition and Decomposition of Derivation Trees

The following easy theory permits us now to argue about the composition and the decomposition of derivation trees.

For composition, let T_1 and T_2 be derivation trees for a common grammar G. Let u be a leaf of T_1 and assume that the labels of u in T_1 and the root ε in T_2 are identical:

$$T_1.\ell(u) = T_2.\ell(\varepsilon). \tag{8}$$

Then, as illustrated in Fig. 79, we can compose trees T_1 and T_2 into a new tree $T = comp(T_1, T_2, u)$ in the following way. Nodes of T are the nodes of T_1 as well as the nodes of T_2 extended by prefix u

$$T.A = T_1.A \cup (\{u\} \circ T_2.A).$$

Labels of nodes in T are imported in the obvious way

$$T.\ell(v) = \begin{cases} T_1.\ell(v), & v \in T_1.A, \\ T_2.\ell(x), & v = u \circ x \end{cases}$$

For $x = \varepsilon$ this is well defined by Equation 8. An easy exercise shows the following.

Lemma 63. $T = comp(T_1, T_2, u)$ *is a derivation tree.*

For decomposition, let T be a derivation tree and $u \in T.A$ be a node of the tree. As illustrated in Fig. 80 we decompose T into the subtree $T_2 = sub(T, u)$ of T with root u and the tree $T_1 = rem(T, u)$ remaining if T_2 is removed from T. Formally

$$T_2.A = \{v \in T.A \mid u \circ v \in T.A\},$$
$$T_1.A = (T.A \setminus (\{u\} \circ T_2.A)) \cup \{u\},$$
$$T_2.\ell(v) = T.\ell(u \circ v),$$
$$T_1.\ell(v) = T.\ell(v).$$

An easy exercise shows the following.

Lemma 64. $T_1 = rem(T,u)$ *and* $T_2 = sub(T,u)$ *are both derivation trees.*

Another easy exercise shows that if we decompose a tree T at node u and then compose the trees $T_1 = rem(T,u)$ and $T_2 = sub(T,u)$ again at u we get back — words of wisdom — the original tree.

Lemma 65.
$$T = comp(rem(T,u), sub(T,u), u).$$

Figures 79 and 80 differ only in the way we draw T_1. In Fig. 79 we stress that u is a leaf of T_1. In Fig. 80 (where u is also a leaf of T_1) we stress that it is some node in the tree T.

10.1.6 Generated Languages

The language generated by nonterminal n in grammar G consists of all words in the terminal alphabet that are derivable from n in G. It is abbreviated by $L_G(n)$:

$$L_G(n) = \{w \mid n \to_G^* w\} \cap G.T^*.$$

An easy exercise shows for the grammar G of the introductory example

$$L_G(\langle BS \rangle) = \mathbb{B}^+,$$
$$L_G(V) = X\mathbb{B}^+ \cup \mathbb{B},$$

where $X\mathbb{B}^+$ is a shorthand for $\{X\} \circ \mathbb{B}^+$. Observe that we could drop the brackets of symbol $\langle BS \rangle$ because the arguments of function L_G are single symbols. The *language* generated by grammar G is the language $L_G(G.S)$ generated by G from the start symbol. A language is called *context-free* if it is generated by some context-free grammar.

10.2 Grammars for Expressions

10.2.1 Syntax of Boolean Expressions

In the hardware chapters we have introduced Boolean expressions by analogy to arithmetic expressions, i.e., we have not given a precise definition. We will now fill this gap by presenting a context-free grammar that generates these expressions.

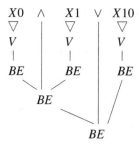

Fig. 81. A derivation tree reflecting the usual priority rules

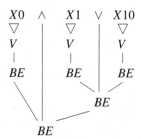

Fig. 82. A derivation tree not reflecting the usual priority rules

We will consider variables in $X\mathbb{B}^+$ and constants in $\mathbb{B}$, and we use the productions of the grammar from the introductory example to produce such variables or constants from symbol V. Boolean expressions are derived from symbol $\langle BE \rangle$ by the productions

$$\langle BE \rangle \rightarrow V \mid \neg \langle BE \rangle \mid (\langle BE \rangle) \mid \langle BE \rangle \wedge \langle BE \rangle \mid \langle BE \rangle \vee \langle BE \rangle \mid \langle BE \rangle \oplus \langle BE \rangle.$$

Note that here we are using the negation symbol '$\neg$' instead of overlining variables. We make $\langle BE \rangle$ the start symbol of the extended grammar G. A derivation tree for the expression $X0 \wedge X1 \vee X10$ in G is shown — without the details for the derivation of the variables from symbol V — in Fig. 81.

If we were to evaluate expressions as suggested by the shape of the tree, this tree would nicely reflect the usual priority among the Boolean operators, where $\wedge$ binds more strongly than $\vee$. However, the simple grammar G also permits a second derivation tree of the same Boolean expression, which is shown in Fig. 82. Although grammar G nicely represents the syntax of expressions, it is obviously not helpful for the definition of expression evaluation. For this purpose we need grammars G generating expressions with two additional properties:

- the grammars are *unambiguous* in the sense that for every word w in the language generated by G there is exactly one derivation tree for G which derives w from the start symbol,
- this unique derivation tree should reflect the usual priorities between operators.

In the next subsection we exhibit such a grammar for arithmetic expressions.

10.2.2 Grammar for Arithmetic Expressions with Priorities

Evaluation of arithmetic expressions is more interesting than meets the eye. People tend to believe that they have learnt expression evaluation in high school. Readers feeling this way are encouraged to solve the following.

Exercise

Evaluate $2/2/2/2$ and $2----2$.

We give two hints: i) first evaluate $2-2-2-2$. This gives you an idea about the order of operators. Then evaluate $2-(-2)$. You will recognize that there are two minus signs: a unary operator $-_1$ and a binary operator $-_2$. So you are really evaluating $2 -_2 (-_12)$. As in Boolean Algebra, unary operators tend to have higher priority than binary operators. This will help you to evaluate $2--2$ and solve the exercise. We postpone the problem of how to identify unary and binary minus signs to a series of exercises in Sect. 10.2.4.

For the time being we do not bother with how constants or variables are generated and temporarily take V as a terminal symbol. A grammar G taking care of the above problems has the following terminal symbols:

$$G.T = \{+, -_2, *, /, -_1, (,), V\}.$$

For each of the priority levels we have a nonterminal symbol

$$G.N = \{F, T, A\}.$$

F like 'factor' stands for the highest priority level of the unary minus or an expression in brackets

$$F \rightarrow V \mid -_1F \mid (A).$$

T like 'term' stands for the priority level of multiplication and division

$$T \rightarrow F \mid T*F \mid T/F.$$

A like 'arithmetic expression' is the start symbol and stands for the priority level of addition and subtraction

$$A \rightarrow T \mid A+T \mid A -_2 T.$$

We illustrate this grammar by a few examples: in Fig. 83 we have a derivation tree for $V/V/V/V$. This tree suggests evaluation of the expression from left to right. For our first exercise above we would evaluate

$$2/2/2/2 = ((2/2)/2)/2 = 1/4.$$

In Fig. 84 we have a derivation tree for $V -_2 -_1 -_1 -_1 V$. According to this tree we would evaluate

$$2----2 = 2-(-(-(-2))) = 4.$$

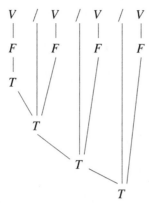

Fig. 83. The grammar determines the order in which division signs are applied

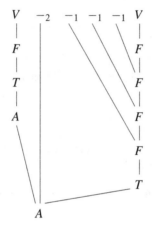

Fig. 84. A derivation tree reflecting the usual priority rules

In Fig. 85 a) we have a tree for $V * V + V$ suggesting that $*$ binds more strongly than $+$. An attempt to start the construction of a derivation tree as in Fig. 85 b) apparently leads to a dead end. Indeed one can show the following for the above grammar G.

Lemma 66. *G is unambiguous.*

This lemma is nothing less than fundamental, because it explains why expression evaluation is well defined. The interested reader can find the proof in the next subsection.

10.2.3 Proof of Lemma 66

We proceed in four fairly obvious steps, showing successively that derivation trees are unique for

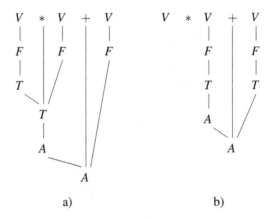

a)   b)

Fig. 85. Construction of a derivation tree not reflecting the usual priority rules

Fig. 86. Deriving V from F

1. factors without brackets,
2. terms without brackets,
3. arithmetic expressions without brackets, and
4. arithmetic expressions with brackets.

Lemma 67. *Let $F \to^* w$ and $w \in \{V, -_1\}^+$. Then the derivation tree U of w from F is unique.*

Proof. First observe that production $F \to (A)$ cannot be used, because w contains no brackets, productions $T \to T \circ F$ and $A \to A \circ T$ cannot be used, because $*, /, +, -_2$ do not occur, and we are left with productions

$$F \to V \mid -_1 F,$$
$$T \to F,$$
$$A \to T.$$

Now we prove the lemma by induction over the number n of occurrences of the unary minus. For $n = 0$ we have $w = V$ and only the production $F \to V$ can be used to derive the V as shown in Fig. 86. The symbol F shown in that figure must be at the root of the tree. Otherwise there would be an ancestor of F which is labeled F. The production $F \to -_1 F$ would produce a unary minus, thus it cannot be used. Thus with the productions available ancestors of a node labeled F can only be labeled with T and A.

Fig. 87. Derivation of a border word with $n \geq 1$ unary minus signs from F

Fig. 88. Deriving a border word without multiplication and division signs from T

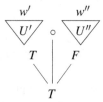

Fig. 89. Deriving a border word with $n \geq 1$ multiplication and division signs from T

For the induction step let $n > 0$ and let U be a derivation tree for $-_1^n V$. Consider the *first* occurrence of symbol $-_1$ in w. It can only be derived by production $F \rightarrow -_1 F$ as shown in Fig. 87. We claim that the node labeled with F at the bottom of this figure is the root. If it had a father labeled F, then the $-_1$ in the figure would not be the leftmost one. Other labels of ancestors can only be T and A.

Decompose the derivation tree U into $U' = sub(U, 1)$ and the rest (that we have just determined) as shown in Fig. 87. U' is unique by induction hypothesis. Hence the composition is unique. By Lemma 65 this composition is U.

Lemma 68. *Let* $T \rightarrow^* w$ *and* $w \in \{V, -_1, *, /\}^+$. *Then the derivation tree* U *of* w *from* T *is unique.*

Proof. We can only use the productions

$$F \rightarrow V \mid -_1 F,$$
$$T \rightarrow F \mid T * F \mid T/F,$$
$$A \rightarrow T,$$

because the other productions produce terminals not in the border word of U.

We prove the lemma by induction over the number n of occurrences of symbols $*$ and $/$. For $n = 0$ the derivation must start at the root labeled T with an application

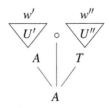

Fig. 90. Deriving a border word without addition and subtraction signs from A

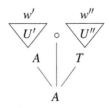

Fig. 91. Deriving a border word with $n \geq 1$ addition and subtraction signs from A

of production $T \to F$ as shown in Fig. 88, because other productions would produce terminals $*$ or $/$ not in the border word. The remaining part U' of the tree is unique by Lemma 67.

For $n > 0$ consider the *last* occurrence of a multiplication or division sign $\circ \in \{*, /\}$. Then $w = w' \circ w''$. It can only be generated by production $T \to T \circ F$ as shown in Fig. 89. We claim that the lower node in the figure labeled T is the root. If it had a father labeled T, then $\circ$ in the picture would not be the rightmost one in the border word. Thus ancestors can only be labeled with A. Subtree U'' is unique by Lemma 67. Subtree U' is unique by induction hypothesis.

Lemma 69. *Let $T \to^* w$ and $w \in \{V, -_1, *, /, +, -_2\}^+$. Then the derivation tree U of w from T is unique.*

Proof. We can only use the productions

$$F \to V \mid -_1 F,$$
$$T \to F \mid T * F \mid T/F,$$
$$A \to T \mid A + T \mid A -_2 T.$$

We prove the lemma by induction over the number n of occurrences of symbols $+$ and $-_2$. For $n = 0$ the derivation must start at the root labeled A with application of production $A \to T$ as shown in Fig. 90 because other productions would produce terminals $+$ or $-_2$ not in the border word. The remaining part U' of the tree is unique by Lemma 68.

For $n > 0$ consider the *last* occurrence of an addition or binary minus sign $\circ \in \{+, -_2\}$. Then $w = w' \circ w''$. It can only be generated by production $A \to A \circ T$ as shown in Fig. 91. We claim that the lower node in the figure labeled A is the root. If it had a father labeled A, then $\circ$ in the picture would not be the rightmost one in

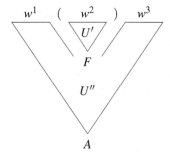

Fig. 92. Deriving a border word $w^1(w^2)w^3$ with an innermost pair of brackets around w^2 from A

the border word. Thus there are no ancestors. Subtree U'' is unique by Lemma 68. Subtree U' is unique by induction hypothesis.

Now we prove Lemma 66 by induction on the number n of bracket pairs. For $n = 0$ we get the uniqueness of the derivation tree from Lemma 69. For $n > 0$ decompose

$$w = w^1(w^2)w^3,$$

where the pair of brackets shown is an innermost bracket pair, i.e., w^2 contains no brackets. The bracket pair shown is generated by production $F \to (A)$ as shown in Fig. 92. Subtree U' of the derivation tree is unique by Lemma 69. Assume there are two different derivation trees U_1'' and U_2'' deriving $w^1 F w^3$ from A. Applying production $F \to V$ in both trees to the leaf F shown gives two derivation trees for $w^1 V w^3$ which has $n-1$ pairs of brackets. This is impossible by induction hypothesis.

10.2.4 Distinguishing Unary and Binary Minus

The theory of this chapter cannot yet be applied directly to ordinary arithmetic expressions, simply because in ordinary expressions one does not write $-_1$ or $-_2$. One simply writes $-$. Fortunately there is a very simple recipe to identify a binary minus: it only stands to the right of symbols V or $)$, whereas the unary minus never stands to the right of these symbols. The reader can justify this rule by proving the following lemma.

Lemma 70.

1. If $a \in L(F) \cup L(T) \cup L(A)$ is any expression, then a ends with symbol $)$ or V.
2. If $u -_2 v \in L(A)$ is an arithmetic expression, then u ends with $)$ or V.
3. If $u -_1 v \in L(A)$ is an arithmetic expression, then u ends with $-_1, +, -_2, *$, or $/$.

10.3 Final Remarks

In our view, the key technical points to remember are

- about tree regions: names of nodes are standardized
  - the root is named $\varepsilon \in \mathbb{N}^*$,
  - son i of node u is $u \circ i \in \mathbb{N}^*$.
- about grammars
  - obviously, the definition of grammars $G = (T, N, P, S)$,
  - the quick and dirty definition of derivation trees,
  - the formal definition of derivation trees as $T = (A, \ell)$; in particular, if $s = outdeg(u, g(A))$, then

$$\ell(u) \rightarrow_G \ell(u \circ 0) \ldots \ell(u \circ s - 1).$$

- about unambiguous grammars for expression evaluation
  - one nonterminal for each priority level: F, T, A,
  - top priority is obvious: $F \rightarrow V \mid -_1 F \mid (A)$,
  - $T \rightarrow F \mid F * T \mid F / T$ or was it $T * F \mid T / F$...? Too hard to memorize. Memorize instead that evaluation is from left to right. One of the two alternatives will do it, the other will not. Same trick with A and $\{+, -_2\}$.
- about the uniqueness proof
  - obvious order of lemmas. Successively introduce i) $-_1$ ii) $*, /$ iii) $+, -_2$ iv) brackets,
  - for proof of lemmas consider i) leftmost $-_1$ ii)-iii) rightmost $\circ$ iv) innermost pair of brackets.
- for parties: $2/2/2/2 =?$ and $2 - - - -2 =?$

10.4 Exercises

1. Which of the following are tree regions? For the tree regions A, draw the trees $g(A)$.
   a) $\{\varepsilon\}$,
   b) $\{\varepsilon, 0, 1, 2\}$,
   c) $\{\varepsilon, (2, 0), (2, 1), (2, 2), (2, 3)\}$,
   d) $\{\varepsilon, 0, 1\}$,
   e) $\{\varepsilon, 2, (2, 0), (2, 1), (2, 2), (2, 3)\}$,
   f) $\{\varepsilon, 0, 1, 2, (2, 0), (2, 1), (2, 2), (2, 3)\}$.
2. Specify a context-free grammar G such that
   a) $L(G) = \{0^n 1^n \mid n \in \mathbb{N}\}$,
   b) $L(G) = \{a^i \mid i > 0\}$,
   c) $L(G) = \{a^i b^j c^j \mid i, j > 0\}$,
   and prove by induction that your solution is correct.
3. Consider the following grammar $G = (N, T, P, S)$ with $N = \{S, A, B\}$, $T = \{a, b\}$ and the productions

$$S \rightarrow aB \mid bA,$$
$$A \rightarrow a \mid aS \mid bAA,$$
$$B \rightarrow b \mid bS \mid aBB.$$

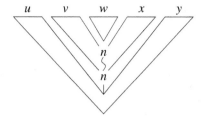

Fig. 93. Derivation tree with a path where a nonterminal n is repeated

For $x \in \{a, b\}$ we denote by $\#_x(w)$ the number of occurrences of symbol x in word w. Show for all $w \in T^+$:

$$S \to^* w \leftrightarrow \#_a(w) = \#_b(w),$$
$$A \to^* w \leftrightarrow \#_a(w) = \#_b(w) + 1,$$
$$B \to^* w \leftrightarrow \#_b(w) = \#_a(w) + 1.$$

4. Prove Lemma 70.
5. For the grammar studied in Sect. 10.2.2 write down the derivation trees for
   a) $(a + b + c * d)/e$,
   b) $(a/b) * (c + d) * e$.
6. Specify a grammar for pure Boolean expressions that reflects the usual priorities between the operators $\neg$, $\wedge$, and $\vee$.
7. We change the grammar of Sect. 10.2.2 to

$$F \to V \mid -_1 V \mid (A),$$
$$T \to F \mid F * T \mid F/T,$$
$$A \to T \mid T + A \mid T -_2 A.$$

   a) What is the derivation tree of $3/3/3$ with this grammar?
   b) What is the value of the expression, if the evaluation of the expression is guided by this derivation tree?
8. Test for any high-level language running on your computer how the compiler or interpreter handles exponentiation '$\wedge$'. To this end:
   a) Write a program which evaluates expressions like $x^\wedge a^\wedge b^\wedge c$ with appropriate constants and decide how the compiler or interpreter fills in brackets.
   b) Produce an extended grammar for arithmetic expressions which reflects this.
9. $uvwxy$-theorem.

   Let G be any context-free grammar and let k be the largest number of symbols in the right-hand side of a production. Let $T = (A, \ell)$ be a derivation tree for G. Let m be the number of nonterminal symbols in G.
   a) Show
      i. for any node u in T

$$outdeg(u, g(A)) \leq k.$$

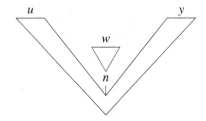

Fig. 94. Smaller derivation tree where the portion of the path between the repeated nonterminal is removed

    ii. if tree T has depth d, then it has at most k^d leaves.

    iii. if the border word of the tree is longer than k^m, then its depth is greater than m.

b) Let $w \in L(G)$ and $|w| > k^m$. Prove that there is a path in the derivation tree on which a nonterminal $n \in N$ is repeated.

    Hint: use the pigeonhole principle.

c) Decompose the tree as shown in Fig. 93. If portions v and x are both the empty word, remove the portion between the repeated nonterminal from the tree as shown in Fig. 94. The remaining tree has still a path longer than m. Why?

    Hint: you can repeat the argument until $v \neq \varepsilon \vee x \neq \varepsilon$.

d) Show that if $w \in L(G)$ and $|w| > k^m$, then w can be decomposed as

$$w = uvwxy$$

with

$$v \neq \varepsilon \vee x \neq \varepsilon,$$

such that for all $i \geq 0$

$$uv^i wx^i y \in L(G).$$

10. Show that

$$L = \{a^i b^i c^i \mid i \in \mathbb{N}\}$$

is not a context-free language.

Hint: use the $uvwxy$-theorem.

The Language $C0$

We present the syntax and semantics of an imperative programming language called $C0$. In a nutshell it is Pascal [HW73] with C syntax. Although $C0$ is powerful enough to do some serious programming work, its context-free grammar G shown in Table 12 fits on a single page.

We proceed in the obvious order. In Sect. 11.1 we present the grammar and give an informal introduction to the language. As an early programming example we indicate in Sect. 11.1.8 how to represent derivation trees as $C0$ data structures. This is meant as an aid for anybody who wishes to implement algorithms which process derivation trees like the compiler from Chap. 12. In Sect. 11.1.9 we develop some basic machinery for navigating in all kinds of sequences (statement sequences, declaration sequences, parameter sequences, etc.) occurring in derivation trees of grammar G.

Then we formally define the declared types in Sect. 11.2, configurations in Sect. 11.3, expression evaluation in Sect. 11.4, and statement execution in Sect. 11.5.

The start symbol of the grammar is

$$G.S = \langle prog \rangle.$$

The language $L(prog)$ generated by the grammar is a superset of all $C0$ programs. Numerous restrictions on the syntax are defined when we define the semantics, i.e., the meaning of $C0$ programs. In the context of program semantics the restrictions will appear most natural: they are just the conditions which happen to make the definition of the semantics work. In the course of formulating these restrictions and studying their consequences a simple theory of type correctness evolves.

In Sect. 11.6 we illustrate the use of the $C0$ semantics with the correctness proofs of a few example programs. For the remainder of this chapter and the next chapter this grammar stays fixed. Thus we drop all subscripts G.

In lectures we take our time at two places: i) explaining that the range of types cannot be defined by the obvious straightforward induction, because *struct* types t can contain components pointing to variables of type t; the standard example is linked lists; ii) introducing the components of $C0$ configurations. Once this is done, expression evaluation and statement execution can be treated with amazing speed in the classroom. Most lemmas belonging to the theory of type correctness are just stated in the classroom, because their proofs boil down to bookkeeping. The only exception is in Sect. 11.4.7, where the absence of dangling pointers is derived from the software condition that 'address of' is never taken for variables (or subvariables) on the stack.

11.1 Grammar of $C0$

In order to give the reader some overview of the language we proceed first with a brief, preliminary, and quite informal discussion of the grammar. The language uses well-known keywords like int, typedef, etc. We treat these as single symbols of the terminal alphabet. Similarly, certain mathematical operators such as $! =$ or $\&\&$ are treated like single symbols of the terminal alphabet.

11.1.1 Names and Constants

Symbol $\langle Di \rangle$ generates decimal digits

$$L(Di) = \{0, \ldots, 9\}$$

and symbol $\langle DiS \rangle$ generates sequences of digits

$$L(DiS) = \{0, \ldots, 9\}^+.$$

Symbol $\langle Le \rangle$ generates small and capital letters as well as the symbol '_'

$$L(Le) = \{a, \ldots, z, A, \ldots, Z, \_\}.$$

Symbol $\langle DiLe \rangle$ generates digits, small, and capital letters including the symbol '_'

$$L(DiLe) = \{0, \ldots, 9, a, \ldots, z, A, \ldots, Z, \_\}.$$

Symbol $\langle DiLeS \rangle$ generates sequences of such symbols

$$L(DiLeS) = \{0, \ldots, 9, a, \ldots, z, A, \ldots, Z, \_\}^+.$$

Names are generated by symbol $\langle Na \rangle$. They are sequences of digits and letters starting with a letter

$$L(Na) = L(Le) \circ \{0, \ldots, 9, a, \ldots, z, A, \ldots, Z, \_\}^*.$$

Constants are generated by symbol $\langle C \rangle$. They are decimal representations of numbers, possibly followed by the letter u, or the symbolic value null representing the content of uninitialized pointer variables (null pointer constant)

$$L(C) = L(Di)^+ \cup (L(Di)^+ \circ \{u\}) \cup \{\text{null}\}.$$

$\langle Di \rangle$	$\longrightarrow 0 \mid \ldots \mid 9$	digit
$\langle DiS \rangle$	$\longrightarrow \langle Di \rangle \mid \langle Di \rangle \langle DiS \rangle$	digit sequence
$\langle Le \rangle$	$\longrightarrow a \mid \ldots \mid z \mid A \mid \ldots \mid Z \mid \_$	letter
$\langle DiLe \rangle$	$\longrightarrow \langle Le \rangle \mid \langle Di \rangle$	alphanumeric symbol
$\langle DiLeS \rangle$	$\longrightarrow \langle DiLe \rangle \mid \langle DiLe \rangle \langle DiLeS \rangle$	sequence of symbols
$\langle Na \rangle$	$\longrightarrow \langle Le \rangle \mid \langle Le \rangle \langle DiLeS \rangle$	name
$\langle C \rangle$	$\longrightarrow \langle DiS \rangle \mid \langle DiS \rangle u \mid$ null	int/uint/null pointer constant
$\langle CC \rangle$	$\longrightarrow \ '\_' \mid \ldots \mid \ '\sim'$	*char*-constant with ASCII code
$\langle BC \rangle$	$\longrightarrow$ true $\mid$ false	*bool*-constant
$\langle id \rangle$	$\longrightarrow \langle Na \rangle \mid \langle id \rangle.\langle Na \rangle \mid \langle id \rangle[\langle E \rangle] \mid$	identifier
	$\langle id \rangle * \mid \langle id \rangle \&$	

$\langle F \rangle$	$\longrightarrow \langle id \rangle \mid -_1 \langle F \rangle \mid (\langle E \rangle) \mid \langle C \rangle$	factor
$\langle T \rangle$	$\longrightarrow \langle F \rangle \mid \langle T \rangle * \langle F \rangle \mid \langle T \rangle / \langle F \rangle$	term
$\langle E \rangle$	$\longrightarrow \langle T \rangle \mid \langle E \rangle + \langle T \rangle \mid \langle E \rangle - \langle T \rangle$	expression
$\langle Atom \rangle$	$\longrightarrow \langle E \rangle > \langle E \rangle \mid \langle E \rangle >= \langle E \rangle \mid$	atom
	$\langle E \rangle < \langle E \rangle \mid \langle E \rangle <= \langle E \rangle \mid$	
	$\langle E \rangle == \langle E \rangle \mid \langle E \rangle \ ! = \langle E \rangle \mid \langle BC \rangle$	

$\langle BF \rangle$	$\longrightarrow \langle id \rangle \mid \langle Atom \rangle \mid !\langle BF \rangle \mid (\langle BE \rangle)$	Boolean factor
$\langle BT \rangle$	$\longrightarrow \langle BF \rangle \mid \langle BT \rangle \&\& \langle BF \rangle$	Boolean term
$\langle BE \rangle$	$\longrightarrow \langle BT \rangle \mid \langle BE \rangle \ || \ \langle BT \rangle$	Boolean expression

$\langle St \rangle$	$\longrightarrow \langle id \rangle = \langle E \rangle \mid \langle id \rangle = \langle BE \rangle \mid \langle id \rangle = \langle CC \rangle \mid$	assignment statement
	if $\langle BE \rangle \ \{\langle StS \rangle\} \mid$	conditional statement (if-then)
	if $\langle BE \rangle \ \{\langle StS \rangle\}$ else $\{\langle StS \rangle\} \mid$	conditional statement (if-then-else)
	while $\langle BE \rangle \ \{\langle StS \rangle\} \mid$	loop
	$\langle id \rangle = \langle Na \rangle(\langle PaS \rangle) \mid \langle id \rangle = \langle Na \rangle() \mid$	function call
	$\langle id \rangle =$ new $\langle Na \rangle *$	allocation of memory
$\langle rSt \rangle$	$\longrightarrow$ return $\langle E \rangle \mid$ return $\langle BE \rangle \mid$ return $\langle CC \rangle$	return statement
$\langle Pa \rangle$	$\longrightarrow \langle E \rangle \mid \langle BE \rangle \mid \langle CC \rangle$	parameter
$\langle PaS \rangle$	$\longrightarrow \langle Pa \rangle \mid \langle Pa \rangle, \langle PaS \rangle$	parameter sequence
$\langle StS \rangle$	$\longrightarrow \langle St \rangle \mid \langle St \rangle ; \langle StS \rangle$	statement sequence

$\langle prog \rangle$	$\longrightarrow \langle TyDS \rangle ; \langle VaDS \rangle ; \langle FuDS \rangle \mid$	C0-program
	$\langle VaDS \rangle ; \langle FuDS \rangle \mid$	no type declaration
	$\langle TyDS \rangle ; \langle FuDS \rangle \mid$	no variable declaration
	$\langle FuDS \rangle$	only function declarations

$\langle TyDS \rangle$	$\longrightarrow \langle TyD \rangle \mid \langle TyD \rangle ; \langle TyDS \rangle$	type declaration sequence
$\langle TyD \rangle$	$\longrightarrow$ typedef $\langle TE \rangle \ \langle Na \rangle$	type declaration
$\langle Ty \rangle$	$\longrightarrow$ int $\mid$ bool $\mid$ char $\mid$ uint $\mid \langle Na \rangle$	type names
$\langle TE \rangle$	$\longrightarrow \langle Ty \rangle[\langle DiS \rangle] \mid$	type expression, array
	$\langle Ty \rangle * \mid$ struct $\{\langle VaDS \rangle\}$	pointer, struct
$\langle VaDS \rangle$	$\longrightarrow \langle VaD \rangle \mid \langle VaD \rangle ; \langle VaDS \rangle$	variable declaration sequence
$\langle VaD \rangle$	$\longrightarrow \langle Ty \rangle \ \langle Na \rangle$	variable declaration

$\langle FuDS \rangle$	$\longrightarrow \langle FuD \rangle \mid \langle FuD \rangle ; \langle FuDS \rangle$	function declaration sequence
$\langle FuD \rangle$	$\longrightarrow \langle Ty \rangle \ \langle Na \rangle \ (\langle PaDS \rangle) \{ \langle VaDS \rangle ; \langle body \rangle \} \mid$	function declaration
	$\langle Ty \rangle \ \langle Na \rangle \ (\langle PaDS \rangle) \{ \langle body \rangle \} \mid$	no local variables
	$\langle Ty \rangle \ \langle Na \rangle \ () \{ \langle VaDS \rangle ; \langle body \rangle \} \mid$	no parameters
	$\langle Ty \rangle \ \langle Na \rangle \ () \{ \langle body \rangle \}$	only body
$\langle PaDS \rangle$	$\longrightarrow \langle VaD \rangle \mid \langle VaD \rangle, \langle PaDS \rangle$	parameter declaration sequence
$\langle body \rangle$	$\longrightarrow \langle rSt \rangle \mid \langle StS \rangle ; \langle rSt \rangle$	function body

Table 12. Grammar of C0

Without the letter u they specify so-called integer constants representable by two's complement numbers of appropriate length. With the letter u they specify so-called unsigned integer constants representable by binary numbers of appropriate length. Constants of character type are generated by symbol $\langle CC \rangle$. They are digits or letters enclosed by quotation marks. The two Boolean constants generated by symbol $\langle BC \rangle$ are *true* and *false*. Names are required to be different from keywords.

Printable symbols with an ASCII-code are

```
!"#$%&'()*+,-./0123456789:;<=>?
@ABCDEFGHIJKLMNOPQRSTUVWXYZ[\]^_
`abcdefghijklmnopqrstuvwxyz{|}~
```

Using three dots we sometimes write the set of printable ASCII symbols as

$$\{ {}_{\llcorner}, \ldots, \tilde{\ } \}.$$

We have used this in the grammar for the definition of character symbols. We denote the set of *all* ASCII symbols as *ASC* and thus have

$$L(CC) \subset \{'\alpha' \mid \alpha \in ASC\}.$$

By

$$ascii : ASC \rightarrow \mathbb{B}^8$$

we denote the function which maps every ASCII symbol $\alpha \in ASC$ to its ASCII code $ascii(\alpha)$. For later reference we remark that the non-printable symbol NUL has ASCII code 0^8

$$ascii(NUL) = 0^8.$$

11.1.2 Identifiers

Variables and subvariables are specified by identifiers generated by symbol $\langle id \rangle$. Identifiers come in the following five flavors:

- plain variable names x,
- struct component accesses of the form $x.n$ where x is an identifier and n is the name of a struct component,
- array accesses of the form $x[e]$ where x is an identifier and e is an arithmetic expression,
- memory access $x*$ performed by dereferencing an identifier x, and
- taking address of $x\&$, which results in a pointer to the variable or subvariable specified by identifier x.

The names $.n$ of struct component accesses and the indices $[e]$ of array accesses select subvariables of variables or, if applied repeatedly, of subvariables. We stress the fact that we have *not* defined variables or subvariables yet. Thus far, we have only specified names and identifiers for variables and subvariables.

11.1.3 Arithmetic and Boolean Expressions

The productions for the arithmetic expressions were basically discussed in the previous chapter. Factors can now be identifiers and constants. Also we use the same sign now for both unary and binary minus.

The productions for Boolean expressions have exactly the same structure. Boolean factors are identifiers, Boolean expressions in brackets, and atoms; atoms are either Boolean constants or they have the form $a \circ b$ where a and b are arithmetic expressions and $\circ$ is a comparison operator.

11.1.4 Statements

There are six kinds of statements:

- *assignment statements.* They have the form

$$id = e$$

  where identifier *id* on the left-hand side specifies a variable or subvariable to be updated. On the right-hand side *e* can be an arithmetic expression, a Boolean expression, a character constant, or an address-of operation. The value of the variable or subvariable specified by the current value of *id* is updated by the current value of *e*. The reader should recall that at this point in the discussion we have not defined yet what variables or subvariables are. Thus we clearly don't have a formal concept of their current values yet.
- *conditional statements.* They can have the form

$$\text{if} \quad e \quad \{\text{\textit{if-part}}\}$$

  or

$$\text{if} \quad e \quad \{\textit{if-part}\} \quad \text{else} \quad \{\textit{else-part}\}$$

  where *e* is a Boolean expression and *if-part* as well as *else-part* are sequences of statements without return statements. If expression *e* evaluates to *true* the *if-part* is executed. If *e* evaluates to *false* then the *else-part* is executed if it exists.
- *while statements.* They have the form

$$\text{while} \quad e \quad \{\textit{loopbody}\}$$

  where *e* is a Boolean expression and {*loopbody*} is a sequence of statements without return statements. Expression *e* is evaluated. If it evaluates to *false* the loop body is skipped and execution of the while loop is complete. If it evaluates to *true*, the loop body is executed and the while loop is executed again. Thus the body is repeatedly executed while expression *e* evaluates to *true* (which might be forever).

- *function calls.* They have the form

$$id = f(e_1, \ldots, e_p)$$

where f is the name of a declared function. Function f is called with parameters given by the current values of expressions e_i. The sequence of parameters can be empty. The value returned by the (return statement of the) function is assigned to the variable or subvariable specified by the current value of identifier id on the left-hand side.
- *return statements.* They have the form

$$\text{return} \quad e$$

where e is either an expression (arithmetic or Boolean) or a character constant. Always executed at the end of the execution of a function, a return statement returns the current value of expression e.
- *new statements.* They allocate new variables in a memory region called the *heap* and have the form

$$id = \text{new} \quad t*$$

where t is the name of an elementary or declared type. The newly generated variable has type t. A pointer to that variable of type $t*$ is assigned to the variable or subvariable specified by the current value of identifier id. See Sect. 11.1.6 for more information on types.

11.1.5 Programs

Programs have three parts:

- a sequence of type declarations $\in L(TyDS)$,
- a sequence of declarations of so-called global variables $\in L(VaDS)$, and
- a sequence of function declarations $\in L(FuDS)$.

Type declarations and declarations of global variables can be absent.

11.1.6 Type and Variable Declarations

A type $x \in L(Ty)$ either belongs to the set ET of elementary types

$$ET = \{int, uint, bool, char\}$$

or it is a (declared) type name $x \in L(Na)$. Types are declared by means of type declaration sequences $\in L(TyDS)$, which are sequences of type declarations $\in L(TyD)$. A type declaration has the form

$$\text{typedef} \quad te \quad x$$

where $x \in L(Na)$ is the name of the declared type and $te \in L(TE)$ is a *type expression* specifying the declared type. Type expressions come in three flavors:

- array type expressions of the form

$$t[z]$$

specify array types. An array variable X of this type consists of array elements $X[i]$ of type t and its length is specified by decimal number z.
- pointer type expression of the form

$$t*$$

specify pointer types. Variables of type $t*$ are pointers to variables or subvariables of type t.
- struct type expressions of the form

$$\text{struct} \quad \{ t_1 \quad n_1 \ ; \ \dots \ ; \ t_s \quad n_s \}$$

specify struct types. A struct variable X of this type consists of components $X.n$ with names $n \in \{n_1, \dots, n_s\}$. The type of component $X.n_i$ is t_i.

In a well-formed $C0$ program one would like to require that type declarations are *non-circular*, i.e., that any type declaration uses only type names declared earlier in the type declaration sequence (in addition to the elementary types, which may be used freely). The only exception to this rule will be pointer type declarations which may reference types that are declared later in the program. This feature allows us to define complex data types like linked lists and trees (cf. Sect. 11.1.8).

Variable declaration sequences $\in L(VaDS)$ are sequences of variable declarations $\in L(VaD)$ separated by semicolons. A variable declaration has the form

$$t \quad x$$

where t is the type of the declared variable and x is its name. We will see that variable declaration sequences occur in three places: i) in the declarations of the global variables of a program, ii) in the declaration of the local variables of a function, and iii) in the declarations of struct components. Later we will be able to treat the first two cases as a special case of the third case. Note that, in a well-formed $C0$ program, every non-elementary type name used in a variable declaration is declared previously by a corresponding type declaration.

11.1.7 Function Declarations

Function declaration sequences $\in L(FuDS)$ are sequences of function declarations $\in L(FuD)$ separated by semicolons. A function declaration has the form

$$t \quad f \quad (t_1 \quad x_1, \ \dots \ , t_p \quad x_p) \{ t_{p+1} \quad x_{p+1} \ ; \ \dots \ ; \ t_s \quad x_s \ ; \ body \}$$

where

- t is the type of the result returned by the declared function,

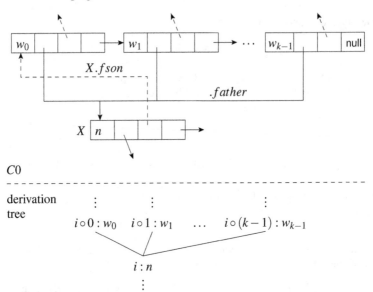

Fig. 95. Representing a node in a derivation tree and its sons by structs of type DTE

- f is the name of the declared function,
- $t_1\, x_1, \ldots, t_p\, x_p$ is the sequence of parameter declarations. It differs from variable declaration sequences in the profound aspect that parameter declarations are separated by commas instead of colons,
- $t_{p+1}\, x_{p+1}; \ldots; t_s\, x_s$ is the sequence of declarations of local parameters,
- $body \in L(body)$ is a sequence of statements $\in L(StS)$ followed by a return statement $\in L(rSt)$.

11.1.8 Representing and Processing Derivation Trees in $C0$

Later in this text we will present numerous algorithms that work on derivation trees T for the above $C0$ grammar G. In case one wishes to implement these algorithms in $C0$, one has to assume that the tree T is represented as a $C0$ data structure. We describe such a representation. To do that, we switch to the clean representation of derivation trees from Sect. 10.1.4. There, we defined a derivation tree as a pair $T.A$ and $T.\ell$, where

- $T.A$ is the set of nodes of the tree,
- $T.\ell : T.A \to N \cup T$ is a mapping that assigns to each node $i \in T.A$ of the tree its label $T.\ell(i)$. In drawings we represent a node i with its label $n = T.\ell(i)$ as $i : n$.

Also recall that the names for nodes $i \in T.A$ are taken from the sequences of integers $(i_1, \ldots, i_s) \in \mathbb{N}^*$, while edges are implicitly coded in the names of the nodes. If node i has k sons, they are included in $T.A$ as $i \circ 0, \ldots, i \circ (k-1)$, where the last index specifies the order of the sons. Thus, if $i \circ x$ is one of these nodes with $x < k - 1$, its brother to the right is $i \circ (x+1)$.

Nodes i of a derivation tree T will be stored as struct variables $X(i)$ of type DTE (for derivation tree element). Type DTE and the type of pointers to DTE are declared in the following way

```
typedef DTE* DTEp;
typedef struct {
   uint label;
   DTEp father;
   DTEp fson;
   DTEp bro
} DTE;
```

Figure 95 illustrates how a node i with nonterminal n and its sons with k nonterminals or terminals $w[0 : k - 1]$ are represented by structs of type DTE.

The interpretation of the components of a struct variable $X(i)$ of type DTE representing node $i = i_1 \ldots i_s$ in the tree is as follows:

- $X(i).label$ contains label $T.\ell(i)$ coded as a string in $\mathbb{B}^{32}$.
- $X(i).father$ points to the struct variable $X(i_1 \ldots i_{s-1})$ representing the father of node i in $T.A$. If i is the root, we make $X(i).father$ the null pointer.
- $X(i).fson$ points to the struct variable representing $X(i \circ 0)$, i.e., to the *first* son of i. If i is a leaf we make $x.fson$ the null pointer.
- $X(i).bro$ points to the next brother $X(i_1 \ldots i_{s-1} \circ (i_s + 1))$ to the right of i in the tree if there is such a brother. Otherwise we make $X(i).bro$ the null pointer.

As a first programming example we specify a function *nson* with two parameters:

- a pointer p to a variable $X(i)$ of type DTE; we assume that $X(i)$ represents a node $i \in T.A$ of a derivation tree T with at least one son, and
- a number $j \in \mathbb{N}$ representing a possibly existing son $i \circ j$ of i in T.

The function is supposed to return a pointer to $X(i \circ j)$, i.e., to the struct variable representing son number j of i, if such a son $i \circ j$ exists, and the null pointer otherwise. In the implementation the 'first son' component $X.fson$ of X is dereferenced in order to let p point to (a struct representing) the first son of X. In the while loop 'brother pointers' are chased and j is decremented as long as there is a brother and $j > 0$. If the loop ends with $j = 0$ then, for the initial j, son $i \circ j$ exists and p points to the struct variable $X(i \circ j)$ representing this son. Otherwise the null pointer is returned.

```
DTEp nson(DTEp p, uint j)
{
   DTEp result;
   p = p*.fson;
   while p*.bro!=null && j!=0
   {
```

```
      p = p*.bro;
      j = j - 1
   };
   if j==0 {result = p};
   return result
}
```

Our second programming example uses data type *LEL* for linked lists of unsigned integers which is specified by:

```
typedef LEL* LELp;
typedef struct {uint label; LELp next} LEL;
```

We specify a function border word *BW*, which has two parameters:

- a pointer p to a variable $X(i)$ of type *DTE* representing a node i in the tree, and
- a pointer q pointing to a variable U of type *LEL*.

Function *BW* appends to list element U a linked list of elements, whose *.label* components code the border word $bw(T, i)$ of the subtree with root i. Pointer q always points to the last list element appended so far (see Fig. 96). The implementation quite literally follows the recursive definition of border words for derivation trees from Sect. 10.1.4, where we defined:

- if i is a leaf of tree T, then its border word is its label:

$$bw(i, T) = T.\ell(i),$$

- the border word of a node i with k sons is obtained by concatenating the border words of its sons

$$bw(i, T) = bw(i \circ 0, T) \circ \ldots \circ bw(i \circ (k-1), T).$$

Thus new list elements are appended whenever the function is called with a parameter p pointing to a leaf X, i.e., when $p * .fson =$ null. If X is not a leaf, its sons are processed recursively from left to right until one arrives at the last son Y, which has no brother. The result returned by the function is a pointer to the last list element inserted so far.

```
LELp BW (DTEp p, LELp q)
{
   if p*.fson==null
   {
      q*.next = new LEL;
      q = q*.next;
      q*.label = p*.label
   }
```

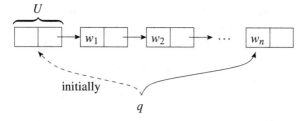

Fig. 96. Appending a list to a list element U. Pointer q always points to the end of the list

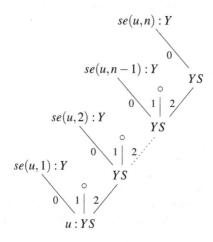

Fig. 97. Deriving a sequence of elements Y at nodes $se(u,i)$ from YS at node u

```
else
{
   p = p*.fson;
   while p!=null
   {
      q = BW(p,q);
      p = p*.bro
   }
}
return q
}
```

11.1.9 Sequence Elements and Flattened Sequences in the *C*0 Grammar

The *C*0 grammar has at many places productions for Y-sequences of the form

$$\langle YS \rangle \to Y \mid Y \circ \langle YS \rangle$$

where $\circ$ is a comma or a colon separating the sequence elements Y. Figure 97 shows a derivation tree with a node u labeled YS and n nodes labeled Y. For these nodes we introduce the notation $se(u,i)$ with $i \in [1:n]$. An easy exercise shows for all i

$$se(u,i) = u2^{i-1}0,$$

where 2^i is a sequence of i twos. The *flattened sequence* of the nodes $se(u,i)$ is denoted by

$$fseq(u) = se(u,1) \circ \ldots \circ se(u,n).$$

If we want to advance a pointer p from $Y(se(u,i))$ to $Y(se(u,i+1))$ we first follow the father pointer, then select son number 2 and finally advance to son number 0:

```
q = nson(p*.father,2);
p = nson(q,0)
```

11.2 Declarations

We split the presentation of the $C0$ semantics into two parts: static and dynamic semantics. In the section on static semantics we only treat concepts which can be defined by looking at the program without running it. In the section on dynamic semantics we continue the definition of semantics by defining $C0$ configurations c and a transition function δ_C, which computes for $C0$ configuration c the next configuration

$$c' = \delta_C(c).$$

As usual one then defines $C0$ computations as sequences (c^i) of configurations satisfying

$$c^{i+1} = \delta_C(c^i).$$

11.2.1 Type Tables

Programs begin with type declaration sequences

$$\text{typedef} \quad te_1 \quad na_1 \; ; \ldots ; \text{typedef} \quad te_s \quad na_s$$

where the $te_i \in L(Ty)$ are type expressions and the na_i are the names of the declared types. We collect the set of names of declared types into the set TN

$$TN = \{na_i \mid i \in [1:s]\}.$$

Type expressions $te \in L(Ty)$ can be of the form $t'[z]$ or $t'*$ where t' is a type name or of the form $struct\{t_1 \; n_1; \ldots; t_u \; n_u\}$ where the t_i are type names. We transform such expressions te into a mathematically more convenient form. For declarations of struct types we delete the keyword $struct$. For array type declarations we replace decimal numbers $z[k-1:0]$ by the natural number

$$\langle z \rangle_Z = \sum_{i=0}^{k-1} z_i \cdot Z^i$$

where $Z = 9 + 1$[1].

$$\tau(te) = \begin{cases} \{t_1 \ n_1; \ldots; t_u \ n_u\}, & te = struct\{t_1 \ n_1; \ldots; t_u \ n_u\}, \\ t'[\langle z \rangle_Z], & te = t'[z], \\ t' *, & te = t' * . \end{cases}$$

The type table tt simply maps the names na_i to the encoding $\tau(te_i)$ of the defining type expressions:

$$tt(na_i) = \tau(te_i).$$

It is extremely tempting to abbreviate $tt(na_i) = \tau(te_i)$ to $na_i = \tau(te_i)$. Thus the type declaration for list elements and their pointers in an earlier example can be written as

$$LELp = LEL*,$$
$$LEL = \{uint \ label; LELp \ next\}.$$

We will give in to this temptation because it simplifies notation. *However*, the only equations we will ever write are of the form

$$e = e'$$

where the left-hand side is an expression evaluating to a type name $t \in ET \cup TN$ and the right-hand side is a type table entry. The equations $e = e'$ of this form are simply abbreviations defined by

$$e = e' \leftrightarrow tt(e) = e'.$$

We will *not* state or attempt to justify any equations of the form

$$LEL = \{uint \ label; LEL* \ next\}$$

or

$$LEL = \{uint \ label; \{uint \ label; LEL* \ next\}* \ next\}.$$

We call a type $t = t'*$ a pointer type; we call a type t *simple* if it is elementary or a pointer type:

$$pointer(t) \leftrightarrow \exists t' : t = t'*,$$
$$simple(t) \leftrightarrow t \in ET \lor pointer(t).$$

Otherwise, i.e., if it is an array or struct type, we call it *composite*. An obvious context condition requires types to be uniquely defined.

---

[1] This definition was already presented in a preliminary and more informal form in Definition 4.

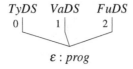

$$\varepsilon : prog$$

Fig. 98. Labels *TyDS*, *VaDS*, and *FuDS* are at nodes 0, 1, and 2 of the derivation tree

Context Condition 1

$$i \neq j \rightarrow na_i \neq na_j.$$

Moreover in struct types the names selecting the struct components should be distinct.

Context Condition 2

$$na = \{t_1 \, n_1; \ldots; t_u \, n_u\} \wedge i \neq j \rightarrow n_i \neq n_j.$$

A further context condition requires that declarations of composite types only use elementary or previously defined types.

Context Condition 3

For array types we require the type of elements to be elementary or to be previously defined.

$$na_i = t'[n] \rightarrow t' \in ET \vee \exists j < i : na_j = t'.$$

For struct types we require the types of components to be elementary or to be previously defined.

$$na_i = \{t_1 \, n_1; \ldots; t_u \, n_u\} \rightarrow \forall k : (t_k \in ET \vee \exists j < i : na_j = t_k).$$

Pointers must point to elementary types or to types declared elsewhere but not necessarily previously.

$$na_i = t' * \rightarrow t' \in ET \vee \exists j : na_j = t'.$$

Indeed in all our examples we declared type *DTE* after *DTEp* and *LEL* after *LELp*. We explicitly encourage readers to feel slightly worried about future consequences of this definition. This is not an ordinary recursive definition where we define new types strictly from known types, and this will indeed produce certain technical complications. Unfortunately, such a definition cannot be avoided, because we wish to have struct types X that have components $X.n$ pointing to variables of type X.

Implementation Hints

A programmer wishing to extract the names na_i of declared types from the derivation tree should observe that label *TyDS* is at node 0 of the tree (see Fig. 98). Sequence element $j \in [1:n]$ with label *TyD* is at node $se(0, j)$; the name na_j can be found as the

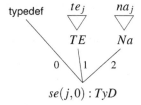

Fig. 99. Name na_j and type expression te_j of the j'th declared type are the border words of nodes $se(0, j)2$ and $se(0, j)1$ resp.

border word of son 2 of $se(0, j)$ (see Fig. 99) and the corresponding type expression te_j as the border word of son 1 of $se(0, j)$. Thus

$$na_j = bw(se(0, j)2),$$
$$te_j = bw(se(0, j)1).$$

Computing a list of these names is an easy exercise with the small portions of $C0$ programs given so far. Inspection of the labels of son 1 of node $se(0, j)1$ allows one to test what kind of a type is defined: a struct if $\ell(se(0, j)11) = \{$, an array if $\ell(se(0, j)11) = [$, and a pointer if $\ell(se(0, j)11) = *$. Finally, if one were to program a data structure for the type table one would store the binary representation of $\langle z \rangle z$; thus one would have to write a program to convert from decimal to binary numbers.

11.2.2 Global Variables

In programs the type declaration sequence is followed by a variable declaration sequence

$$t_1 \quad x_1 \; ; \; \ldots \; ; \; t_r \quad x_r$$

for the global variables. Obviously, used type names should be declared

Context Condition 4

$$t_i \in ET \cup TN$$

and declared variable names should be distinct.

Context Condition 5

$$i \neq j \to x_i \neq x_j.$$

We collect the names of the global variables in a set

$$VN = \{x_i \mid i \in [1 : r]\}.$$

Except for the missing keyword *struct* and the brackets this looks very much like the declaration of a struct type. Indeed, it turns out that both in semantics and in compiler construction, global variables x are treated *exactly* like components $gm.x$ of a struct variable gm, which stands for the *global memory*. We therefore introduce for the type of this variable gm a new name $\$gm$, extend the range of the type table to include gm, and define

$$tt(\$gm) = \{t_1\ x_1; \ldots; t_r\ x_r\}$$

resp. for short

$$\$gm = \{t_1\ x_1; \ldots; t_r\ x_r\}.$$

11.2.3 Function Tables

The sequence of function declarations of a program has the form

$$t_1 \quad f_1 \quad d_1 ; \ldots ; t_k \quad f_k \quad d_k$$

where for all i: t_i is the type of the returned result, f_i is the name of the declared function, and d_i is everything else.

Obviously, types of returned results should be declared

Context Condition 6

$$t_i \in ET \cup TN$$

and declared function names should be distinct.

Context Condition 7

$$i \neq j \rightarrow f_i \neq f_j.$$

We collect the names of declared functions into a set

$$FN = \{f_i \mid i \in [1:k]\}$$

and create a function table ft, mapping function names $f \in FN$ to *function descriptors* $ft(f)$, which consist of several components $ft(f).y$. In order to specify these components and what is recorded there we expand a function declaration to

$$t \quad f\ (t_1 \quad x_1\ , \ldots,\ t_p \quad x_p\ )\ \{\ t_{p+1} \quad x_{p+1} ; \ldots ; t_r \quad x_r\ ;\ body\ \}.$$

Obviously, parameters and local variables should have declared types

Context Condition 8

$$t_i \in ET \cup TN$$

and distinct names.

Context Condition 9

$$i \neq j \rightarrow x_i \neq x_j.$$

We record in the function table $ft(f)$ the following information:

- the type of the result

$$ft(f).t = t,$$

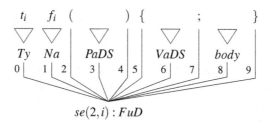

$$se(2,i) : FuD$$

Fig. 100. The body of a function with a nonempty variable declaration sequence is generated by son number 8

- the set of declared parameters and local variables

$$ft(f).VN = \{x_1, \ldots, x_r\},$$

- the number of parameters

$$ft(f).p = p,$$

- the body of function f

$$ft(f).body = body.$$

Implementation Hints

Although quite intuitive, this is not yet a complete definition. Making it precise is very slightly tedious. The function declaration sequence is generated by node 2 (Fig. 98) of the entire derivation tree. If $f = f_i$ then its declaration is generated by node $se(2,i)$. For functions with parameters and local variables Fig. 100 shows that the body is generated by son 8 of $se(2,i)$. Drawing the analogous figures for declarations of functions without parameters, without local variables, and without both, one finds

$$nbody(f_i) = \begin{cases} se(2,i)8, & outdeg(se(2,i)) = 10, \\ se(2,i)7, & outdeg(se(2,i)) = 9, \\ se(2,i)6, & outdeg(se(2,i)) = 8, \\ se(2,i)5, & outdeg(se(2,i)) = 7, \end{cases}$$

and the body of f is the border word of this node

$$body(f_i) = bw(nbody(f_i)).$$

Basically we are using the edge labels of the derivation tree as a table of contents of the program text. When people read in a table of contents that the body of function f_i is found in chapter 2, section i, subsection 8 they tend to find nothing wrong with this. Readers who nevertheless find this 'too technical' are strongly encouraged to look for a *precise* definition of the body of the i'th function of a program in less than five short lines.

In the same spirit, the name f_i of the i'th declared function can be obtained from the derivation tree as

$$f_i = bw(se(2,i)1).$$

When programs start running, they start by executing the body of a parameterless function *main*. In order for this to work function *main* obviously has to be declared.

Context Condition 10

$$main \in FN \quad and \quad f_i = main \rightarrow \ell(se(2,i)3) =).$$

During program execution, every call of function f will create a so-called *function frame* for f containing a new copy of the parameters and local variables of the function. We proceed as we did in the case of global memory and treat all parameters and local variables x_i of a function frame for f as components of a single struct variable. Thus we record in the type table

$$\$f = \{t_1\ x_1; \ldots; t_p\ x_p; t_{p+1}\ x_{p+1}; \ldots; t_r\ x_r\}.$$

11.2.4 Variables and Subvariables of all $C0$ Configurations

Recall that — for the purpose of defining semantics — we treat global variables as components of the global memory and we treat parameters and local variables of functions as components of function frames. The only other $C0$ variables are created by the *new* statement; they belong to the so-called *heap* and have no name. Therefore the set V of variables of all $C0$ programs contains at the top level only three kinds of variables (see Fig. 102):

- the global memory frame *gm*. Thus

$$gm \in V,$$

- the heap variables. They have no name, so we simply number them with elements $j \in \mathbb{N}$. Thus

$$\mathbb{N} \subset V,$$

- function frames for declared functions f; they belong to the so-called *stack*. Because functions can be called recursively, several function frames (f,i) can be present for the same function f. Observing that f is a name $\in L(Na)$ and using indices $i \in \mathbb{N}$ we get

$$L(Na) \times \mathbb{N} \subset V.$$

Other variables will not occur. Thus we define the set of all possible $C0$ variables as

$$V = \{gm\} \cup \mathbb{N} \cup L(Na) \times \mathbb{N}.$$

Next, we define the set SV of all possible subvariables in all $C0$ programs. Variables of struct or array type can have subvariables. Subvariables of structs are selected by *selectors* of the form $.n$ where n is a name $\in L(Na)$. Subvariables of arrays are selected by selectors of the form $[n]$ with $n \in \mathbb{N}$. We define the set S of all selectors of all $C0$ programs as

$$S = \{.n \mid n \in L(Na)\} \cup \{[n] \mid n \in \mathbb{N}\}.$$

Subvariables can be of struct or array type too. Thus selectors can be applied repeatedly (to variables of appropriate type). The set of finite sequences of selectors is S^+ and we get

$$V \circ S^+ \subset SV.$$

Allowing empty selector sequences ε we consider each variable X as the subvariable

$$X = X \circ \varepsilon.$$

Thus we define

$$SV = V \circ S^*.$$

11.2.5 Range of Types and Default Values

For elementary or declared types t we define the *range* of t, i.e., the set of values that variables of type t can assume. As the reader will expect, the definition is recursive. The obvious base case is the elementary types

$$ra(t) = \begin{cases} B_{32}, & t = uint, \\ T_{32}, & t = int, \\ ASC \cup \{NUL\}, & t = char, \\ \mathbb{B}, & t = bool. \end{cases}$$

It is quite a common practice to define in textbooks $ra(int) = \mathbb{Z}$ and $ra(uint) = \mathbb{N}$. This has the advantage of being elegant but also a crucial disadvantage: for real programs running on real computers it is simply wrong. Real computers are finite and can only represent finitely many numbers. In an n-bit processor unsigned integer arithmetic is usually modulo 2^n, therefore we get

$$(2^n - 1) + 1 = 0,$$

which violates the Peano axiom that 0 is not the successor of any natural number.

For arrays and struct types t we define the set $sel(t)$ of their selectors as

$$sel(t) = \begin{cases} [0 : n - 1], & t = t'[n], \\ \{n_1, \ldots, n_s\}, & t = \{t_1 \, n_1; \ldots; t_s \, n_s\}. \end{cases}$$

A value of an array of type $t'[n]$ is then defined as a mapping f assigning to selectors $i \in sel(t)$ values in the range of t'. Values of struct types are mappings from $sel(t)$ to the union of the $ra(t_i)$, with the restriction that a component $.n_i$ must be mapped into the range of t_i.

$$\begin{pmatrix} a_{0,0} & \cdots & a_{0,j} & \cdots & a_{0,m-1} \\ \vdots & & \vdots & & \vdots \\ a_{i,0} & \cdots & a_{i,j} & \cdots & a_{i,m-1} \\ \vdots & & \vdots & & \vdots \\ a_{n-1,0} & \cdots & a_{n-1,j} & \cdots & a_{n-1,m-1} \end{pmatrix}$$

Fig. 101. Matrices in *C0* are declared as arrays of rows $a_{i,0}, \ldots, a_{i,m-1}$

$$ra(t) = \begin{cases} \{f \mid f : sel(t) \to ra(t')\}, & t = t'[n], \\ \{f \mid f : sel(t) \to \bigcup_i ra(t_i) \wedge \forall i : f(n_i) \in ra(t_i)\}, & t = \{t_1 \, n_1; \ldots; t_s \, n_s\}. \end{cases}$$

Example: We can declare the type of an *n*-by-*m* matrix of integers such that we get from the type table

$$row = int[m],$$
$$matrix = row[n].$$

If $g \in ra(matrix)$ is a value of type *matrix*, then for $i \in [0 : n-1]$

$$g(i) = f_i$$

is a value of type *row*. In order to represent row *i* of the matrix in Fig. 101, we set

$$f_i(j) = a_{i,j}$$

for $j \in [0 : m-1]$. Then we get for all *i* and *j*

$$g(i)(j) = f_i(j)$$
$$= a_{i,j},$$

which fortunately matches our intuition.

Pointers are the null pointer or they point to subvariables. Thus we define the range of pointer types *t* as

$$ra(t) = SV \cup \{null\} \quad \text{if} \quad pointer(t)$$

and can show the following.

Lemma 71. *Let $t_1, \ldots, t_s$ be the sequence of declared types $t_i \in TN$ in the order of their declaration. Then for all i the range $ra(t_i)$ is well defined.*

Proof. For elementary types and for pointer types, the definitions are not recursive, thus there is nothing to show. For array and struct types the lemma follows by induction on *i*.

If $t_i = t'[n]$, then we require t' to be elementary, in which case $ra(t')$ is obviously well defined, or to be previously defined, i.e., $t' = t_j$ with $j < n$, in which case $ra(t_j)$ is well defined by induction. Similarly, if $t_i = \{r_1 \, n_1; \ldots; r_v \, n_v\}$ then we require component type r_u to be elementary, in which case $ra(r_u)$ is obviously well defined, or to be previously defined, i.e., $r_u = t_k$ with $k < i$, in which case $ra(r_u)$ is well defined by induction.

For every simple type t we define in an obvious way a default value $dft(t) \in ra(t)$. This default value $dft(t)$ is later used to initialize subvariables of type t that are created without an immediate assignment of their value. This will concern all variables except parameters of functions.

$$dft(t) = \begin{cases} 0, & t \in \{int, uint\}, \\ NUL, & t = char, \\ 0, & t = bool, \\ null, & t = t' *. \end{cases}$$

The definition of default values is extended to composite types in an obvious way. The default value of array type $t'[n]$ maps every index i to the default value $dft(t')$ of the array element's type:

$$t = t'[n] \wedge i \in [0 : n - 1] \rightarrow dft(t)(i) = dft(t').$$

The default value of a struct type $\{t_1 \, n_1; \ldots; t_s \, n_s\}$ maps every selector n_i to the default value $dft(t_i)$ of the corresponding component type:

$$t = \{t_1 \, n_1; \ldots; t_s \, n_s\} \wedge i \in [1 : s] \rightarrow dft(t)(n_i) = dft(t_i).$$

11.3 *C*0 Configurations

All previous definitions are implicitly part of the definitions made in this section, but as they don't change in the course of computations, we don't list them explicitly in the mathematical formalism. As announced previously we will keep introducing context conditions at the places where they help in an obvious way to make certain mechanisms work, although they can be checked statically.

11.3.1 Variables, Subvariables, and Their Type in *C*0 Configurations c

In order to provide intuition for the following formal definitions we sketch the use of the stack and the concept of recursion depth in the future definitions: computations start with recursion depth 0 and a single function frame $(main, 0)$ on the stack. If a function f is called in a configuration of recursion depth rd, then frame $(f, rd + 1)$ is put on the stack and $rd + 1$ is made the new recursion depth. Conversely, if a return statement is executed in a configuration of recursion depth rd, then the top frame (with second component rd) is removed from the stack and the recursion depth is decreased by 1. A frame (f, i) on the stack is called a frame at recursion depth i.

*C*0 configurations have among others the following components (see Fig. 102):

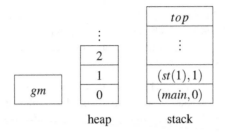

Fig. 102. Top-level variables in the $C0$ semantics: the global memory frame gm, the heap variables numbered $0, 1, \ldots$, and the function frames $(st(i), i)$ of the stack. The bottom frame of the stack belongs to function *main*

- the recursion depth

$$c.rd \in \mathbb{N}.$$

It counts the number of functions that have been called and that have not returned yet. While function *main* is being executed we have $c.rd = 0$.
- a stack function

$$c.st : [0 : c.rd] \rightarrow FN$$

specifying for each recursion depth $i \in [0 : c.rd]$ the name of the function $c.st(i) \in FN$ such that stack frame

$$ST(c, i) = (c.st(i), i)$$

is the frame with recursion depth i on the stack. We will see later that we always have $c.st(0) = main$, i.e., the bottom frame of the stack is always $(main, 0)$. The top frame, i.e., the frame with maximal recursion depth on the stack is denoted by

$$top(c) = ST(c, c.rd)$$

and the function $c.st(c.rd)$ belonging to the maximal recursion depth on the stack is called the current function $cf(c)$ of configuration c

$$cf(c) = c.st(c.rd).$$

- the number

$$c.nh \in \mathbb{N}$$

of variables on the heap. Recall that we refer to variables on the heap simply by numbers. Thus the set of heap variables in configuration c is $[0 : c.nh - 1]$.
- a function

$$c.ht : [0 : c.nh - 1] \rightarrow ET \cup TN$$

assigning to each heap variable j its type $c.ht(j)$, which is either elementary or declared.

With these definitions we can define the set $V(c)$ of (top-level) variables of a configuration c as the global memory frame, the heap variables, and the stack frames present in c:

$$V(c) = \{gm\} \cup [0 : c.nh - 1] \cup \{ST(c,i) \mid i \in [0 : c.rd]\}.$$

The *variable type* $vtype(x,c)$ of variable x in configuration c is defined in the obvious way. The type of the global memory frame gm is $\$gm$ as defined in the type table. Similarly, the type of a function frame (f,i) is $\$f$ as defined in the type table. The type of a heap variable j is defined by the heap type function as $c.ht(j)$.

$$vtype(x,c) = \begin{cases} \$gm, & x = gm, \\ \$f, & x = ST(i,c) \wedge c.st(i) = f, \\ c.ht(x), & x \in [0 : c.nh - 1]. \end{cases}$$

We define the set of subvariables $SV(c)$ of configuration c and the *variable type* $vtype(x,c)$ of such subvariables $x \in SV(c)$ by the obvious recursive definitions:

- variables x of configuration c are subvariables of configuration c and their variable type $vtype(x,c)$ is already defined:

$$V(c) \subseteq SV(c).$$

- if $x \in SV(c)$ is a subvariable of array type, then elements $x[i]$ are subvariables with the corresponding element type:

$$vtype(x,c) = t'[n] \rightarrow \forall i \in [0 : n-1] : (x[i] \in SV(c) \wedge vtype(x[i],c) = t').$$

- if $x \in SV(c)$ is a subvariable of struct type, then components $x.n_i$ are subvariables with the corresponding element type:

$$vtype(x,c) = \{t_1 \, n_1; \ldots; t_s \, n_s\} \rightarrow \forall i \in [1 : s] : (x.n_i \in SV(c) \wedge vtype(x.n_i,c) = t_i).$$

A subvariable $x \in SV(c)$ is called *simple* if its variable type is simple; it is called a pointer if its variable type is a pointer type:

$$simple(x,c) \equiv simple(vtype(x,c)),$$
$$pointer(x,c) \equiv pointer(vtype(x,c)).$$

Subvariables $x \in SV(c)$ have the form $x = ys$ where $y \in V(c)$ is a variable and $s \in S^*$ is a selector sequence. Subvariables ys can belong to the global memory, the heap, or the stack iff the corresponding variable y belongs to the global memory, heap, or stack. Thus for $x \in SV(c)$ we define predicates:

$$ingm(x,c) \equiv \exists s \in S^* : x = gm \circ s,$$
$$onheap(x,c) \equiv \exists i \in \mathbb{N}, s \in S^* : x = i \circ s,$$
$$onstack(x,c) \equiv \exists i \in \mathbb{N}, s \in S^* : x = ST(i,c) \circ s.$$

11.3.2 Value of Variables, Type Correctness, and Invariants

Variables and subvariables have types

$$t \in ET \cup TN.$$

Subvariables of type t should assume values in $ra(t)$. Thus the set

$$VA = \bigcup_{t \in ET \cup TN} ra(t)$$

should include all possible values that variables and subvariables can assume.

The value of variables $x \in V(c)$ is specified by a new 'memory content' component of $C0$ configurations

$$c.m : V(c) \rightarrow VA$$

simply mapping variables x of type t to a value $c.m(x)$. We call a configuration c *type correct for variables* and write $tc\text{-}V(c)$ if for all variables x of the configuration their value $c.m(x)$ lies in the range of $vtype(x,c)$:

$$tc\text{-}V(c) \equiv \forall x \in V(c) : (vtype(x,c) = t \rightarrow c.m(x) \in ra(t)).$$

Function $c.m$ is extended to subvariables x by the obvious definitions. If $x \in SV(c)$ is a subvariable of array type $t = t'[n]$, then its value $c.m(x) \in ra(t)$ is a mapping

$$c.m(x) : [0 : n-1] \rightarrow ra(t').$$

For the subvariables $x[i]$ we define

$$c.m(x[i]) = c.m(x)(i).$$

Similarly, if $x \in SV(c)$ is a subvariable of struct type $t = \{t_1\ n_1; \ldots; t_s\ n_s\}$, then its value $c.m(x) \in ra(t)$ is a mapping

$$c.m(x) : \{n_1, \ldots, n_s\} \rightarrow \bigcup_{i=1}^{s} ra(t_i),$$

s.t. for every i we have $c.m(x)(n_i) \in ra(t_i)$.

For the subvariables $x.n_i$ we define

$$c.m(x.n_i) = c.m(x)(n_i).$$

Type correctness can now be extended to subvariables. We define

$$tc\text{-}SV(c) \equiv \forall x \in SV(c) : (vtype(x,c) = t \rightarrow c.m(x) \in ra(t))$$

and get by a trivial induction that type correctness — if it holds — is extended to subvariables by the above definition.

Lemma 72.

$$tc\text{-}V(c) \leftrightarrow tc\text{-}SV(c).$$

Conversely, we can also conclude type correctness of all subvariables from the type correctness of all simple subvariables. We define

$$tc\text{-}s(c) \equiv \forall x \in SV(c) : (vtype(x,c) = t \land simple(t) \to c.m(x) \in ra(t))$$

and conclude as follows.

Lemma 73.

$$tc\text{-}s(c) \leftrightarrow tc\text{-}SV(c)$$

Proof. As type correctness for simple subvariables is a special case of type correctness, the direction from right to left is trivial. The proof in the other direction is intuitively a straightforward induction, but a little bit of technical work has to be invested in defining the induction scheme used. For this purpose we define the *height* $h(t)$ of a type in the following way: the height of simple types is 1:

$$simple(t) \to h(t) = 1.$$

The height of an array type is the height of the element type plus one:

$$t = t'[n] \to h(t) = h(t') + 1.$$

The height of a struct type is the *maximum* of its component types plus one:

$$t = \{t_1\ n_1; \ldots; t_s\ n_s\} \to h(t) = \max\{h(t_i) \mid i \in [1:s]\} + 1.$$

Now the obvious induction on height i shows that type correctness holds for subvariables x with $h(vtype(x,c)) = i$.

For variables of pointer type t we were forced to define the range $ra(t) = SV \cup \{null\}$ in a rather broad way in order to make the definition of $ra(t)$ well defined. Now we are able to formulate the requirement that pointers of type $t'*$ which are not *null* should point to subvariables of type t'. More precisely: subvariables x of variable type $t'*$ should point to (have as value) a subvariable in $SV(c)$ of variable type t'. We call configurations satisfying this requirement *type correct for pointers*:

$$tc\text{-}p(c) \equiv \forall x \in SV(c) :$$
$$(vtype(x,c) = t'* \land c.m(x) \neq null \to c.m(x) \in SV(c) \land vtype(c.m(x),c) = t').$$

A pointer variable x with a value $c.m(x) \notin SV(c)$, i.e., outside the subvariables of configuration c, is called a dangling pointer and violates type correctness for pointers. In order to show the absence of dangling pointers we will later define (quite restrictive) context conditions which permit us to show that pointers only point to the global memory or the heap. We define the corresponding predicate:

$$p\text{-}targets(c) \equiv \forall x \in SV(c) :$$
$$(pointer(x,c) \land c.m(x) \neq null \to ingm(c.m(x),c) \lor onheap(c.m(x),c)).$$

We define a configuration c to be *type correct* if it is type correct for subvariables and in the strengthened sense for pointers:

$$tc(c) \equiv tc\text{-}s(c) \wedge tc\text{-}p(c).$$

We will define the transition function δ_C and add context conditions such that in the end we will be able to show that all C0 computations are type correct.

Lemma 74. *Let (c^i) be a C0 computation. Then*

$$\forall i : tc(c^i).$$

The proof will obviously proceed by induction on i showing i) that the initial configuration c^0 is type correct and ii) — once the transition function δ_C for C0 is defined — that c^{i+1} is type correct if c^i is type correct. Properties which hold for all configurations c of a computation are called *invariants*, and Lemma 74 can be reformulated as follows.

Invariant 1
$$tc(c).$$

We will also show the following.

Invariant 2
$$p\text{-}targets(c).$$

Obviously, invariants are simply parts of induction hypotheses about computations (c^i) where the index i has been dropped.

11.3.3 Expressions and Statements in Function Bodies

We wish to define when an expression e or a statement s occurs in the body of a function f. Overloading the $\in$-sign we will denote this by

$$e \in f \quad \text{resp.} \quad s \in f.$$

Intuitively we consider the subtree T' of the derivation tree whose border word is the body of f, and we say that e or s occurs in f if there is a subtree T'' of T' with border word e or s resp. This is illustrated in Fig. 103. For the definition of semantics this is precise enough. But later, when we argue about compiler correctness, we need to be able to distinguish between possibly multiple *occurrences* of the same statement or expression (see, e.g., Fig. 104). In the full model of derivation trees this is easily done via the roots of the aforementioned subtrees T''.

We call a node $i \in T.A$ of the derivation tree a *statement node* if it is labeled with St or rSt, and we collect the statement nodes of the tree into the set

$$stn = \{i \in T.A \mid T.\ell(i) \in \{St, rSt\}\}.$$

Similarly we define the set *en* of *expression nodes* as

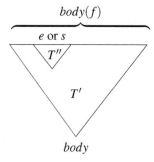

Fig. 103. Statement *s* or expression *e* in the body of function *f*

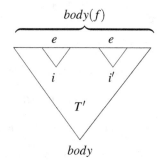

Fig. 104. Multiple occurrences of the same expression in a function body

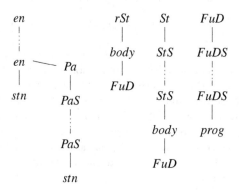

Fig. 105. Sequences of labels on paths in the derivation tree. Every expression or statement node *i* has a unique ancestor $FuD(i)$ labeled FuD, which produces the corresponding function declaration

$$en = \{i \in T.A \mid T.\ell(i) \in \{CC, BC, id, F, T, E, Atom, BF, BT, BE\}\}.$$

From now on we abbreviate $T.\ell$ by ℓ where it is unambiguous. We show that every occurrence of an expression is part of an occurrence of a statement, and that every

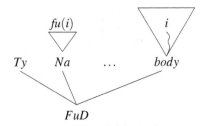

Fig. 106. Every expression or statement node i belongs to a function body of some declared function $fu(i)$

occurrence of a statement is part of a function body which has been declared in the function declaration sequence.

Lemma 75.

- *Every expression node i has an ancestor (prefix) j which is a statement node:*

$$i \in en \rightarrow \exists j \in stn : prefix(j, i).$$

- *Every statement node j has an ancestor k which is labeled with FuD:*

$$j \in stn \rightarrow \exists k : \ell(k) = FuD \wedge prefix(k, j).$$

- *Ancestors r of nodes k labeled FuD are not labeled FuD:*

$$\ell(k) = FuD \wedge prefix(r, k) \rightarrow \ell(r) \neq FuD.$$

Proof. Inspection of the grammar shows that fathers of expression nodes are either expression nodes or statement nodes, except for nonterminals E, BE, and CC, which can also be derived from PaS via Pa (see Fig. 105). Such a parameter sequence, however, can only be the son of a function call statement. This proves the first statement. Fathers of statement nodes are labeled StS or $body$; fathers of nodes labeled StS are labeled StS or $body$. Fathers of nodes labeled $body$ are labeled FuD. This proves the second statement. Fathers of nodes labeled FuD are labeled $FuDS$. Fathers of nodes labeled $FuDS$ are labeled $FuDS$ or $prog$ and a node labeled $prog$ has no father. This proves the third statement.

Let i be an expression or statement node as shown in Fig. 106. By Lemma 75 we can define the unique ancestor $FuD(i)$ of i which is labeled FuD as

$$FuD(i) = \in \{r \mid \ell(r) = FuD \wedge prefix(r, i)\}.$$

The name $fu(i)$ of the function to which node i belongs is the border word of son 1 of $FuD(i)$:

$$fu(i) = bw(FuD(i)1).$$

For expressions e we define formally that e occurs in f if it is the border word of an expression node i with $fu(i) = f$:

$$e \in f \equiv \exists i \in en : fu(i) = f \wedge bw(i) = e.$$

Similarly, we define for statements s:

$$s \in f \equiv \exists i \in stn : fu(i) = f \wedge bw(i) = s.$$

Different occurrences of the same expression or statement are border words of different subtrees.

11.3.4 Program Rest

As the next component of C0 configurations c we introduce the *program rest* or *continuation $c.pr$* of configuration c. Program rests are sequences of statements separated by semicolons, i.e., they have the form

$$c.pr = c.pr[1 : n] = c.pr[1]; \ldots; c.pr[n],$$

where

$$\forall i : c.pr[i] \in L(St) \cup L(rSt).$$

The *length n* of the program rest is denoted by $|c.pr|$. Also we overload functions hd and $tail$ to be used on program rests with similar semantics to plain sequences, i.e., $hd(c.pr) = c.pr[1]$ and $tail(c.pr) = c.pr[2]; \ldots; c.pr[n]$. The last element of the program rest is denoted by $last(c.pr)$.

Intuitively speaking this is the sequence of statements and return statements (separated by semicolons) that is yet to be executed. The statement to be executed in configuration c is the first statement $hd(c.pr)$ of the program rest. Recall that the top function name $c.st(c.rd)$ on the stack is called the current function $cf(c)$ of configuration c:

$$cf(c) = c.st(c.rd).$$

Stack $c.st$ and program rest $c.pr$ are coupled by the following crucial result.

Lemma 76. *The statement $hd(c.pr)$ executed in configuration c belongs to the current function $cf(c)$ of configuration c*

$$hd(c.pr) \in cf(c).$$

Once the definition of the semantics is complete we can prove this lemma by induction on the steps i of the computation (c^i). In order to be able to perform the induction step we have to strengthen the induction hypothesis considerably. We first have to show that the number of return statements in the program rest equals $c.rd + 1$; intuitively speaking every function frame of the stack belongs to a function whose execution has started and which has not returned yet. Moreover the last statement in a program rest is always a return statement, i.e.,

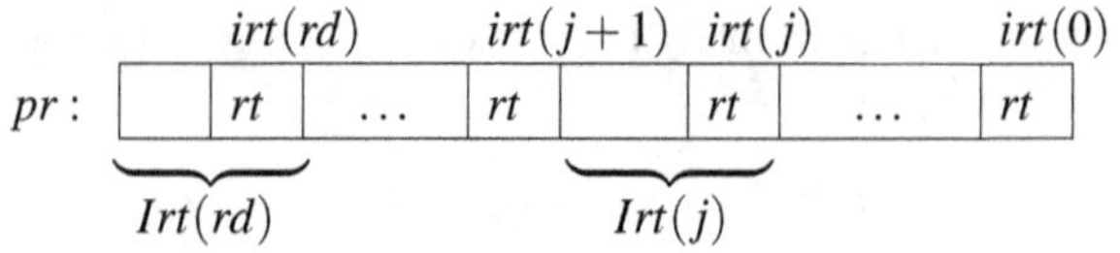

Fig. 107. Decomposing the indices of the program rest into intervals $Irt(j)$ ending with return statements rt. A statement node for $pr[k]$ with $k \in Irt(j)$ belongs to function $st(j)$. Argument c is omitted in this figure

$$\#\{i \mid c.pr[i] \in L(rSt)\} = c.rd + 1$$

and

$$last(c.pr) \in L(rSt).$$

We now locate *from right to left* the indices $irt(j,c)$ such that the root node of a derivation tree with border word $c.pr[irt(j,c)]$ is labeled with rSt (see Fig. 107):

$$irt(0,c) = |c.pr|,$$
$$irt(j+1,c) = \max\{i \mid i < irt(j,c) \wedge c.pr[i] \in L(rSt)\},$$

and divide the indices of program rest elements into intervals $Irt(j,c)$ by

$$Irt(j,c) = \begin{cases} [irt(j+1,c)+1 : irt(j,c)], & j < c.rd, \\ [1 : irt(j,c)], & j = c.rd, \end{cases}$$

i.e., interval $Irt(j,c)$ extends to the left of return statement j until the next return statement if there is such a statement and otherwise to the head of the program rest. Now we simply formulate an invariant stating that statements $c.pr[k]$ with index k in interval $Irt(j,c)$ belong to the function of stack frame $c.st(j)$:

$$j \in [0 : c.rd] \wedge k \in Irt(j,c) \rightarrow c.pr[k] \in c.st(j).$$

We collect the above conditions into an invariant about the program rest.

Invariant 3 *We say that inv-pr(c) holds if the following conditions are fulfilled:*

1. $\qquad\qquad\qquad \#\{i \mid c.pr[i] \in L(rSt)\} = c.rd + 1,$
2. $\qquad\qquad\qquad last(c.pr) \in L(rSt),$
3. $\qquad\qquad j \in [0 : c.rd] \wedge k \in Irt(j,c) \rightarrow c.pr[k] \in c.st(j).$

Proving that statement execution maintains this invariant will be very easy once the definition of the semantics is completed.

11.3.5 Result Destination Stack

As the last component of *C0*-configurations c we introduce a stack for storing the destination of results to be returned by function calls.. The result destination stack

$$c.rds : [1 : c.rd] \rightarrow SV(c)$$

stores for all recursion depths $i \in [1 : c.rd]$ the subvariable where the result of function $c.st(i)$ will be returned.

Clearly, we want the type of the subvariable $c.rds(i)$, where function $f = c.st(i)$ returns its result, to be the type of this result. It is recorded in the function table at $ft(f).t$:

$$vtype(c.rds(i),c) = ft(c.st(i)).t.$$

Also it is highly desirable that the result of the function $c.st(i)$ at recursion depth i is returned to a subvariable that still exists after the function has returned. Indeed we will show that entries $x = c.rds(i)$ on the result destination stack are either i) on the heap, or ii) in the global memory, or iii) in a subvariable (specified by a selector sequence $s \in S^*$) of function frame $ST(j,c)$ below frame $ST(j,c)$:

$$c.rds(i) = x \rightarrow onheap(x,c) \vee ingm(x,c) \vee \exists s \in S^*, j < i : x = ST(j,c)s.$$

We collect these two conditions into an invariant for the result destination stack.

Invariant 4 *We say that inv-rds*(c) *holds if for all* $i \in [1 : c.rd]$ *the following conditions are fulfilled:*

1. $$vtype(c.rds(i),c) = ft(c.st(i)).t,$$
2. $$c.rds(i) = x \rightarrow onheap(x,c) \vee ingm(x,c) \vee \exists s \in S^*, j < i : x = ST(j,c)s.$$

We collect all invariants concerning the well-formedness of *C0*-configurations into a single invariant.

Invariant 5

$$inv\text{-}conf(c) \equiv tc(c) \wedge p\text{-}targets(c) \wedge inv\text{-}pr(c) \wedge inv\text{-}rds(c).$$

In what follows we aim at a definition of the successor configuration $c' = \delta_C(c)$ of c that permits us to show

$$inv\text{-}conf(c) \rightarrow inv\text{-}conf(c').$$

11.3.6 Initial Configuration

Initial configurations c^0 are defined in the following way: the recursion depth is zero

$$c^0.rd = 0,$$

the only stack frame present is a frame for function *main*

Fig. 108. Initial configurations c^0 have only two top-level variables: the global memory frame gm and the function frame $ST(0,c) = (main, 0)$. The heap is initially empty

$$c^0.st(0) = main,$$

the heap is empty

$$c^0.nh = 0,$$

thus there are no arguments for function $c^0.ht$ and we have

$$V(c^0) = \{gm, (main, 0)\}$$

as shown in Fig. 108. Simple subvariables are initialized with their default values:

$$x \in SV(c^0) \wedge simple(x, c^0) \;\rightarrow\; c^0.m(x) = dft(vtype(x, c^0)).$$

Hence configuration c^0 is type correct for simple subvariables and invariants $tc(c^0)$ and $p\text{-}targets(c^0)$ hold initially. The program rest of c^0 is the body of function *main*:

$$c^0.pr = ft(main).body.$$

This sequence has a single return statement at the last node, thus the number of returns equals the number of stack frames which equals the recursion depth increased by one. Also, all nodes $c^0.pr[k]$ of the program rest belong to current function *main*:

$$c^0.st(0) = main \wedge \forall k : c^0.pr[k] \in main.$$

Thus invariant $inv\text{-}pr(c)$ holds initially. Finally, invariant $inv\text{-}rds(c)$ holds trivially, because stack indices $i \geq 1$ are not present. We summarize as follows.

Lemma 77.

$$inv\text{-}conf(c^0).$$

11.4 Expression Evaluation

In this chapter we define how to evaluate expressions e in configurations c which satisfy invariant $inv\text{-}conf(c)$; among other things this will involve determining the type $etype(e, c)$ of expressions e in configurations c. Expression evaluation is part of the semantics of statement execution, and later we will define the first statement $hd(c.pr)$ of the program rest as the statement to be executed in configuration c. By invariant $inv\text{-}pr(c)$ this statement belongs to the function $cf(c)$ currently executed in configuration c:

$$hd(c.pr) \in cf(c).$$

Thus at first glance it suffices to define expression evaluation for expressions $e \in cf(c)$, and if all we want to do is define the semantics of programs, this is indeed good enough. Also evaluation of e will certainly hinge on the derivation tree of e, particularly in order to handle priorities of operators, and this tree is already a subtree of the entire derivation tree T of the program as long as $e \in f$ for any function $f \in FN$. However defining a semantics without doing something with it is like building a car and not driving it. With our semantics we want to do two things:

- show that the compiler we construct is correct. Indeed we will use the $C0$ semantics as a guideline for the construction of the compiler.
- prove that programs written in $C0$ do what they are supposed to do.

For both applications of the semantics it turns out that it is better to define expression evaluation in a slightly more general way:

- the code generated by a compiler for an expression depends on the type of the expression. But at compile time we do not know the configurations c in which e will be evaluated. All we know is the function f to which e belongs. Thus we had better be able to define the type of an expression e in function f as a function $etype(e, f)$.
- when we want to argue about the correctness of programs, it is often convenient to consider expressions e which are formed with the variable names x and identifiers of the program but which are *not* part of the program. Fortunately we can refer to the unique derivation trees T' of arbitrary expressions even if they are not subtrees of the derivation tree T of the program.

Let e' be an expression with derivation tree T', and let e'' be an expression. We define that e'' is a *subexpression* of e' and write $e'' \in e'$ if in the derivation tree T' of e' there is an expression node i such that e'' is the border word of i in T':

$$e'' \in e' \equiv \exists i \in en \wedge bw(i, T') = e''.$$

A small example illustrates that this definition of subexpressions respects the priorities of operators. We have

$$1 \in 1+2*3, \qquad 2*3 \in 1+2*3,$$

but

$$1+2 \notin 1+2*3$$

because the derivation tree for $1+2*3$ does not contain a derivation tree for $1+2$ as a subtree.

The following definitions are quite obvious; however we will need a somewhat strong context condition to guarantee the absence of dangling pointers.

11.4.1 Type, Right Value, and Left Value of Expressions

For expressions $e \in f$, functions $f \in FN$, and $C0$ configurations c satisfying invariant $inv\text{-}conf(c)$ we will simultaneously define three values:

- the *expression type* $etype(e, f)$ of the expression e evaluated in function f.
- the *right value* $va(e, c)$ of e in configuration c. This is a value in the intuitive sense.
- the *left value* $lv(e, c)$ when e is an identifier. For identifiers $e \in L(id)$ either the left value $lv(e, c)$ is the subvariable specified by e in configuration c or the left value is undefined. In the latter case we write $lv(e, c) = \bot$. Thus we always have

$$lv(e, c) \in SV(c) \cup \{\bot\}.$$

The names 'left value' and 'right value' are motivated by the semantics of assignment statements $x = e$ (where $x \in L(id)$ and $e \in L(E) \cup L(BE) \cup L(CC)$) which will be defined in Sect. 11.5. The 'ordinary' value of the *right*-hand side (expression e) is assigned to the subvariable specified by the *left*-hand side (expression x).

Left values $lv(e, c)$ and right values $va(e, c)$ of expressions $e \in f$ will only be defined for configurations c where statements from f are executed, i.e., where

$$cf(c) = f,$$

in which case the top frame has the format

$$top(c) = (f, c.rd).$$

Whenever the left value $lv(e, c) \in SV(c)$ of an expression e is defined, its right value is *always* determined by the current memory content of subvariable $lv(e, c)$:

$$va(e, c) = c.m(lv(e, c)).$$

As we proceed with the definitions we show (by induction on the height of the derivation tree of e) that the following invariant $inv\text{-}expr(e, c)$ concerning the type correctness and pointer targets of expression e hold for all expressions e under consideration.

Invariant 6 *We say that $inv\text{-}expr(e, c)$ holds if the following conditions are fulfilled:*

1. *Left values $lv(e, c)$ — if they exist — are subvariables of the current configuration:*
$$lv(e, c) \in SV(c).$$

2. *The variable type of the left value of an expression is the expression type of the expression:*
$$vtype(lv(e, c), c) = etype(e, f).$$

3. *The value of an expression is in the range of its expression type:*

$$va(e, c) \in ra(etype(e, f)).$$

4. *Expressions with pointer type $t'*$ and a non-null value point to a subvariables of c which have type t':*

$$etype(e,f) = t'* \wedge va(e,c) \neq null \rightarrow va(e,c) \in SV(c) \wedge vtype(va(e,c),c) = t'.$$

5. *Non-null pointers in expression evaluation point to the global memory or the heap:*

$$etype(e,f) = t'* \wedge va(e,c) \neq null \rightarrow ingm(va(e,c),c) \vee onheap(va(e,c),c).$$

If an expression e has a left value $lv(e,c)$, then one only has to check parts 1 and 2 of *inv-expr(c)*. Parts 3–5 follow.

Lemma 78. *Let e have a left value $lv(e,c) \in SV(c)$, and assume $vtype(lv(e,c),c) = etype(e,f)$ as well as inv-conf(c). Then*

1. $$va(e,c) \in ra(etype(e,f)),$$
2. $etype(e,f) = t'* \wedge va(e,c) \neq null \rightarrow va(e,c) \in SV(c) \wedge vtype(va(e,c),c) = t'$,
3. $etype(e,f) = t'* \wedge va(e,c) \neq null \rightarrow ingm(va(e,c),c) \vee onheap(va(e,c),c)$.

Proof. Let $x = lv(e,c)$.

1. By definition of $va(e,c)$ and invariant $tc(c)$ we have

$$va(e,c) = c.m(x) \in ra(vtype(x,c)).$$

By hypothesis we have

$$vtype(x,c) = etype(e,f),$$

hence

$$va(e,c) \in ra(etype(e,f)).$$

2. Let $etype(e,f) = t'*$ and assume $va(e,c) \neq null$. We conclude

$$\begin{aligned}
t'* &= etype(e,f) \\
&= vtype(x,c) \quad \text{(hypothesis)}, \\
va(e,c) &= c.m(x) \quad \text{(definition of } va(e,c)) \\
&\in SV(c) \quad \text{(invariant } tc\text{-}p(c)), \\
vtype(va(e,c),c) &= vtype(c.m(x),c) \\
&= t' \quad \text{(invariant } tc\text{-}p(c)).
\end{aligned}$$

3. Let $etype(e,f) = t'*$ and $va(e,c) = c.m(x) \neq null$. Invariant p-$targets(c)$ gives

$$ingm(c.m(x),c) \vee onheap(c.m(x),c),$$

i.e.,

$$ingm(va(e,c),c) \vee onheap(va(e,c),c).$$

11.4.2 Constants

Evaluation of constants e is independent of the current configuration c and function f. Left values of constants are not defined. There are four kinds of constants e:

- Boolean constants $\in \{true, false\}$. We define

$$etype(true, f) = etype(false, f) = bool$$

and

$$va(false, c) = 0,$$
$$va(true, c) = 1.$$

- Character constants 'a' where a is a digit or a letter. We define

$$etype('a', f) = char$$

and

$$va('a', c) = a.$$

- Integer constants $e \in [0:9]^+$. We define

$$etype(e, z) = int$$

and identify e with a decimal number, which we evaluate tmod 2^{32}

$$va(e, c) = (\langle e \rangle_Z \text{ tmod } 2^{32}).$$

Due to the modulo arithmetic, this leaves $\langle e \rangle_Z$ unchanged only if we have

$$\langle e \rangle_Z \in [0:2^{31} - 1].$$

In contrast, for $z = 2147483648$ we have $\langle z \rangle_Z = 2^{31}$ and thus as an integer constant it evaluates to the negative number

$$va(2147483648, c) = (2^{31} \text{ tmod } 2^{32}) = -2^{31}.$$

- Unsigned integer constants $e = e'u \in ([0:9]^+ \circ \{u\})$. We set

$$etype(e, z) = uint$$

and identify e' with a decimal number, which we evaluate mod 2^{32}

$$va(e, c) = (\langle e' \rangle_Z \text{ mod } 2^{32}).$$

This leaves $\langle e' \rangle_Z$ unchanged only if we have

$$\langle e' \rangle_Z \in [0:2^{32} - 1].$$

In particular we have

$$va(2147483648u, c) = (2^{31} \text{ mod } 2^{32}) = 2^{31}.$$

- The null pointer $e = null$ has to receive a somewhat special treatment, because it lies in the range of *all* pointer types. It gets its own expression type, which we denote by $\perp$. Thus we define

$$etype(null, f) = \perp,$$
$$va(null, c) = null.$$

Invariant 6.3 obviously holds for e. Parts 1 and 2 of the invariant concern left values and do not apply. Parts 4 and 5 of the invariant concern pointers and don't apply either.

11.4.3 Variable Binding

We want to evaluate variable names X in functions f resp. configurations c. Variable names $X \in f$ occurring in the body of function f have to be declared either as global variables $\in VN$ or as parameters or local variables $\in ft(f).VN$.

Context Condition 11

$$X \in f \rightarrow X \in VN \cup ft(f).VN.$$

For variable names $X \in e$ occurring in expressions $e \notin f$, which we want to evaluate for function f even if they do not occur in the body of f, the same context condition applies. Let

$$\$f = \{t_1\ n_1, \ldots, t_s\ n_s\},$$
$$\$gm = \{t'_1\ n'_1, \ldots, t'_r\ n'_r\}.$$

We take the expression type $etype(X, f)$ of X from the type table for $\$f$ if X is a parameter or local variable of function f. Otherwise we take it from the type table of $\$gm$.

$$etype(X, f) = \begin{cases} t_i, & X = n_i \wedge X \in ft(f).VN, \\ t'_i, & X = n'_i \wedge X \in VN \setminus ft(f).VN. \end{cases}$$

We *bind* variable name $X \in f$ to a subvariable $lv(X, c) \in SV(c)$ of configurations c satisfying $top(c) = (f, c.rd)$, i.e., we have

$$(f, c.rd) \in SV(c).$$

This permits us to bind variable name X to a local (sub)variable $top(c).X$ of $top(c)$, if X is the name of a local variable of function $f = fu(c.rd)$. Otherwise we bind X to global (sub)variable $gm.X$. This is illustrated in Fig. 109.

$$lv(X, c) = \begin{cases} top(c).X, & X \in ft(f).VN, \\ gm.X, & \text{otherwise.} \end{cases} \tag{9}$$

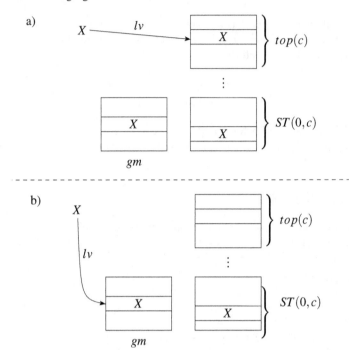

Fig. 109. Variable name X is bound by function lv to a local variable or parameter $top(c).X$ in the top frame if a local variable or parameter with that name exists, as illustrated in a. Otherwise it is bound to component $gm.X$ in global memory — see (b)

Equation 9 is called the *visibility rule* of $C0$ (and of C) because it defines how variables are 'visible' from a function body: the local (sub)variable or parameter $top(c).X$ obscures, if it exists, the global variable $gm.X$.

We have $lv(X,c) \in SV(c)$, thus Invariant 6.1 holds for X. If $X \in ft(f).VN$, let $X = n_i$; for the variable type of $lv(X,c)$ we get

$$\begin{aligned} vtype(lv(X,c),c) &= vtype(top(c).X,c) \\ &= vtype((f,c.rd).n_i,c) \\ &= t_i \\ &= etype(X,f). \end{aligned}$$

Otherwise, let $X = n_i'$. We get

$$\begin{aligned} vtype(lv(X,c),c) &= vtype(gm.X,c) \\ &= vtype(gm.n_i',c) \\ &= t_i' \\ &= etype(X,f). \end{aligned}$$

Thus Invariant 6.2 holds for X. Parts 3–5 follow from Lemma 78.

11.4.4 Pointer Dereferencing

Let $e = e'*$ be an expression in the body of function f, where subexpression $e' \in e$ has an expression type $etype(e', f)$. We require that this is a pointer type and we do not dereference null pointers.

Context Condition 12 *There is a type $t' \in ET \cup TN$ such that:*

$$etype(e', f) = t'*,$$
$$va(e', c) \neq null.$$

Note that the second condition cannot be tested statically. We set

$$lv(e, c) = va(e', c),$$
$$etype(e, f) = t'.$$

Because $va(e', c) \neq null$, we get from Invariant 6.4 for e'

$$lv(e, c) = va(e', c) \in SV(c).$$

This is Invariant 6.1 for e. Again using Invariant 6.4 for e' we get

$$vtype(lv(e, c), c) = vtype(va(e', c), c)$$
$$= t'$$
$$= etype(e, c).$$

This is Invariant 6.2 for e. Parts 3–5 of the invariant follow from Lemma 78.

11.4.5 Struct Components

Let $e = e'.n$ be an expression in the body of function f, where subexpression $e' \in e$ has an expression type $etype(e', f)$. We require that this is a struct type and that n is one of the declared selectors for this type.

Context Condition 13 *There are types $t_1, \ldots, t_s \in ET \cup TN$ such that:*

$$etype(e', f) = \{t_1 \ n_1; \ldots; t_s \ n_s\},$$
$$\exists j : n = n_j.$$

We define

$$lv(e, c) = lv(e', c).n,$$
$$etype(e, f) = t_j.$$

From Invariant 6.1 for e' we know $lv(e', c) \in SV(c)$, hence

$$lv(e, c) = lv(e', c).n \in SV(c),$$

i.e., we have Invariant 6.1 for e. Using Invariant 6.2 for e' we get

$$
\begin{aligned}
vtype(lv(e,c),c) &= vtype(lv(e',c).n,c) \\
&= t_j \\
&= etype(e,f).
\end{aligned}
$$

This is Invariant 6.2 for e. Parts 3–5 of the invariant follow from Lemma 78.

11.4.6 Array Elements

Let $e = e'[e'']$ be an expression in the body of function f, where subexpressions $e',e'' \in e$ have expression types $etype(e',f)$ resp. $etype(e'',f)$. We require that $etype(e',f)$ is an array type and that $etype(e'',f)$ is either *int* or *uint*.

Context Condition 14 *There is a type* $t' \in ET \cup TN$ *such that:*

$$
\begin{aligned}
etype(e',f) &= t'[n], \\
etype(e'',f) &\in \{int, uint\}.
\end{aligned}
$$

Moreover we require that the current value of e'' specifies an index within the array bounds.

Context Condition 15
$$
va(e'',c) \in [0:n-1].
$$

Note that there is no general method to check the latter condition at compile time. We define

$$
\begin{aligned}
lv(e,c) &= lv(e',c)[va(e'',c)], \\
etype(e,f) &= t'.
\end{aligned}
$$

From Invariant 6.1 for e' we know $lv(e',c) \in SV(c)$, hence

$$
lv(e,c) = lv(e',c)[va(e'',c)] \in SV(c),
$$

i.e., we have Invariant 6.1 for e. Using Invariant 6.2 for e' we get

$$
\begin{aligned}
vtype(lv(e,c),c) &= vtype(lv(e',c)[va(e'',c)],c) \\
&= t' \\
&= etype(e,f).
\end{aligned}
$$

This is Invariant 6.2 for e. Parts 3–5 of the invariant follow from Lemma 78.

11.4.7 'Address of'

Let $e = e'\&$ be an expression in the body of function f, where subexpression $e' \in L(id)$. We require the left value $lv(e',c)$ to be defined and to not lie on the stack.

Context Condition 16

$$lv(e',c) \neq \bot \wedge (ingm(lv(e',c),c) \vee onheap(lv(e',c))).$$

We define only

$$va(e,c) = lv(e',c),$$
$$etype(e,f) = etype(e',f)*.$$

The left value $lv(e,c)$ is not defined. Thus Invariants 6.1 and 6.2 do not apply to e. Since for any pointer type $t'*$

$$SV(c) \subseteq ra(t'*),$$

we have by Invariant 6.1 for e'

$$va(e,c) = lv(e',c)$$
$$\in SV(c)$$
$$\subseteq ra(etype(e,f)).$$

This is Invariant 6.3 for e. By Invariant 6.2 for e' we have

$$vtype(va(e,c),c) = vtype(lv(e',c),c)$$
$$= etype(e',f).$$

Thus we have Invariant 6.4 for e. Substituting $va(e,c) = lv(e',c)$ into context condition 16 immediately gives

$$ingm(va(e,c),c) \vee onheap(va(e,c),c),$$

i.e., we have Invariant 6.5 for e.

11.4.8 Unary Operators

From now on, neither new left values nor new pointers are introduced. Thus we have $etype(e,f) \in \{int, uint, bool\}$ and only Invariant 6.3 is relevant. The trivial proof that it holds is omitted in all remaining cases. We also treat bracketed expressions here.

Let $e = \circ e'$ be an expression in the body of function f, where $e' \in e$ and $\circ$ is an unary operator. There are two subcases:

- unary minus: $\circ = -_1$. We require

Context Condition 17

$$etype(e', f) \in \{int, uint\}$$

and set

$$etype(e, f) = etype(e', f),$$

$$va(e, c) = \begin{cases} (-va(e', c) \text{ tmod } 2^{32}), & etype(e', f) = int, \\ (-va(e', c) \text{ mod } 2^{32}), & etype(e', f) = uint. \end{cases}$$

This definition looks 'natural', but if one has no experience with modulo arithmetic one might be in for a surprise. Let

$$etype(e', f) = int,$$
$$va(e', c) = -2147483648$$
$$= -2^{31},$$

then

$$etype(-e', f) = int,$$
$$va(-e', c) = (2^{31} \text{ tmod } 2^{32})$$
$$= -2^{31}$$
$$= va(e', c).$$

- negation: $\circ = !$. We require

Context Condition 18

$$etype(e', f) = bool$$

and set

$$etype(e, f) = etype(e', f),$$
$$va(e, c) = \neg va(e', c).$$

Let $e = (e')$ be an expression in the body of function f, where $e' \in e$. Expression e' represents an arithmetic or Boolean expression, hence we require

Context Condition 19

$$etype(e', f) \in \{int, uint, bool\}$$

and set

$$etype(e, f) = etype(e', f),$$
$$va(e, c) = va(e', c).$$

11.4.9 Binary Operators

Let $e = e' \circ e''$ be an expression in the body of function f, where $e', e'' \in e$ and $\circ$ is a binary operator. Recall that subexpressions were defined via subtrees of the derivation tree of e, thus the decomposition of e into e' and e'' reflects the priorities of the operators. Because we want to avoid type casting we require here that subexpressions e' and e'' have the same type. We allow tests for equality or inequality for all simple types. For Boolean operations we require the type to be *bool*. For arithmetic operations and comparisons involving order we require this type to be *int* or *uint*. Testing for the null pointer requires a special case. For $etype(e', f) = t'$ we require the following.

Context Condition 20

$$t' = etype(e'', f) \vee pointer(t') \wedge e'' = null,$$

$$t' \in \begin{cases} \{t \mid simple(t)\}, & \circ \in \{==, !\,=\}, \\ \{bool\}, & \circ \in \{\&\&, \|\}, \\ \{int, uint\}, & otherwise. \end{cases}$$

The type of the result is the type of the operands for arithmetic operations and *bool* for comparisons or Boolean operators:

$$etype(e, f) = \begin{cases} etype(e', f), & \circ \in \{+, -, *, /\}, \\ bool, & otherwise. \end{cases}$$

Now we make the obvious case split:

- addition, subtraction, and multiplication: $\circ \in \{+, -, *\}$. We define

$$va(e, c) = \begin{cases} (va(e', c) \circ va(e'', c) \text{ tmod } 2^{32}), & etype(e', f) = int, \\ (va(e', c) \circ va(e'', c) \text{ mod } 2^{32}), & etype(e', f) = uint. \end{cases}$$

We have mentioned it before and we stress it again: this is modulo arithmetic, which is very different from arithmetic in $\mathbb{N}$. For instance if

$$etype(e, f) = uint,$$
$$va(e', c) = 2^{32} - 1,$$
$$va(e'', c) = 1,$$
$$\circ = +,$$

we compute for $va(e, c)$ the result of the addition modulo 2^{32}:

$$va(e, c) = (2^{32} - 1 + 1 \text{ mod } 2^{32})$$
$$= 0.$$

- division: $\circ = /$. Of course we forbid division by zero.

Context Condition 21

$$\circ = / \;\rightarrow\; va(e'',c) \neq 0.$$

Let

$$y' = va(e',c),$$
$$y'' = va(e'',c).$$

Using definitions from Sect. 9.3, we determine the result in the usual way:

$$va(e,c) = \begin{cases} (divt(y',y'') \bmod 2^{32}), & etype(e',f) = int, \\ ((y'/y'') \bmod 2^{32}), & etype(e',f) = uint. \end{cases}$$

Note that integer division *can* overflow (in a nasty way) due to the non-symmetric range T_n of two's complement numbers. We already know the relevant example:

$$e' = -2^{31},$$
$$e'' = -1,$$
$$y = y'/y'' = 2^{31},$$
$$(y \bmod 2^{32}) = -2^{31}.$$

- computation of atoms: $\circ \in \{==, !=, <, >, <=, >=\}$ is a comparison operator. We convert the operator $\circ$ in the obvious way into an operator

$$\circ' \in \{=, \neq, <, >, \leq, \geq\}$$

and define

$$va(e,c) = \begin{cases} 1, & va(e',c) \circ' va(e'',c), \\ 0, & \text{otherwise.} \end{cases}$$

- Boolean operators: $\circ \in \{\&\&, ||\}$. We define

$$va(e,c) = \begin{cases} va(e',c) \wedge va(e'',c), & \circ = \&\&, \\ va(e',c) \vee va(e'',c), & \circ = ||. \end{cases}$$

11.5 Statement Execution

The semantics of statement execution is defined by a case split on the first element $hd(c.pr)$ of the program rest of configuration c. The current configuration is denoted by c; the new configuration is denoted by c'. We assume invariant $inv\text{-}conf(c)$ which implies invariant $inv\text{-}expr(e,c)$ for the expressions $e \in hd(c.pr)$ evaluated in configuration c. Let $f = cf(c)$ be the name of the function executed in configuration c. Definitions will be fairly straightforward. As there are six kinds of statements we get the obvious cases.

11.5.1 Assignment

The head of the program rest has the form

$$hd(c.pr) : e = e'.$$

We require that the left value $lv(e,c)$ exists and that the types of the left-hand side and right-hand side of the assignment statement match *and* that these types are simple[2]. Assigning the null pointer, whose type we left undefined, again requires a special case. For $etype(e,f) = t$ we require the following.

Context Condition 22

$$lv(e,c) \neq \bot \wedge (simple(t) \wedge t = etype(e',f) \vee$$
$$pointer(t) \wedge e' = null).$$

The execution of the statement does not create or delete heap variables and it does not manipulate the stack function nor the result destination stack. Formally

$$X \in \{rd, st, rds, nh, ht\} \rightarrow c'.X = c.X,$$

which implies

$$SV(c') = SV(c).$$

The memory content $c.m$ of the single simple subvariable $lv(e,c) \in SV(c)$ specified by the left-hand side is updated by assigning it the value $va(e',c)$ specified by the right-hand side. All other simple subvariables keep their previous value. The assignment statement is dropped from the program rest.

$$\forall x \in SV(c) : simple(x,c) \rightarrow c'.m(x) = \begin{cases} va(e',c), & x = lv(e,c), \\ c.m(x), & \text{otherwise,} \end{cases}$$
$$c'.pr = tail(c.pr).$$

A single simple subvariable $lv(e,c)$ is updated. For this subvariable $tc\text{-}s(c')$ follows from Invariant 6.3 for the right-hand side e' of the assignment. If it is a pointer, $tc\text{-}p(c')$ follows from Invariant 6.4 for e' and $p\text{-}targets(c')$ follows from Invariant 6.5 for e'.

For all other simple subvariables the memory content stays the same:

$$\forall x \in SV(c) \setminus \{lv(e,c)\} : simple(x,c) \rightarrow c'.m(x) = c.m(x).$$

Thus for these subvariables $tc\text{-}s(c')$, $tc\text{-}p(c')$ and $p\text{-}targets(c')$ follows by induction from the corresponding invariants for c.

No returns are added to or deleted from the program rest, the recursion depth stays the same, and rds is not manipulated. Thus invariant $inv\text{-}rds$ is preserved. If the first statement of a program rest is not a return statement, dropping it from the program rest preserves invariant $inv\text{-}pr$. Hence, $inf\text{-}conf(c')$ holds.

---

[2] This is a usual restriction in programming languages; relaxing this restriction is a fairly simple exercise.

11.5.2 Conditional Statement

The head of the program rest can have two forms: if-statements

$$hd(c.pr) : \text{if } e \ \{\textit{if-part}\}$$

or if-then-else statements

$$hd(c.pr) : \text{if } e \ \{\textit{if-part}\} \text{ else } \{\textit{else-part}\}.$$

Conditional statements only change the program rest

$$X \neq pr \ \rightarrow \ c'.X = c.X.$$

Condition e is evaluated. We require it to be of type $bool$.

Context Condition 23

$$etype(e, f) = bool.$$

- if-statements: If the condition evaluates to 1, the conditional statement is replaced in program rest by the if-part. Otherwise the statement is simply dropped from the program rest.

$$c'.pr = \begin{cases} \textit{if-part};tail(c.pr), & va(e,c) = 1, \\ tail(c.pr), & va(e,c) = 0. \end{cases}$$

- if-then-else-statements. If the condition evaluates to 1, the conditional statement is replaced in the program rest by the if-part. Otherwise it is replaced by the else-part.

$$c'.pr = \begin{cases} \textit{if-part};tail(c.pr), & va(e,c) = 1, \\ \textit{else-part};tail(c.pr), & va(e,c) = 0. \end{cases}$$

Thus invariants tc, p-$targets$, and inv-rds are obviously preserved. Statements s from if-parts or else-parts which are added to the program rest belong to function f, i.e., $s \in cf(c)$. Moreover, in the $C0$ grammar we see that if-parts and else-parts do not contain return statements. Hence, invariant inv-pr is preserved.

11.5.3 While Loop

The head of the program rest has the form

$$hd(c.pr) : \text{while } e \ \{\textit{body}\}.$$

It changes only the program rest

$$X \neq pr \ \rightarrow \ c'.X = c.X.$$

The loop condition e is evaluated. We require it to be of type $bool$.

Context Condition 24

$$etype(e,f) = bool.$$

If it evaluates to 0 the while statement is dropped from the program rest. Otherwise the flattened loop body is put in front of the program rest:[3]

$$c'.pr = \begin{cases} tail(c.pr), & va(e,c) = 0, \\ body;c.pr, & va(e,c) = 1. \end{cases}$$

Preservation of invariants is shown as for conditional statements.

11.5.4 'New' Statement

The head of the program rest has the form

$$hd(c.pr) : e = \text{new } t *.$$

We require t to be an elementary or declared type. Moreover, the subvariable $lv(e,c)$ should be defined and have type $t*$.

Context Condition 25

$$t \in TN \cup ET \ \wedge lv(e,v) \neq \bot \wedge \ etype(e,f) = t *.$$

Execution of a 'new' statement does not change recursion depth, stack, and result destination stack:

$$X \in \{rd, st, rds\} \ \rightarrow \ c'.X = c.X.$$

The other components of the configuration change in the following way (which is illustrated in Fig. 110):

- the number of heap variables in increased by 1:

$$c'.nh = c.nh + 1.$$

This creates the new heap variable $c.nh$ and we have

$$V(c') = V(c) \cup \{c.nh\}.$$

- the type of this new heap variable is fixed to t. The types of the previously existing heap variables stay the same.

$$c'.ht(x) = \begin{cases} t, & x = c.nh, \\ c.ht(x), & x < c.nh. \end{cases}$$

---

[3] There is a certain analogy between this rule and the use of a (potentially limitless) role of toilet paper. One tests whether one is done and if so one can forget about the entire roll. Otherwise one unrolls a single piece of toilet paper (the loop body) and iterates.

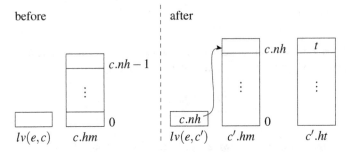

Fig. 110. A 'new' statement creates a new nameless variable $c'.nh - 1 = c.nh$ on the heap, records its type t at $c'.ht(c.nh)$ and stores $c.nh$ in the pointer variable $lv(e,c)$ designated by e. This makes $lv(e,c')$ point to the newly created variable $c.nh$

- the new heap variable is initialized with its default value:

$$c'.hm(c.nh) = dft(t).$$

- the pointer variable $lv(e,c)$ specified by the left-hand side is made to point to the newly created variable. The content of all other simple subvariables in $SV(c)$ remains unchanged.

$$\forall x \in SV(c) \setminus \{c.nh\} : simple(x,c) \ \rightarrow \ c'.m(x) = \begin{cases} c.nh, & x = lv(e,c), \\ c.m(x), & \text{otherwise.} \end{cases}$$

- the 'new' statement is dropped from the program rest.

$$c'.pr = tail(c.pr).$$

The new heap variable $c.nh$ is initialized with its default value. Subvariables of the new variable are initialized with default values as well. Thus invariants $tc(c')$ and $p\text{-}targets(c')$ hold trivially for the new variable.

By Invariant 6.2, the pointer subvariable $x = lv(e,c)$ has type

$$\begin{aligned} vtype(x,c) &= vtype(lv(e,c),c) \\ &= etype(e,f) \\ &= t* \end{aligned}$$

and gets assigned the value

$$\begin{aligned} c'.m(x) &= c'.m(lv(e,c)) \\ &= c.nh \end{aligned}$$

of type

$$vtype(c'.m(x), c') = vtype(c.nh, c')$$
$$= c'.ht(c.nh)$$
$$= t.$$

Because

$$c'.m(x) \in SV(c') \wedge onheap(c'.m(x), c'),$$

we have tc-$p(c')$ and p-$targets(c')$ for subvariable x. All other subvariables of c stay unchanged, thus their invariants are maintained. Only the first statement is dropped from the program rest. Thus invariants inv-pr and inv-rds are maintained by the known arguments.

11.5.5 Function Call

The head of the program rest has the form

$$hd(c.pr) : \begin{cases} e = g(e_1, \ldots, e_p), & p \geq 1, \\ e = g(), & p = 0, \end{cases}$$

where g is the name of the called function and p is the number of its parameters specified in the function table as

$$p = ft(g).p.$$

Obviously, we require this function name to be declared.

Context Condition 26

$$g \in FN.$$

The type of function frames for g is specified in the type table as

$$tt(\$g) = \{t_1\ x_1; \ldots; t_s\ x_s\}.$$

The type t of the result returned by g is specified in the function table as

$$t = ft(g).t.$$

There are several more requirements:

- the left value $lv(e, c)$ is defined.
- the expression type of the left-hand side is simple and matches the type t of the function's return value.
- for $j \in [1 : p]$ the type of the j'th parameter is simple and matches the type t_j of the j'th parameter x_j in the function declaration. A null pointer as parameter requires the usual special case.

Context Condition 27

$$lv(e, c) \neq \bot \wedge (simple(t) \wedge t = etype(e, f)) \wedge$$
$$\forall j \in [1 : p]: \quad (simple(t_j) \wedge t_j = etype(e_j, f) \vee$$
$$pointer(t_j) \wedge e_j = null).$$

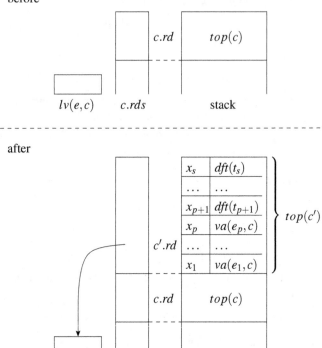

Fig. 111. A function call increases recursion depth creating a new stack frame $top(c')$ and a new entry on the return destination stack $c'.rds$. It passes parameter values $va(e_j,c)$ to parameters x_j of the new top function frame and initializes local variables of the new frame to their default values. A pointer to the result destination $lv(e,c)$ of the call is stored at the new entry $c'.rds(c'.rd)$ of the result destination stack

The execution of a function call has several effects as illustrated in Fig. 111:

- the heap stays the same. The recursion depth is increased to $c.rd + 1$ and a new function frame for g is created at the top of the stack.

$$c'.nh = c.nh,$$
$$c'.ht = c.ht,$$
$$c'.rd = c.rd + 1,$$
$$c'.st(x) = \begin{cases} g, & x = c'.rd, \\ c.st(x), & x \leq c.rd. \end{cases}$$

Thus the new top frame of the stack is

$$top(c') = (g, c'.rd).$$

The variables of c' are the old variables together with the new top frame

$$V(c') = V(c) \cup \{top(c')\}.$$

- parameters are initialized with their values, whereas local variables are initialized with their default values:

$$c'.m(top(c').x_j) = \begin{cases} va(e_j,c), & j \leq p, \\ dft(t_j), & j > p. \end{cases}$$

- the subvariable $lv(e,c)$, where the function's result is returned, is recorded in the new entry $c'.rds(c'.rd)$ of the return destination stack. Old parts of this stack stay unchanged.

$$c'.rds(x) = \begin{cases} lv(e,c), & x = c'.rd, \\ c.rds(x), & x \leq c.rd. \end{cases}$$

- in the program rest the function call is replaced by the body of g, which is recorded in the function table.

$$c'.pr = ft(g).body; tail(c.pr).$$

The new variable $top(c')$ is initialized. Arguing that invariants are maintained involves two different cases:

- parameters: one argues that $tc\text{-}SV(c')$, $tc\text{-}p(c')$, and $p\text{-}targets(c')$ hold as in the case of assignment statements.
- local variables: they are initialized with their default values. One argues that their invariants hold as in the case of the new heap variable in a 'new' statement.

For the new entry $c'.rds(x)$ at argument

$$x = c'.rd$$

of the return destination stack we argue:

$$\begin{aligned} vtype(c'.rds(x),c) &= vtype(lv(e,c),c) \\ &= etype(e,f) \quad \text{(part 2 of } inv\text{-}expr(e,c)) \\ &= t \quad \text{(context condition)} \\ &= ft(g).t \\ &= ft(c'.st(x)).t. \end{aligned}$$

This shows part 1 of $inv\text{-}rds(c')$. The value $lv(e,c)$ assigned to $c'.rds(x)$ by part 1 of invariant $inv\text{-}expr(e,c)$ refers to the old subvariables:

$$\begin{aligned} c'.rds(x) &= lv(e,c), \\ lv(e,c) &\in SV(c). \end{aligned}$$

Hence in c' it lies on the heap, in global memory, or in the stack under the new top frame. This shows part 2 of *inv-rds*(c').

Old subvariables are not changed, thus their invariants are preserved. The number of returns in the program rest and recursion depth are both increased by 1. The new front portion of the program rest until the first return from the left (which is the new last return from the right) comes from statements s belonging to the function g of the new top frame

$$s \in g,$$
$$g = cf(c'),$$

thus invariant *inv-pr* is preserved.

11.5.6 Return

The head of the program rest has the form

$$hd(c.pr) : \text{return } e.$$

We require that the type $etype(e,f)$ of the expression in the return statement matches the type t of the return result of the function f containing the return statement. As usual, a null pointer as returned value requires a special treatment. For $ft(f).t = t$ we require the following.

Context Condition 28

$$simple(t) \wedge t = etype(e,f) \vee$$
$$pointer(t) \wedge e = null.$$

Execution of the return statement has the following effects as illustrated in Fig. 112:

- the value $va(e,c)$ is assigned to the subvariable

$$y = c.rds(c.rd),$$

which is the destination of the return result:

$$c'.m(y) = va(e,c).$$

By *inv-rds*(c) subvariable y has (simple) type t:

$$vtype(y,c) = t.$$

- the top frame is deleted from the stack, and the heap is unchanged. For $x \in [0 : c.rd - 1]$:

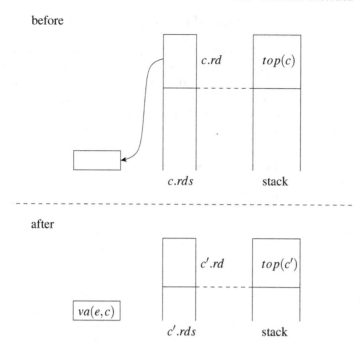

Fig. 112. A return statement assigns the return value $va(e,c)$ to the subvariable pointed to by the top entry $c.rds(c.rd)$. Then it decreases recursion depth, thereby dropping the old top frame and the old top entry in the return destination stack $c.rds$

$$c'.rd = c.rd - 1,$$
$$c'.st(x) = c.st(x),$$
$$c'.nh = c.nh,$$
$$c'.ht = c.ht,$$

thus

$$V(c') = V(c) \setminus \{top(c)\}.$$

- subvariables of c other than the return destination do not change. For $x \in SV(c')$:

$$simple(x,c) \wedge x \neq y \rightarrow c'.m(x) = c.m(x).$$

- the return statement is dropped from the program rest:

$$c'.pr = tail(c.pr).$$

The old top frame disappears. The simple subvariable y is assigned value $va(e,c)$. Now $tc(c')$ and — in case t is a pointer type — p-$targets(c')$ are concluded for y as in the case of assignments. The number of returns in the program rest and the recursion depth both decrease by 1. The return statement which is dropped from the program rest was the last statement belonging to the function f of the old top frame. Thus invariant inv-$pr(c')$ follows from inv-$pr(c)$.

11.6 Proving the Correctness of $C0$ Programs

We use the $C0$-semantics presented so far to analyze a few small programs. These examples serve several purposes: i) they simply illustrate the constructs of the semantics; ii) they show that properties of $C0$ computations, i.e., of program runs, can be rigorously proven as mathematical theorems; iii) they illustrate that when using computer arithmetic, one *always* has to keep in mind that it is finite. Several examples were chosen to highlight this fact.

11.6.1 Assignment and Conditional Statement

Consider the following program

```
int x;
int main()
{
   x = 3;
   x = x + 1;
   if x>0 {x = 1} else {x = 2};
   return -1
}
```

It produces a finite computation (c^i). There are no function calls and no new statements. Thus for all c^i we have

$$c^i.rd = 0,$$
$$c^i.nh = 0,$$
$$c^i.st(0) = main,$$
$$V(c^i) = \{gm, (main, 0)\}.$$

Function *main* has no parameters and no local variables. A single global variable x is declared. Thus, global memory and the function frame for *main* have types

$$\$gm = \{int\ x\},$$
$$\$main = \{\ \}.$$

As illustrated in Fig. 113, in the body of function *main* variable name x is always bound to

$$lv(x, c) = gm.x$$

and its value and expression type are

$$va(x, c) = c.m(gm.x),$$
$$etype(x, main) = int.$$

We evaluate $va(x, c^i)$ and $c^i.pr$ for $i \in [0 : 4]$:

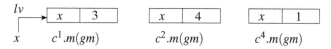

Fig. 113. Value of $gm.x$ in various configurations

0. initially we have
$$va(x, c^0) = dft(int) = 0$$
and the program rest
$$c^0.pr = ft(main).body$$

1. after execution of the first assignment we have
$$va(x, c^1) = 3$$
and the program rest $c^1.pr$ is

```
x = x + 1;
if x>0 {x = 1} else {x = 2};
return -1
```

2. after the next step we have
$$va(x, c^2) = 4$$
and the program rest $c^2.pr$ is

```
if x>0 {x = 1} else {x = 2};
return -1
```

3. execution of the conditional statement only changes the program rest. Expression $x > 0$ is evaluated to
$$4 > 0$$
which is true, thus the program rest $c^3.pr$ is

```
x = 1;
return -1
```

4. after the assignment statement we have
$$va(c^4.x) = 1$$
and the program rest is

```
return -1
```

5. execution of the return statement of function main at recursion depth 0 has no semantics yet. In theory we might say the program halts. In practice it should return to the operating system. In later sections we will see how this can be implemented.

11.6.2 Computer Arithmetic

In the previous example we change the first assignment in the body of function *main*
to

```
x = 2147483647
```

Clearly, in ordinary arithmetic the test $x > 0$ in configuration c^2 would evaluate to
true, because

$$2147483647 + 1 > 0.$$

However in two's complement arithmetic we get

$$va(x, c^1) = 2147483647$$
$$= 2^{31} - 1,$$
$$va(x, c^2) = (2^{31} - 1 + 1 \text{ tmod } 2^{32})$$
$$= -2^{31}$$
$$< 0.$$

11.6.3 While Loop

We consider the following example program:

```
int n;
int res;
int main()
{
  n = 32768;
  res = 0;
  while n>0
  {
    res = res + n;
    n = n - 1
  }
  return -1
}
```

We would hope to show that after a certain number of steps T the program ends with
a program rest

$$c^T.pr : \text{return} -1$$

and for the result the following holds:

$$va(res, c^T) = \sum_{i=1}^{32768} i.$$

res	0
n	32768

$c^2.m(gm)$

res	b
n	a

$c.m(gm)$

res	$b+a$
n	$a-1$

$d.m(gm)$

Fig. 114. Global memory in configurations of the execution of a while loop

Variable names n and res are bound in all configurations to global memory components

$$lv(n,c) = gm.n,$$
$$lv(res,c) = gm.res.$$

After the first two steps we have

$$va(n,c^2) = 32768,$$
$$va(res,c^2) = 0,$$

and the program rest is

```
while n>0
{
   res = res + n;
   n = n - 1
}
return -1
```

Now let c be a configuration where n is strictly positive

$$va(n,c) > 0$$

and the program rest is the same as above

$$c.pr = c^2.pr.$$

Let

$$d = \delta_C^3(c)$$

be the configuration obtained after three more C0 steps. Analysis along the lines of Sect. 11.6.1 gives

$$d.pr = c.pr,$$
$$va(n,d) = (va(n,c) - 1 \text{ tmod } 2^{32}),$$
$$va(res,d) = (va(res,c) + va(n,c) \text{ tmod } 2^{32}).$$

This is illustrated in Fig. 114.
By induction on $j \in [0 : 32768]$ one shows

$$c^{2+3\cdot j}.pr = c^2.pr,$$
$$va(n, c^{2+3\cdot j}) = 32768 - j,$$
$$va(res, c^{2+3\cdot j}) = \sum_{i=32768-j+1}^{32768} i.$$

The proof is not completely trivial: for the last line one needs to use the formula on arithmetic sums of Sect. 3.3.2 to show the absence of overflows:

$$\sum_{i=1}^{32768} i = 32768 \cdot (32768 + 1)/2$$
$$\leq 32768^2$$
$$= 2^{30}$$
$$< 2^{31}.$$

For $j = 32768$ we have

$$va(n, c^{2+3\cdot j}) = 0,$$

the condition of the while loop is false and with $T = 2 + 3 \cdot 32768 + 1$ we get the desired result. If we changed the initial assignment of n to a number with well-known bad properties (2147483648), we would of course never execute the loop body, because

$$va(n, c^1) = (2^{31} \operatorname{tmod} 2^{32})$$
$$= -2^{31}$$
$$< 0.$$

11.6.4 Linked Lists

We consider programs with the following sequence of type declarations:

```
typedef LEL* u;
typedef struct {int content; u next} LEL;
```

Type *LEL* for list elements and the type $u = LEL*$ of pointers to list elements were already informally studied in the introductory examples of Sect. 11.1.8. Now let

$$x[0 : n - 1] \in [0 : c.nh - 1]^n$$

be a sequence of n heap variables of configuration c. We say that $x[0 : n - 1]$ is a linked list and write

$$llist(x[0 : n - 1], c)$$

if the following conditions hold:

1. the x_i are distinct

$$i \neq j \to x_i \neq x_j.$$

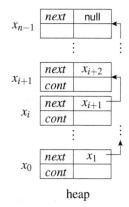

heap

Fig. 115. Linked list on the heap of configuration c

2. the x_i have type *LEL*

$$vtype(x_i,c) = c.ht(x_i) = LEL.$$

3. for $i < n-1$ the next component of x_i points to x_{i+1} and the next component of x_{n-1} is the null pointer

$$c.m(x_i.next) = \begin{cases} x_{i+1}, & i < n-1, \\ null, & i = n-1. \end{cases}$$

This is illustrated in Fig. 115.

The following program initializes variable n to a natural number N. We assume $0 < N < 2^{31}$. We show that the program creates a linked list of length N.

```
typedef LEL* u;
typedef struct {int content; u next} LEL;
u first;
u last;
int n;
int main()
{
  n = N;
  first = new LEL*;
  last = first;
  n = n - 1;
  while n>0
  {
    last*.next = new LEL*;
    last = last*.next;
    n = n - 1
  }
```

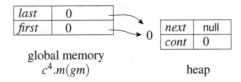

Fig. 116. In configuration c^4 the heap contains a list of length 1. Pointers *first* and *last* point to the single element of this list

```
    return -1

}
```

After four steps of the C0 semantics variable n has value $N - 1$, and there is a single variable of type *LEL* on the heap; it is initialized with default values. Both pointers point to this variable.

$$va(n, c^4) = N - 1,$$
$$c^4.nh = 1,$$
$$c^4.ht(0) = LEL,$$
$$c^4.m(0.content) = 0,$$
$$c^4.m(0.next) = null,$$
$$va(first, c^4) = 0,$$
$$va(last, c^4) = 0.$$

The program rest $c^4.pr$ is

```
while n>0
{
   last*.next = new LEL*;
   last = last*.next;
   n = n - 1
}
return -1
```

As illustrated in Fig. 116 there is a linked list with the single element $x[0] = 0$ in c, i.e.,

$$llist(x[0:0], c^4).$$

We define the list x of length N by

$$x[0:N-1] = (0, 1, \ldots, N-1)$$

and prove a lemma that is illustrated in Fig. 117.

Lemma 79. *For $i \in [0:N-1]$ let*

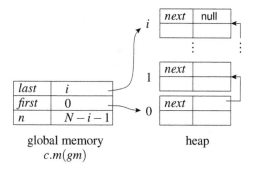

Fig. 117. In configuration c after i iterations of the loop. The heap contains a list of length $i+1$. Pointers *first* and *last* point to the first and last elements of this list, respectively

$$c = c^{4+4\cdot i}$$

be the configuration after i iterations of the loop. Then:

1. *c has the above program rest*

$$c.pr = c^4.pr.$$

2. *c has i+1 heap variables*

$$c.nh = i+1.$$

3. *x[0 : i] is a linked list in c*

$$llist(x[0 : i], c).$$

4. *pointers first and last point to x_1 and x_i resp.*

$$va(first, c) = 0,$$
$$va(last, c) = i.$$

5. *variable n has value $N - i - 1$*

$$va(n, c) = N - i - 1.$$

Proof. For $i = 0$ and $c = c^4$ the statement of the lemma was shown above. The induction step is illustrated in Fig. 118. Assume the lemma holds for $c = c^{4+4\cdot i}$ for $i < N - 1$.

1. Since $va(n, c) = N - i - 1 > 0$, the loop condition evaluates to 1 and thus the next program rest $\delta_C(c).pr$ is

```
last*.next=new LEL*; last=last*.next; n=n-1; c.pr
```

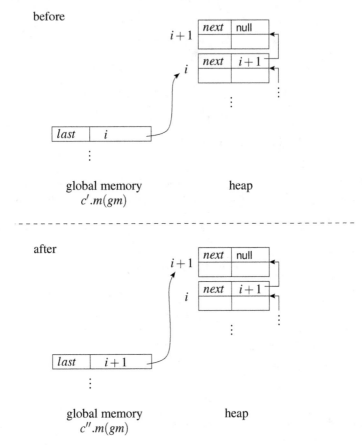

Fig. 118. In configuration c' the heap contains a list of length $i+2$. In configuration c'' *last* points to the last element of this list

2. Let

$$c' = \delta_C^2(c)$$

be the configuration after execution of the new statement. Execution of the new statement increases the number of heap variables

$$c'.nh = c.nh + 1$$
$$= i + 2.$$

The new heap variable $x[i+1]$ is distinct from the elements of the existing linked list $x[1:i]$

$$x[i+1] = i+1$$
$$\notin [0:i]$$
$$= x[0:i].$$

It has type *LEL* and is initialized to its default value

$$vtype(i+1,c') = LEL,$$
$$va((i+1).content, c') = c'.m(i+1)(content) = 0,$$
$$va((i+1).next, c') = c'.m(i+1)(next) = null.$$

Applying rules of expression evaluation and applying the induction hypothesis we find that the left-hand side of the new statement has left value

$$lv(last*.next, \delta_C(c)) = lv(last*, \delta_C(c)).next$$
$$= va(last, \delta_C(c)).next$$
$$= i.next.$$

The value assigned to this subvariable of pointer type is the new heap variable $i+1$ and we get

$$c'.m(i.next) = i+1.$$

Thus we have a linked list of length $i+2$ in c'

$$llist(x[0:i+1], c')$$

and the program rest $c'.pr$ is

```
last=last*.next; n=n-1; c.pr
```

3. Let

$$c'' = \delta_C(c') = \delta_C^3(c) = c^{4+4\cdot i+3}$$

be the configuration after execution of the first assignment. Expression evaluation gives

$$va(last*.next, c') = c'.m(lv(last*.next, c'))$$
$$= c'.m(lv(last*, c'))(next)$$
$$= c'.m(va(last, c'))(next)$$
$$= c'.m(i)(next)$$
$$= c'.m(i.next)$$
$$= i+1.$$

Execution of the assignment gives

$$c''.m(last) = i+1.$$

4. One instruction later n is decremented:

$$va(n, c^{4+4\cdot(i+1)}) = va(n, c) - 1$$
$$= N - (i+1) - 1$$

and the program rest is

$$c^{4+4\cdot(i+1)}.pr = c.pr$$
$$= c^4.pr.$$

Thus the lemma holds for $i + 1$.

For $i = N - 1$ we get

$$llist(x[0 : N - 1], c^{4N}) \wedge va(n, c^{4N}) = 0,$$

therefore the loop condition evaluates to 0 and the loop is dropped from the next program rest:

$$c^{4N+1}.pr : \text{return} -1.$$

11.6.5 Recursion

Next we analyze a program with recursive function calls. For natural numbers n the Fibonacci numbers $fib(n)$ are defined by

$$fib(0) = 0,$$
$$fib(1) = 1,$$
$$fib(n) = fib(n - 2) + fib(n - 1).$$

A trivial induction gives

$$fib(n) < 2^n.$$

This implies that Fibonacci numbers for $n \le 31$ are in the range of $C0$ integer variables and in the calculations below modulo arithmetic coincides with ordinary arithmetic.

For the computation of Fibonacci numbers we consider the following program:

```
int x;
int Fib(int n)
{
   int res;
   int f1;
   int f2;
   if n<2 {res = n} else
   {
      f1 = Fib(n-2);
      f2 = Fib(n-1);
      res = f1 + f2
   };
   return res
};
int main()
{
   x = Fib(31);
   return -1
}
```

We want to show

$$\exists T : va(x, c^T) = fib(31) \wedge (c^T.pr : \text{return} -1).$$

In configurations c for this program, global memory gm and frames for function Fib have types:

$$\$gm = \{int\ x\},$$
$$\$Fib = \{int\ n; int\ res; int\ f1; int\ f2\}.$$

Function $main$ has no local variables or parameters. For $z \in [0 : 31]$ we prove by induction on z the following result.

Lemma 80. *Let*

$$hd(c.pr) : y = Fib(e)$$

be a call of function Fib with parameter e and result destination y, and let

$$z = va(e, c) \leq 31.$$

Then there is a step number T such that for the configuration T steps later

$$c' = \delta_C^T(c)$$

the following holds:

1. only configuration components m and pr have changed:

$$X \notin \{m, pr\} \rightarrow c'.X = c.X.$$

2. the function call has been dropped from the program rest

$$c'.pr = tail(c.pr).$$

3. simple variable $lv(y,c)$ is updated with Fibonacci number $fib(z)$. Other simple variables keep the old value.

$$\forall x \in SV(c) : simple(x, c) \rightarrow c'.m(x) = \begin{cases} fib(z), & x = lv(y, c), \\ c.m(x), & otherwise. \end{cases}$$

The effect of calling function Fib as specified by the lemma is illustrated in Fig. 119.

Proof. Execution of the function call gets to a configuration d illustrated in Fig. 120 with

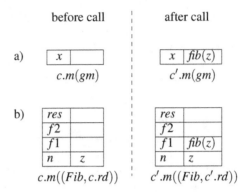

Fig. 119. Effect of calling function *Fib* in configuration c. a) If *Fib* is called from function *main* with left-hand side x, then $gm.x$ is updated. b) If *Fib* is called recursively from function *Fib* with left-hand side $f1$, then the instance of local variable $f1$ in the top frame is updated

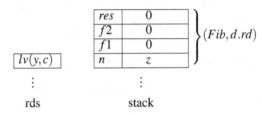

Fig. 120. Configuration d after the call. The parameter value z is passed to the instance of parameter n in the new top frame. If the call was from function *main* with $y = x$, we have $lv(y,c) = gm.x$. If the call was recursive with $y = f1$ resp. $y = f2$ we have $lv(y,c) = (Fib,c.rd).f1$ resp. $lv(y,c) = (Fib,c.rd).f2$, which is in the previous top frame

$$d.rd = c.rd + 1,$$
$$c.st(c.rd) = Fib,$$
$$cf(d) = Fib,$$
$$V(d) = V(c) \cup \{top(d)\},$$
$$top(d) = (Fib,d.rd),$$
$$\forall a \in \{n, f1, f2, res\} : lv(a,d) = top(d).a,$$
$$lv(y,c) \in SV(c),$$
$$va(n,d) = d.m(va(top(d)).n)$$
$$= va(e,c)$$
$$= z,$$
$$d.rds(d.rd) = lv(y,c),$$
$$d.pr = ft(Fib).body;tail(c.pr).$$

The base case of the induction occurs with

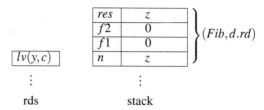

rds   stack

Fig. 121. Configuration h before the return in the base case. Local variable *res* in the top frame contains the result $z = fib(z)$. The variable to be updated by the following return statement is $lv(y,c)$, which was stored in the top entry $d.rds(d.rd)$ of the result destination stack

$$z = va(n,d) < 2.$$

After execution of the conditional statement we are in configuration g with program rest:

$$g.pr : \texttt{res=n; return res;}\ tail(c.pr)$$

Execution of the assignment leads to configuration h with

$$va(res,h) = n = fib(n).$$

which is illustrated in Fig. 121.

Execution of the return statement gives a configuration k with

$$
\begin{aligned}
k.rd &= d.rd - 1 \\
&= c.rd, \\
SV(k) &= SV(h) \setminus \{top(d)\} \\
&= SV(c), \\
va(y,k) &= k.m(lv(y,c)) \\
&= k.m(h.rds(h.rd)) \\
&= va(res,h) \\
&= fib(z), \\
k.pr &= tail(c.pr).
\end{aligned}
$$

This is the situation specified on the right-hand side of Fig. 119.

The case

$$z = va(e,c) \geq 2$$

is handled in the induction step. After execution of the conditional statement one gets to configuration g with program rest:

$$g.pr : \texttt{f1=Fib(n-2); f2=Fib(n-1); res=f1+f2; return res;}\ tail(c.pr)$$

Two function calls have parameters with values

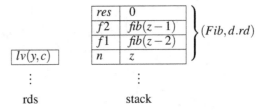

rds   stack

Fig. 122. Configuration h after two recursive calls of function Fib. Local variables $f1$ and $f2$ of the top frame have been updated as illustrated in Fig. 119 b) with $fib(z-1)$ and $fib(z-2)$

$$va(n-1,g) = va(n,c) - 1,$$
$$va(n-2,g) = va(n,c) - 2.$$

Both are smaller than $z = va(n,c)$, thus we can apply the lemma inductively for $z-1$ and $z-2$. We conclude for the configuration h after two calls, which is illustrated in Fig. 122:

$$va(f1,h) = fib(va(n,c) - 1),$$
$$va(f2,h) = fib(va(n,c) - 2),$$

and the program rest $h.pr$ is

$$\texttt{res=f1+f2; return res; } tail(c.pr)$$

Execution of the assignment statement gives a configuration u with

$$va(res,u) = va(f1,h) + va(f2,h)$$
$$= fib(va(n,c) - 1) + fib(va(n,c) - 2)$$
$$= fib(va(n,c)).$$

For the return statement one argues exactly as in the base case.

11.7 Final Remarks

- Grammar: first of all, remember that the specification of syntax and semantics of a useful language does not take more than 40 pages, even in a textbook. Do not try to memorize the grammar. Try instead to memorize the key constructs of $C0$
  - types: *int*, *uint*, *char*, *bool*, struct, array, pointer.
  - declarations of variables or struct types: type first, name second.
  - expressions: the obvious stuff. &, ∗, and selectors are *all* placed behind the operands (to save brackets). Arithmetic expressions and Boolean expressions are coupled via atoms.
  - statements: assignment, conditional, while, new, call, return.

Reproducing the grammar from this is basically an exercise, except for Boolean and arithmetic expressions, but the grammar for this you have already memorized in Chap. 10.

- Types: most important to remember is that you cannot always define new types from known types: linked lists are the simplest example. The solution here: leave pointer types temporarily undefined. Now remember for the range of types
  - identify all (top-level) variables of all $C0$ configurations: global memory, stack frames, heap variables; then their subvariables.
  - define in a first step the range of pointers as all possible subvariables.
  - then define the range of arrays and structs by the obvious inductive definition.
  - restrict the range of pointers further by type correctness in a second step.
  - global memory and function frames for function f are treated as variables with struct types $\$gm$ and $\$f$.
  - the ranges of *int* and *uint* are finite.
- Configurations: simply memorize
  - the components: global memory gm, recursion depth rd, stack st, number of heap variables nh (these variables have no name and are just numbered), heap type ht, memory m (content of subvariables), program rest pr, result destination stack rds.
  - the invariant that program rests consist of $rd + 1$ blocks $Irt(i)$ of statements belonging to recursion depth i.
- Expression evaluation: memorize
  - the same expression can occur at different places in the program with different meanings. Therefore one evaluates occurrences of expressions which are specified by the nodes n in the derivation tree which generate them.
  - right value va, left value lv, and expression type $etype$ are defined simultaneously. $etype(e, f)$ depends only on the function f in which expression e occurs.
  - variable names are bound to subvariables of the global memory or the top stack frame by a visibility rule.
  - if 'address of' is not taken for subvariables on the stack (which can disappear when the stack shrinks), one will have no dangling pointers

  Everything else is fairly straightforward.
- Statement execution: left value on the left-hand side of equations, right value on the right-hand side of assignments. Expansion of the program rest is possible at conditional statements, while loops, and function calls.

A C semantics with address arithmetic, which was used in the SeL4 project [KAE$^+$10] is described in [NK14].

Defining the semantics of multi-threaded dialects of C (running on multi-core processors) is surprisingly hard. This is due to the so-called 'weakly consistent memory systems' used in modern microprocessors. These memory systems are only informally defined in the processor manuals of manufacturers. The recent C/C++11 standard [ISO11] makes an effort to define a memory model which fits all current processor families, but in [VBC$^+$15] it was shown that the model still needs repair.

For the TSO model used in x86 processors the situation is better. The formalization — using store buffers — is reasonably straightforward. This allows one to define semantics for multi-threaded C dialects in two ways: i) define semantics for a strong (sequentially consistent [Lam79]) memory model and enforce software conditions which guarantee that the store buffers are invisible, i.e., a strong memory system is simulated [CPS13, Bau14, CS10], ii) directly using a TSO memory model in the multi-threaded C semantics [VN13].

11.8 Exercises

1. Draw the derivation tree for each of the following programs if it exists. Otherwise argue why no derivation tree for the C0 grammar exists.

   a)

   ```
   int a;
   int b;
   int main()
   {
      int d;
      if a==b {b = 5} else {b = 0; d = a};
      return d
   }
   ```

   b)

   ```
   int main() {return 4};
   int function(int x) {return 9*x};
   ```

   c)

   ```
   int a;
   int main() {return 0};
   a = main()
   ```

2. A C0 program starts with

   ```
   typedef uint[32] u;
   typedef uint[8] v;
   typedef struct {u gpr; v spr} pcb;
   typedef pcb[p+1] PCB_t;
   ```

   Specify the corresponding portion of the type table. What components of gm do you know already?

3. Declare the type of a three-dimensional array capable of storing 125 integers. Specify the type table and determine the sizes of the types involved.

4. Specify the type table for the function b, which is given below. Do not forget the $tt(\$b)$.

```
int b(int x)
{
    int result;
    if x==73896 {result = 100}

    else {if x==69695 {result = -100}};

    return result
}
```

---

x	$ft(x).t$	$ft(x).p$	$ft(x).VN$	$ft(x).body$	$tt(x)$
b					

5. Consider the following program (numbers of lines are comments and not part of the program).

```
 0: typedef int* intp;
 1: typedef intp[10] intparr;
 2: typedef intparr* intparrp;
 3: intparr a;
 4: int fak(int x)
 5: {
 6:     int y;
 7:     if x==1 {y = x} else
 8:     {
 9:         y = fak(x-1);
10:         y = x*y
11:     };
12:     return y
13: };
14: int main()
15: {
16:     intparrp b;
17:     int i;
18:     b = a&;
19:     i = 1;
20:     while i<11
21:     {
22:         b*[i-1]* = fak(i);
23:         i = i + 1
24:     };
25:     return 42
26: }
```

---

a) Draw the derivation tree for the type declaration in line 0.

b) Draw the derivation tree for the assignment in line 22.

c) Specify the type table and function table.

6. Recall derivation trees that encode sequences, i.e., derivation trees using production rules $XS \to X|XoXS$, where X is a nonterminal and o is some terminal. Let $u \in \mathbb{N}^*$ be the root of such a derivation tree.

a) Recall, for $i \in \mathbb{N} \setminus \{0\}$, the definition of $se(u,i)$ that gives the i'th sequence element of the tree starting in u. Consider the following (alternative) recursive definition:

$$se(u,1) = u \circ 0,$$
$$se(u,i+1) = se(u \circ 2, i).$$

Prove that $se(u,i) = u \circ 2^{i-1} \circ 0$.

b) Recall the definition of $fseq(u) = se(u,1) \circ \ldots \circ se(u,n)$, which turns such a derivation tree into the list of its sequence elements. Give a formal definition of $fseq(u)$, i.e., a definition that does not use three dots.

7. Give a type declaration sequence, i.e., a word in $L(TyDs)$, where the derivation tree starting with the nonterminal $TyDs$ has three nodes with label $\langle TyD \rangle$. In the following, consider this derivation tree.

a) Give $bw(se(\varepsilon,i) \circ 1)$ and $bw(se(\varepsilon,i) \circ 2)$ for $i \in \{1,2,3\}$.

b) Explain in two or fewer sentences why we use the border words of $se(0,i) \circ 1$ and $se(0,i) \circ 2$ to get the i'th type declaration, and why this does not work in our example.

8. Define formally how to get t_i and x_i, i.e., the i'th global type and variable name, from a derivation tree where the root ε has label $\langle Prog \rangle$.

9. Prove that $ra(t)$ is well defined for $t \in TN$.

10. Give formal definitions of $irt(i)$ and $Irt(i)$ which give the index of the i'th return statement in $c.pr$, as well as the list of statements between the i'th and $(i+1)$'th return statement, including the i'th but not the $(i+1)$'th return statement.

11. Recall that the element symbol $\in$ is overloaded in our text. Give the formal definitions of:

a) $e' \in e$ where $e', e \in L(E)$, i.e., e' is a subexpression of e.

b) $e \in f$ where $e \in L(E), f \in FN$, i.e., e is an expression occurring in the body of f.

c) $s \in f$ where $s \in L(St), f \in FN$, i.e., s is a statement in the body of f.

d) $x \in f$ where $x \in L(id), f \in FN$, i.e., x is a local variable defined in f.

12. Recall the following functions, which are used to define the semantics of statements in a C0-configuration c where e is an expression:

- $lv(e,c)$: the subvariable specified by e (only defined if e has a left value),
- $etype(e,f)$: the type of e,
- $va(e,c)$: the value of e.

Give definitions of $lv(e,c)$, $etype(e,f)$, and $va(e,c)$ for the following expressions:

a) $e = e'.n$ and n is a valid field of struct type $etype(e',f)$,

b) $e = true$,

c) $e = null$.

13. Draw the derivation trees of the following statements and specify their semantics:

    a) $e = e' == e''$,

    b) $e = e' <= e''$.

    Attention: in the second case the semantics depends on the types involved.

14. Specify the semantics of the following assignments:

    a) $e = e' < e'' \mathbin{||} e' == 0$,

    b) $e = (e' - e'') * e'''$,

    c) $e = e' \& *$,

    d) $e = e' * .n\&$,

    e) $e = x[e' + e'']$.

    Assume that the types involved permit the evaluation.

15. Prove that for expressions e in the body of function f the following holds:

$$inv\text{-}conf(c) \wedge lv(e,c) \in SV(c) \wedge vtype(lv(e,c),c) = etype(e,f) \;\to\; inv\text{-}expr(e,c).$$

16. Fill in the following program such that $fc(x)$ returns $(x! \bmod 2^{32})$.

```
int fc(int n)
{
    ...
    return res
}
```

Prove formally that your program is correct.

17. Prove or disprove that $hd(c.pr) \in L(St)$ is an invariant during statement execution.

18. Consider the following program:

```
typedef bool* pb_t;

int a;
int main()
{
    a = 5;
    pb_t pb;
    pb = new bool*;
    pb* = true;
    if pb* {a = a + 1} else {a = a - 1};
    return 1
}
```

    a) Specify the type table and function table.

    b) Specify the initial configuration c^0.

    c) Specify the computation (c^i) of the program. Specify only those components of c^i which change.

19. Repeat the exercise above for the following program:

```
typedef struct {int exp; int val} pow_struct;
int x;
int pow(int base, int n)
{
  pow_struct p;
  p = new pow_struct*;
  p.exp = 0;
  p.val = 1;
  while p.exp<n
  {
    p.exp = p.exp + 1;
    p.val = p.val*p.base
  };
  return p.val
}
int main()
{
  x = pow(2,2);
  return x
}
```

A $C0$-Compiler

Compilers translate source programs p from high-level languages L to target programs $code(p)$ from some instruction set architecture ISA, such that the target program simulates the source program. Of course, here we are interested in $C0$ as the source language and MIPS ISA as the target language. Thus, roughly speaking, our $C0$ compiler computes a translation function

$$code : L(prog) \rightarrow MIPS\ ISA.$$

In Sect. 12.1 we define a simulation relation

$$consis(c,d)$$

coupling $C0$ configurations c with MIPS configurations d. This simulation relation (which we also refer to by the term *compiler consistency*, or for short, *consistency*) formalizes the idea that the low-level MIPS configuration d encodes (more or less) the high-level $C0$ configuration c. Then, we can show a step-by-step simulation of source programs p by target programs $code(p)$. Formally, we compare $C0$ computations (c^i) and MIPS computations (d^i) which start in a consistent pair of configurations, i.e.,

$$consis(c^0, d^0)$$
$$\wedge\ \forall i : c^{i+1} = \delta_C(c^i)$$
$$\wedge\ \forall i : d^{i+1} = \delta_M(c^i).$$

We then show: if $C0$ configuration c and MIPS configuration d are consistent, and in one $C0$ step c progresses to $c' = \delta_C(c)$, then in some number s of MIPS steps d progresses to a configuration $\delta_M^s(d)$ which is consistent with c'.

Lemma 81.

$$consis(c,d) \wedge c' = \delta_C(c) \rightarrow \exists s : consis(c', \delta_M^s(d)).$$

By induction we get a simulation theorem stating the correctness of non-optimizing compilers.

© Springer International Publishing Switzerland 2016   253
W.J. Paul et al., *System Architecture*, DOI 10.1007/978-3-319-43065-2_12

Lemma 82. *There exists a sequence $(s(i))$ of natural numbers such that*

$$\forall i : consis(c^i, d^{s(i)}).$$

Proof. by induction on i. For the induction step from i to $i+1$, we apply Lemma 81 with

$$c = c^i,$$
$$d = d^{s(i)},$$

and define

$$s(i+1) = s(i) + s.$$

We specify code generation for expressions in Sect. 12.2. There, we extend the consistency relation to expressions in a straightforward way and then generate the obvious code. Showing that consistency for expressions is maintained by the generated code is a bookkeeping exercise. To determine the order of evaluation of subexpressions we use a simple and elegant algorithm due to Sethi and Ullman [SU70], which permits one to evaluate expressions with up to 262144 binary operators using not more than 18 processor registers.

Code generation for statements in Sect. 12.3 maintains consistency as required in Lemma 81, which also results in completely intuitive code. The proof of Lemma 81, however, is more than a bookkeeping exercise. This is due to the fact that code generation is defined by induction over the derivation tree of the program, whereas the program rest is a flat sequence of statements. After the code of a statement $hd(c.pr)$ has been executed on the MIPS machine (whatever that means), we need to show that the program counter $\delta^s(d).pc$ points to the code of $hd(c'.pr)$, but this requires finding (the right instance of) statement $hd(c'.pr)$ in the derivation tree.

In the course of the proof we will show the existence of the step numbers $s(i)$ in a very constructive fashion. In Sect. 12.5 we will use this to strengthen the statement of the compiler correctness result in a crucial way.

12.1 Compiler Consistency

12.1.1 Memory Map

The purpose of a memory map is to indicate what is stored where in the memory of a processor. Here, we present a memory map that specifies how C0 configurations c are encoded in MIPS configurations d.

Recall that the variables of C0 configurations are i) the global memory gm, ii) the stack frames $ST(i,c) = (c.st(i),i)$ for $i \in [0 : c.rd]$, and iii) the nameless heap variables $[0 : c.nh - 1]$. Fig. 123 shows how these variables and the translated program are mapped into the memory $d.m$ of MIPS configuration d. Let us briefly discuss the different memory regions of the C0 compiler's memory map. We have

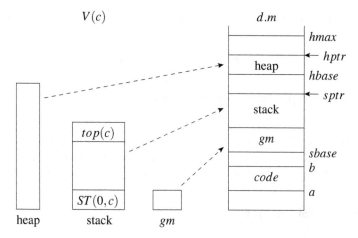

Fig. 123. Memory map for the compiler. Code is stored between addresses a and b. Global memory starts at address *sbase*. The stack is stored on top of global memory. The heap starts at address *hbase*. We keep pointers *sptr* and *hptr* to the first free addresses after the occupied portions of the stack, and the heap, respectively

- the code region: the translated code occupies the address range from address a to address b.
- the global memory region: The encoding of C0's global memory variable *gm* starts at address *sbase*. We maintain a pointer to this address in general-purpose register number 28.[1]

$$sbase = d.gpr(bpt_5) \quad \text{with} \quad bpt = 28.$$

- the stack region: The encodings of the C0 stack frames $ST(0,c)$, $ST(1,c)$, ..., $top(c)$ are stored in order, starting at the first address behind the global memory region. During a computation, the stack grows (by means of function calls) and shrinks (by means of returning from function calls). It must stay below address *hbase*. We maintain a pointer *sptr* to the first free location behind the top stack frame in register number 29.

$$sptr(d) = d.gpr(spt_5) \quad \text{with} \quad spt = 29.$$

- the heap region: The encoding of the C0 heap starts at address *hbase* and should not extend beyond address *hmax*. During the computation, the heap can only

---

[1] While we could consider *sbase* as a given constant, keeping *sbase* in a register means that we never need to do a so-called *relocation* of the global memory. Consider that compilers that use a fixed *sbase* tend to generate code that works exactly for that given *sbase* (by using *sbase* as a constant in the compiled code). If, for some reason, one needs to place the global memory of the program at a different address, all occurrences of *sbase* in the compiled code need to be recomputed, i.e., relocated, in retrospect.

grow (unless we do garbage collection as sketched in Sect. 13.1). We maintain a pointer $hptr$ to the first free location on the heap in register number 30.

$$hptr(d) = d.gpr(hpt_5) \quad \text{with} \quad hpt = 30.$$

12.1.2 Size of Types, Displacement, and Base Address

We compute for each type $t \in ET \cup TN$ the number of bytes $size(t)$ allocated by the compiler to store a value from the range $ra(t)$ of type t. The definition is straightforward: for values with simple types the compiler always allocates four bytes; the size of composite types is the sum of the sizes of their components; empty struct types[2] have size 0.

Definition 21 (Size of Types).

$$size(t) = \begin{cases} 0, & t = \{\ \}, \\ 4, & simple(t), \\ n \cdot size(t'), & t = t'[n], \\ \sum_{i=1}^{s} size(t_i), & t = \{t_1 \ n_1; \ldots; t_s \ n_s\}. \end{cases}$$

Using the $size$ function, we define for each C0-variable $x \in V(c)$ the base address $ba(x, c)$ in the MIPS memory where the value of x is stored.

Definition 22 (Base Address of Variables). *We define base addresses for the different kinds of variables appearing in C0 configurations c:*

- *the global memory variable gm: global memory starts at address sbase.*

$$ba(gm, c) = sbase.$$

- *stack frame variables ST(i, c): behind the global memory, we have eight reserved bytes followed by the encoding of the value of stack frame variable ST(0, c). At the two words below the base address of each stack, we reserve room for auxiliary data for the frame; there, we place the return result destination and the address to which we jump when the function returns. Increasing the base address of the stack frame ST(i, c) by its size size($c.st(i)$) plus 8, one obtains the base address of the next stack frame ST(i + 1, c).*

$$ba(ST(0, c), c) = sbase +_{32} (size(\$gm) + 8)_{32},$$
$$ba(ST(i+1, c), c) = ba(ST(i, c), c) +_{32} (size(\$c.st(i)) + 8)_{32}.$$

This is illustrated in Fig. 124.

---

[2] Empty struct types represent function frames of functions f without parameters and local variables, i.e., $\$f = \{\ \}$.

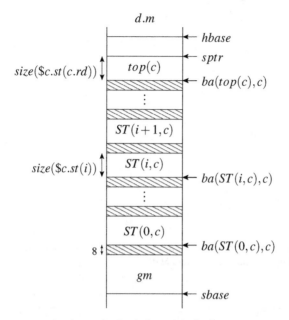

Fig. 124. Memory map for the stack. Stack frames $ST(i,c)$ are stored in the order of their indices behind the global memory frame gm. Below the base address of each frame two words (eight bytes) are reserved for auxiliary data

- *heap variables: nameless heap variable 0 starts at address hbase. Increasing the base address of heap variable i by the size $size(c.ht(i))$ of this variable, we obtain the base address of the next heap variable:*

$$ba(0,c) = hbase,$$
$$ba(i+1,c) = ba(i,c) +_{32} size(c.ht(i))_{32}.$$

This is illustrated in Fig. 125.

For the base addresses of subvariables, we consider the memory layout of composite variables and subvariables as shown in Fig. 126. Assume x is a composite variable of type t with base address $ba(x,c)$ and components $x[i]$ if $t = t'[n]$ is an array type and, respectively, $t.n_i$ if $t = \{t_1\ n_1; \ldots; t_s\ n_s\}$ is a struct type. We store the components $x[i]$ or $x.n_i$ successively in memory, starting at low addresses as shown in Fig. 126. The distance in bytes of the start of component $x[i]$ resp. $x.n_i$ from the base address $ba(x,c)$ is called the *displacement* and denoted by $displ(i,t)$, resp. $displ(n_i,t)$. It equals the sum of the sizes of the preceding components.

Definition 23 (Displacement of a Component in a Composite Variable). *For the two cases, we define the displacement of a component as*

$$displ(i,t) = i \cdot size(t'), \qquad if\ t = t'[n],$$
$$displ(n_i,t) = \sum_{j<i} size(t_i), \qquad if\ t = \{t_1\ n_1; \ldots; t_s\ n_s\}.$$

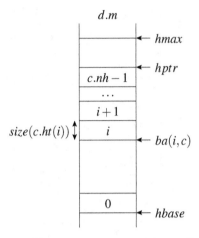

Fig. 125. Memory map for the heap. Nameless variables $i \in [0 : c.nh - 1]$ are stored in the order of their indices starting from address *hbase*

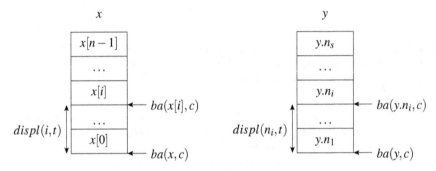

Fig. 126. Layout of composite variables and their subvariables. Right: array variable x with subvariables $x[0], \ldots, x[n-1]$. Left: struct variable y with subvariables $y.n_1, \ldots, y.n_s$

Definition 24 (Type of a Selected Component of a Composite Type).

$$ctype(i,t) = t', \quad if\ t = t'[n],$$
$$ctype(n_i,t) = t_i, \quad if\ t = \{t_1\ n_1; \ldots; t_s\ n_s\}.$$

We extend the definitions of displacements and component types to selector sequences $s = s[1 : r]$.

Definition 25 (Displacement and Type of Components for Selector Sequences).

$$displ(\varepsilon,t) = 0,$$
$$displ(s[1 : r],t) = displ(s_1,t) + displ(s[2 : r], ctype(s_1,t)),$$
$$ctype(\varepsilon,t) = t,$$
$$ctype(s[1 : r],t) = ctype(s[2 : r], ctype(s_1,t)).$$

We define the base addresses of the corresponding subvariables as follows.

Definition 26 (Base Addresses of Subvariables).

$$ba(x[i],c) = ba(x,c) +_{32} displ(i,t)_{32}, \quad \text{if } vtype(x,c) = t = t'[n],$$
$$ba(x.n_i,c) = ba(x,c) +_{32} displ(n_i,t)_{32}, \quad \text{if } vtype(x,c) = t = \{t_1\ n_1;\dots;t_s\ n_s\}.$$

Trivial induction on the length of selector sequences s gives the following.

Lemma 83.

$$ba(xs,c) = ba(x,c) +_{32} displ(s,vtype(x,c))_{32},$$
$$vtype(xs,c) = ctype(s,vtype(x,c)).$$

We extend the interval notation $[a : b]$ to bit strings $a,b \in \mathbb{B}^{32}$ with $\langle a \rangle \le \langle b \rangle$ by defining

$$[a : a] = \{a\},$$
$$[a : b +_{32} 1_{32}] = [a : b] \cup \{b +_{32} 1_{32}\},$$

and for subvariables x we define the address range $ar(x,c)$ as the set of addresses allocated to subvariable x:

$$ar(x,c) = [ba(x,c) : ba(x,c) +_{32} (size(vtype(x,c)) - 1)_{32}].$$

Note that the address range of a simple variable x covers the four bytes of the memory word starting at the base address of the subvariable:

$$simple(x,c) \rightarrow ar(x,c) = [ba(x,c) : ba(x,c) +_{32} 3_{32}].$$

From the definitions, we infer that address ranges of variables are disjoint.

Lemma 84.

$$x,y \in V(c) \rightarrow x \ne y \rightarrow ar(x,c) \cap ar(y,c) = \emptyset.$$

For subvariables we have the following.

Lemma 85. *For a subvariable $x \in SV(c)$ with array type $t = t'[n]$ the address ranges of its subvariables $x[i]$ are disjoint and all lie in the address range of x.*

$$vtype(x,c) = t'[n] \wedge 0 \le i < j < n \rightarrow \begin{aligned} &ar(x[i],c) \cap ar(x[j],c) = \emptyset \\ &\wedge ar(x[j],c) \subseteq ar(x,c). \end{aligned}$$

A similar statement holds if x has a struct type.

Lemma 86.

$$vtype(x,c) = \{t_1\ n_1;\dots;t_s\ n_s\} \wedge 1 \le i < j \le s \rightarrow \begin{aligned} &ar(x.n_i,c) \cap ar(x.n_j,c) = \emptyset \\ &\wedge ar(x.n_j,c) \subseteq ar(x,c). \end{aligned}$$

For subvariables $x, y \in SV(c)$ we say that x is a subvariable of y if $x = ys$ for some selector sequence $s \in S^+$:

$$subvar(x, y) \equiv \exists s \in S^+ : x = ys.$$

An easy induction for subvariables $x, y \in SV(c)$ shows: i) if x is a subvariable of y, then the address range of x is contained in the address range of y; ii) if neither x is a subvariable of y nor y is a subvariable of x, then their address ranges are disjoint.

Lemma 87 (Disjointness of Address Ranges). *Let $x, y \in SV(c)$. Then*

$$subvar(x, y) \rightarrow ar(x, c) \subseteq ar(y, c),$$
$$\neg subvar(x, y) \wedge \neg subvar(y, x) \rightarrow ar(x, c) \cap ar(y, c) = \emptyset.$$

Finally we observe that x and its subvariables xs can share the same base addresses, but in this case they have a different type.

Lemma 88. *Let $x, y \in SV(c)$ and $x \neq y$. Then*

$$ba(x, c) = ba(y, c) \rightarrow vtype(x, c) \neq vtype(y, c).$$

Proof. Subvariables with disjoint address ranges cannot share a base address. Thus, by Lemma 87, x is a proper subvariable of y or vice versa. In this case x and y have different types.

12.1.3 Consistency for Data, Pointers, and the Result Destination Stack

For elementary data types and for the type of the null pointer, i.e., for $t \in ET \cup \{\bot\}$, we define an encoding $enc(va, t) \in \mathbb{B}^{32}$ which encodes values $va \in ra(t)$ in the range of t into machine words $\mathbb{B}^{32}$ in an obvious way: integers are encoded as two's complement numbers; unsigned integers are encoded as binary numbers; Boolean values are zero extended; character constants are encoded in ASCII code and then zero extended.

Definition 27 (Encoding of Values).

$$enc(va, t) = \begin{cases} twoc_{32}(va), & t = int, \\ bin_{32}(va), & t = uint, \\ 0^{31}va, & t = bool, \\ 0^{24}ascii(va), & t = char, \\ 0^{32}, & t = \bot \wedge va = null. \end{cases}$$

We use this to define compiler consistency relations for data, pointers, and the stack. Let c be a C0 configuration and let d be a *MIPS* configuration, then

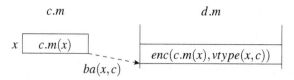

Fig. 127. Illustration of *e-consis*(c,d). The value $c.m(x)$ of elementary variable x is stored in $d.m$ at the memory word starting at the base address $ba(x,c)$

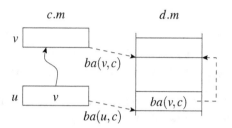

Fig. 128. Illustration of *p-consis*(c,d). Pointer u points in c to subvariable v. Then the base address $ba(v,c)$ is stored in $d.m$ at the memory word starting at the base address $ba(u,c)$

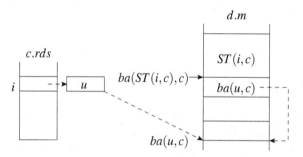

Fig. 129. Illustration of *rds-consis*(c,d). Pointer $c.rds(i)$ points in c to subvariable u. Then, the base address $ba(u,c)$ is stored in $d.m$ at the memory word below the base address $ba(ST(i,c),c)$ of stack frame $ST(i,c)$

- We say that configurations c and d are consistent with respect to the values of elementary subvariables (and the null pointer) and write *e-consis*(c,d) if for all subvariables x with an elementary data type $vtype(x,c) \in ET \cup \{\bot\}$ the obvious encoding of its value $c.m(x)$ is stored in the MIPS memory $d.m$ at the base address of x (see Fig. 127).

$$t = vtype(x,c) \wedge t \in ET \cup \{\bot\} \quad \rightarrow \quad d.m_4(ba(x,c)) = enc(c.m(x),t).$$

- We say that configurations c and d are consistent with respect to (non-null) pointer values and write *p-consis*(c,d) if, for all subvariables u with $pointer(u,c)$, the following holds: if u points in configuration c to subvariable v, i.e., $c.m(u) = v$, then the base address $ba(v,c)$ is stored in the MIPS memory $d.m$ at the base address of u (see Fig. 128):

$$pointer(u,c) \wedge c.m(u) = v \neq null \rightarrow d.m_4(ba(u,c)) = ba(v,c).$$

- We say that c and d are consistent with respect to base, stack, and heap pointers and write $bsp\text{-}consis(c,d)$ if i) the pointers we keep in general-purpose registers bpt, spt, and hpt point to $sbase$, the first free address on the stack, and the first free address on the heap resp. Moreover ii) the pointers should not be out of range (see Figs. 124 and 125):

$$d.gpr(bpt_5) = sbase,$$
$$d.gpr(spt_5) = ba(top(c),c) + size(\$cf(c)) < hbase,$$
$$d.gpr(hpt_5) = ba(c.nh - 1, c) + size(c.ht(c.nh - 1)) < hmax.$$

- Entries $c.rds(i)$ of the return destination stack are treated similarly to pointers: their encoded values are stored in the memory word below the base address $ba(ST(i,c),c)$ of stack frame $ST(i,c)$; the encoded value of subvariable $c.rds(i)$ is its base address, i.e., if $c.rds(i) = u$ then the base address $ba(u,c)$ of u is stored in the MIPS memory at $d.m_4(ba(ST(i,c),c) -_{32} 4_{32})$ (see Fig. 129). Thus, we say that c and d are consistent with respect to the return destination stack and write $rds\text{-}consis(c,d)$ if

$$c.rds(i) = u \rightarrow d.m_4(ba(ST(i,c),c) -_{32} 4_{32}) = ba(u,c).$$

12.1.4 Consistency for the Code

In order to prove that the *MIPS* machine running the compiled code simulates the *C0* machine running the original program, an important prerequisite is that the compiled code of *C0* program is in fact present in the memory of the *MIPS* machine at the correct place. We formalize that the target program of the compilation process is stored in the address range $[a : b]$ and that the code is not changed while the *MIPS* machine is running. Code generation will proceed recursively (from sons to fathers) along the nodes n of the derivation tree T of the program that is translated. For nodes n, by $code(n)$ we define the code generated for the subtree with root n (see Fig. 130). We define the length of the generated code *measured in words* by $|code(n)|$. Then, we have

$$code(n) \in \mathbb{B}^{32 \cdot |code(n)|}.$$

In the initial configuration d^0 of the MIPS computation, $code(n)$ occupies

$$4 \cdot |code(n)|$$

bytes starting from address $start(n)$ to address $end(n)$:

$$d^0.m_{4 \cdot |code(n)|}(start(n)) = code(n),$$
$$end(n) = start(n) +_{32} (4 \cdot |code(n)| - 1)_{32}.$$

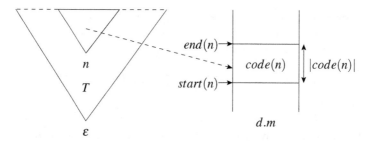

Fig. 130. The code $code(n)$ generated for node n in the derivation tree T is stored at $|code(n)|$ consecutive addresses in $d.m$ from $start(n)$ to $end(n)$

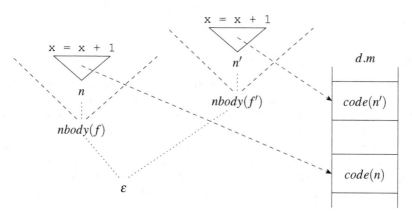

Fig. 131. Different instances of statement $x = x + 1$ are generated in the bodies of functions f and f'. The instances are translated into $code(n)$ and $code(n')$

For the root ε of the derivation tree, we get the target program tp as the code of the translated source program:

$$tp = code(\varepsilon).$$

In the initial configuration d^0, the target program tp lies in the code region $[a : b]$:

$$[start(\varepsilon) : end(\varepsilon)] \subseteq [a : b],$$
$$d^0.m_{4 \cdot |tp|}(start(\varepsilon)) = tp.$$

We say that c and d are *code consistent* and write $code\text{-}consis(c,d)$ if the target program did not modify itself and is still there:

$$d.m_{4 \cdot |tp|}(start(\varepsilon)) = tp.$$

Note that $code(e)$ for expressions e or $code(s)$ for statements s in the source program is in general *not* well defined. Consider several *instances* of the assignment $x = x + 1$ that occur in two different functions of the program: in one function f with local variable x and in another function f' without local variable or parameter x. The statements are executed in configurations c resp. c' with $cf(c) = f$ and $cf(c') = f'$. In these configurations the bindings

$$lv(x,c) = top(c).x \qquad \text{and} \qquad lv(x,c') = gm.x$$

are different, and this will be reflected in the compilation. Thus, $code(x+1)$ is ambiguous. However, the two instances of assignment $x+1$ are border words of some nodes n, and respectively n', of the derivation tree of the program (as shown in Fig. 131). We can refer to their compiled code as $code(n)$ and $code(n')$, respectively.

There is a similar problem with defining $start(e)$ or $start(s)$: even if we had well-defined $code(e)$ and $code(s)$, that code might occur several times in the compiled code of the program.

12.1.5 Consistency for the Program Rest

Before we can define consistency for the program rest, we introduce two so-called *ghost*[3] components of *C*0 configurations. They are not needed to define semantics; obviously we have done it without them. They are also not needed to define the compilation process. We only need them to prove that the compilation is correct, i.e., to establish a simulation theorem between runs of source and target programs. We encode the program rest of a *C*0 configuration by means of the program counter $d.pc$ of the MIPS and by means of a caller stack of return addresses. First, we discuss consistency for the program counter, then we elaborate on the caller stack and its consistency.

Program Counter Consistency

Intuitively, we want to define that c and d are *pc-consistent* if the program counter $d.pc$ points to the start of the code of the head of the program rest, i.e.,

$$d.pc = start(hd(c.pr)).$$

Unfortunately, the $c.pr[i]$ are statements; we have seen that different instances of the same statement can exist in the same program, and thus we have to disambiguate $start(c.pr[i])$ by specifying the nodes

$$c.prn(i) \in T.A$$

which identify for each i the instance of $c.pr[i]$ to whose code the program counter should point. Thus, we define a ghost component

$$c.prn \in T.A^+$$

---

[3] The term *ghost* component or *specification* component describes an auxiliary variable whose introduction does not change the semantics of the original model.

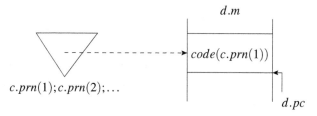

Fig. 132. Illustration of *pc-consis*. The *pc* points to the start of the code generated for the subtree with root $hd(c.prn) = c.prn(1)$

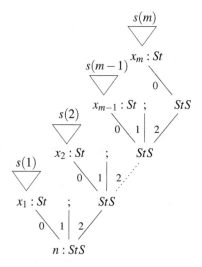

Fig. 133. If n is labeled *StS* then descendant x_i with label *St* is reached following label 2 exactly $i - 1$ times and then label 0. For each i the border word of x_i is statement $s(i)$

of C0 configurations c with these nodes and we call c and d pc-consistent and write $pc\text{-}consis(c,d)$ if the program counter points to the start of the code generated for the subtree with root $hd(c.prn)$ (see Fig. 132):

$$d.pc = start(hd(c.prn)).$$

For the formal definition of this *program rest nodes* component, we proceed in the obvious way: we translate the definition of the program rest $c'.pr$ for the next configuration c' after c into a definition for $c'.prn$ which uses nodes in the derivation tree $T = (A, \ell)$ instead of statements. The details are slightly technical.

Before we do this, we remind the reader of a concept from Sect. 11.1.9. Consider Fig. 133. Let $n \in A$ be a node labeled $\ell(n) = StS$ and its border word

$$bw(n) = s(1);\ldots;s(m) \quad \text{with} \quad s(i) \in L(St).$$

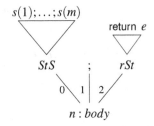

Fig. 134. Node n generates the body $s(1);\ldots;s(m);$ return e of a function. The statement sequence is generated by node $n0$ and the return statement by node $n2$

Let $x[1:m] = se(n, [1:m]) \in A^m$ be the sequence of descendants of n labeled with $\ell(x_i) = St$ such that we have for all i

$$s(i) = bw(x[i]),$$

thus we have

$$bw(n) = bw(x[1]);\ldots;bw(x[m]).$$

We have seen that we get from n in the tree to x_i by following label 2 exactly $i-1$ times and then label 0:

$$x_i = n2^{i-1}0$$

and we called the sequence $x[1:m]$ of these nodes the flattened sequence of n

$$fseq(n) = x[1:m].$$

Now consider the body of function f as shown in Fig. 134, where

$$ft(f).body = s(1);\ldots;s(m);\text{return } e.$$

It is derived from node $n = nbody(f)$ labeled *body*. Following edge 0 we arrive at a node deriving the statement sequence $s[1:m]$ and following edge 2 we arrive at a node deriving the return statement. Thus we define the sequence of statement nodes of the body of f as

$$snodes(f) = fseq(nbody(f)0) \circ nbody(f)2.$$

Initial Configuration

For the initial configuration c^0 we define $c^0.prn$ as the sequence of statement nodes in the body of function *main*

$$c^0.prn = snodes(main).$$

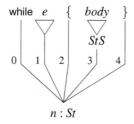

Fig. 135. Node n generates a while statement. The body is generated by node $n3$

Semantics

During program execution, we update the program rest nodes component in such a way that it mimics the behavior of the program rest:

- Assignments, new statements, and return statements: If $hd(c.pr)$ is an assignment statement, a new statement, or a return statement, then the first node in $c.prn$ is dropped:
$$c'.prn = tail(c.prn).$$

- While loops: Let $hd(c.pr)$: while e { *body* }.
  As shown in Fig. 135, this statement is derived by node $n = hd(c.prn)$ labeled St; following edge 3 we arrive at a node labeled StS which derives the loop body. We define
$$c'.prn = \begin{cases} fseq(hd(c.prn)3) \circ c.prn, & va(e,c) = 1, \\ tail(c.prn), & va(e,c) = 0. \end{cases}$$

- If-then-else: Let $hd(c.pr)$: if e { *if-part* } else { *else-part* }.
  As shown in Fig. 136 this statement is derived by node $n = hd(c.prn)$ labeled St; following edge 3 we arrive at a node labeled StS which derives the if-part and following edge 7 we arrive at a node labeled StS which derives the else-part. We define
$$c'.prn = \begin{cases} fseq(hd(c.prn)3) \circ tail(c.prn), & va(e,c) = 1, \\ fseq(hd(c.prn)7) \circ tail(c.prn), & va(e,c) = 0. \end{cases}$$

- If-then: Let $hd(c.pr)$: if e { *if-part* }.
  We define
$$c'.prn = \begin{cases} fseq(hd(c.prn)3) \circ tail(c.prn), & va(e,c) = 1, \\ tail(c.prn), & va(e,c) = 0. \end{cases}$$

- Function call: Let $hd(c.pr)$: $e = f(e_1,\ldots,e_p)$.
  We set
$$c'.prn = snodes(f) \circ tail(c.prn).$$

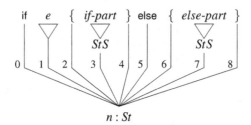

Fig. 136. Node n generates a conditional statement. The if-part is generated by node $n3$ and the else-part is generated by node $n7$

Trivially, we maintain the following invariant.

Invariant 7

$$inv\text{-}prn\text{-}pr(c) \equiv \forall i : c.pr[i] = bw(c.prn(i)).$$

We can also reformulate invariant *inv-pr* for program rest nodes. We identify the indices of program rest nodes that are labeled with rSt:

$$irt'(0,c) = |c.prn|,$$
$$irt'(j+1,c) = \max\{i \mid i < irt'(j,c) \wedge \ell(c.prn(i)) = rSt\}.$$

We divide the indices of program rest nodes into intervals $Irt'(j,t)$ between return statements by

$$Irt'(j,c) = \begin{cases} [irt'(j+1,c)+1 : irt'(j,c)], & j < c.rd, \\ [1 : irt'(j,c)], & j = c.rd \end{cases}$$

and get the following trivial result.

Lemma 89.

$$irt'(j,c) = irt(j,c),$$
$$Irt'(j,c) = Irt(j,c).$$

Invariant *inv-pr* translates into an invariant coupling program rest nodes with function frames.

Invariant 8 *We say that inv-prn(c) holds if the following conditions are fulfilled:*

1. $$\#\{i \mid \ell(c.prn(i)) = rSt\} = c.rd + 1,$$
2. $$\ell(last(c.prn)) = rSt,$$
3. $$j \in [0 : c.rd] \wedge k \in Irt(j,c) \rightarrow fu(c.prn(k)) = c.st(j).$$

Caller Stack Consistency

As a second ghost component of $C0$-configurations c we introduce a caller stack:

$$c.clr : [1 : c.rd] \rightarrow T.A.$$

In entry $c.clr(i) \in T.A$ of this stack, we record the node whose border word is the function call that generated frame i. Formally, we change this stack only at function calls and returns:

- Calls: At function calls, we have $hd(c.pr) : e = f(e_1, \ldots, e_p)$ and $hd(c.pr) = bw(hd(c.prn))$, and we push node $hd(c.prn)$ onto the caller stack:

$$c'.clr(x) = \begin{cases} hd(c.prn), & x = c'.rd, \\ c.clr(x), & x \leq c.rd. \end{cases}$$

- Returns: At the execution of return statements, the recursion depth is decreased and the top entry of the caller stack is dropped:

$$c'.clr(x) = c.clr(x) \quad \text{for} \quad x \leq c'.rd.$$

A function at recursion depth i is always called from the current function of recursion depth $i - 1$. We note this as an invariant.

Invariant 9 *We say that inv-clr(c) holds if for all i*

$$fu(c.clr(i)) = c.st(i - 1).$$

Formally we use Invariant 8 to show this.

We say that c and d are consistent for the caller stack and write *clr-consis*(c, d) if, for all $i > 0$, the second memory word under the base address of $ST(i, c)$ stores the 'return address' where the MIPS program execution should continue after the return of the function. This should be the instruction *behind the end* of the code $code(n)$ of the node n that created the stack frame with the function call $bw(n)$. The node n is recorded in $c.clr(i)$. Thus, we require for all $i > 0$

$$d.m_4(ba(ST(i,c),c) -_{32} 8_{32}) = end(c.clr(i)) +_{32} 1_{32}.$$

Fig. 137 illustrates how this works in a configuration c the next statement of which is a function call.

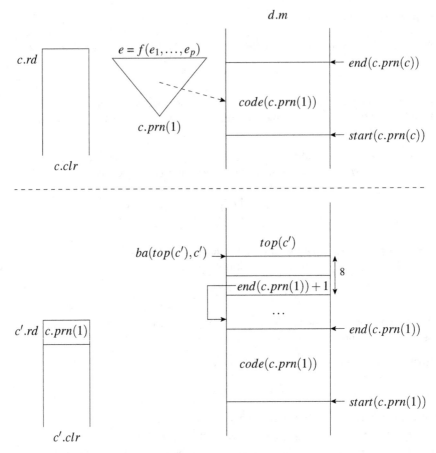

Fig. 137. Maintaining *clr-consis* at a function call generated by node $n = c.prn(1) = hd(c.prn)$. The node n is recorded in the new top entry $c'.clr(c'.rd)$ of the caller stack. A new top frame $top(c')$ is created in the MIPS memory. In the second word below the base address we store the return address for the *pc* after the call. It points one byte behind $end(n)$

When we connect all consistency relations X-*consis*(c,d) with logical *AND*s, we obtain the overall consistency relation

$$consis(c,d) = \bigwedge_X X\text{-}consis(c,d)$$

where $X \in \{e, p, rds, bsp, code, pc, clr\}$.

12.1.6 Relating the Derivation Tree and the Program Rest

In this subsection we show that the entire program rest nodes component $c.prn$ (and hence $c.pr$) of a configuration can be reconstructed from the head $c.prn(1)$ of the

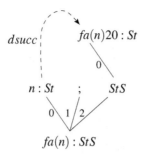

Fig. 138. The direct successor of a node n in a statement sequence is $fa(n)20$. Node n is the last node in a statement sequence if son 2 of its father is absent

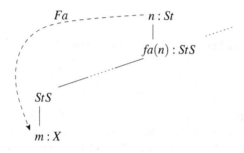

Fig. 139. To find $Fa(n)$ one searches from n upwards in the derivation tree until one finds a node m that is not labeled with StS

program rest nodes and the caller stack $c.clr$ with the help of the derivation tree $T = (A, \ell)$. This technical result will play a surprisingly important role in the sequel. We will first use it in the induction step of the proof of Lemma 82 to show that *pc-consis* is maintained. It will also play a crucial role in the definition of semantics for $C0$ with assembly code.

For nodes n in the tree region A the following definitions are illustrated in Fig. 138. Let n be a statement node in the derivation tree which does not derive a return statement. The father *father*(n) of such a node is always labeled with StS. We abbreviate *father*(n) with $fa(n)$.

$$\ell(n) = St \;\rightarrow\; \ell(fa(n)) = StS.$$

Node n is the last statement node in a statement sequence if it is the only son of its father. Formally we define

$$last\,st(n) \;\equiv\; \ell(n) = St \wedge fa(n)2 \notin A.$$

If n is not the last node of a statement sequence, we define the direct successor of n as

$$dsucc(n) = fa(n)20.$$

The next definition is illustrated in Fig. 139. For nodes n which are labeled with St or StS we search for ancestors until we find a node m that is not labeled with StS. We

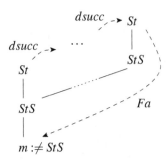

Fig. 140. Successive statement nodes in a statement sequence are direct successors. The father of the node which generated the statement sequence is the Father of every statement node in the sequence

call this node m the Father of n and abbreviate it similarly but using capital F:

$$Fa(n) = \begin{cases} fa(n), & \ell(fa(n)) \in \{St, body\}, \\ Fa(fa(n)), & \text{otherwise.} \end{cases}$$

Figure 140 summarizes the last two definitions. Note that $Fa(n)$ is labeled with St or $body$. In the first case the statement sequence to which n belongs is the body of a while loop or a part (if-part or else-part) of the conditional statement. In the second case it belongs to the statement sequence which is directly derived from a node $nbody(f)$ which generates the body of function $f = fu(n)$. Thus we define predicates

$$inwhile(n) \equiv \ell(Fa(n)0) = while,$$
$$incond(n) \equiv \ell(Fa(n)0) = if,$$
$$inbody(n) \equiv \ell(Fa(n)) = body.$$

For statement nodes which are not labeled with rSt, i.e., which do not derive a return statement, we now define the successor $succ(n)$ of n in the following way: i) take the direct successor if there is one. Otherwise there are three more cases, ii) If n belongs to the body of a while loop, take the node $Fa(n)$ deriving that loop (see Fig. 141), iii) If n belongs to an if-part or an else-part, recursively take the successor of node $Fa(n)$ which derives the conditional statement (see Fig. 142), iv) If n is the last statement of the statement sequence of a function body, take the node $Fa(n)2$ which is labeled rSt and which derives the return statement of the body (see Fig. 143).

$$succ(n) = \begin{cases} dsucc(n), & \neg lastst(n), \\ Fa(n), & lastst(n) \wedge inwhile(n), \\ succ(Fa(n)), & lastst(n) \wedge incond(n), \\ Fa(n)2, & lastst(n) \wedge inbody(n). \end{cases}$$

With this definition we can give a recipe for the successive reconstruction of the program rest nodes until the next return statement.

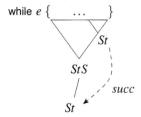

Fig. 141. The successor of the last statement of the body of a while loop is the node generating the loop

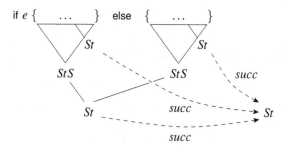

Fig. 142. The successor of the last statement of an if-part or an else-part is the successor of the node generating the conditional statement

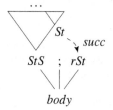

Fig. 143. The successor of the last statement node in the statement sequence of a body is the node deriving the return statement

Lemma 90. *Assume $c.prn(i)$ is not the last node of $c.prn$. Then*

1. *If $c.prn(i)$ does not derive a return statement, then the next program rest node $c.prn(i+1)$ is the successor $succ(c.prn(i))$:*

$$i \neq irt(j,c) \;\rightarrow\; c.prn(i+1) = succ(c.prn(i)).$$

2. *If $c.prn(i)$ derives a return statement for recursion depth j, then the next program rest node is the successor of the node $c.clr(j)$ which produced the call of the function at recursion depth j:*

$$i = irt(j,c) \;\rightarrow\; c.prn(i+1) = succ(c.clr(j)).$$

Proof. by a straightforward induction on the number of steps t in the C0 computation. We examine the program rest in the initial configuration c^0. If $c^0.prn$ contains only one rSt node, there is nothing to show. Otherwise the program rest nodes have the form

$$c^0.prn = snodes(main) = n_1; \ldots; n_x; n_{x+1},$$

where for $y \in [1 : x - 1]$ we have

$$n_{y+1} = dsucc(n_y) = succ(n_y).$$

Node n_x is the last node of the statement sequence and node n_{x+1} derives the return statement. Hence

$$n_{x+1} = succ(n_x).$$

In the induction step we assume that the lemma holds for configuration c and show that it holds in configuration c'. We proceed by an obvious case split on the first node $c.prn(1)$ of the program rest which defines the statement which is executed. There are five cases:

- an assignment statement, a new statement, or a return statement is executed. We have

$$c'.prn = tail(c.prn)$$

and we get the statement of the lemma directly from the induction hypothesis.
- a call of function g is executed. We get a program rest of the form

$$c'.prn = snodes(g); tail(c.prn) = n_1; \ldots; n_r; n_{r+1}; tail(c.prn),$$

where n_{r+1} marks the node that derives the return statement in $snodes(g)$. For the portion $n_1; \ldots; n_r; n_{r+1}$ one argues exactly as in the base case of the induction; for the remaining part one uses induction. This yields the first claim for $c'.prn$. By induction hypothesis and the definition of $c.clr$ we have for $j = c'.rd$:

$$c'.prn(r+2) = c.prn(2) = succ(c.prn(1)) = succ(c.clr(j)).$$

The claim for all return statements in $tail(c.prn)$ follows from the induction hypothesis.
- node $c.prn(1)$ derives a while loop. Let

$$n_1; \ldots; n_x = fseq(c.prn(1)3)$$

be the flattened sequence of the statement nodes of the while loop. We have

$$c'.prn = n_1; \ldots; n_x; c.prn(1); tail(c.prn).$$

For $y < x$ node n_x is the direct successor of node n_{x-1}. Node n_x is the last node of the statement sequence which is derived from

$$c.prn(1) = Fa(n_x) = succ(n_x).$$

The claim for the tail portion follows by induction hypothesis.

• node $c.prn(1)$ derives an if-then-else statement. Let

$$n_1; \ldots; n_x$$

be the flattened sequence of the if-part or the else-part that is executed. The new program rest is

$$c'.prn = n_1; \ldots; n_x; tail(c.prn).$$

For $y < x$ node n_x is the direct successor of node n_{x-1}. The Father $Fa(n_x)$ of n_x is the node $c.prn(1)$ deriving the if-then-else statement that is executed. By induction hypothesis we have

$$c.prn(2) = succ(c.prn(1)) = succ(Fa(n_x)).$$

For the other nodes of the tail the claim follows directly from the induction hypothesis.

• node $c.prn(1)$ derives an if-then statement. Let

$$n_1; \ldots; n_x$$

be the flattened sequence of the if-part. If it is executed we argue as in the previous case. If it is not executed we have

$$c'.prn = tail(c.prn)$$

and obtain the claim directly by induction hypothesis.

For later reference we include Lemma 90 among the invariants which should hold for well-formed $C0$ programs with ghost components.

Invariant 10

We say that invariant inv-prn-clr(c) is fulfilled if Lemma 90 holds for c.

12.2 Translation of Expressions

12.2.1 Sethi-Ullman Algorithm

Expressions e have derivation trees T; the roots n of these trees lie in the set of expression nodes en which was defined in Sect. 11.3.3. Inspection of the $C0$ grammar shows that expression nodes have at most two sons which are expression nodes.[4] We have defined the semantics of expressions e by induction over the structure of their derivation trees T. Code generation will be defined by an induction over these trees, too. In general, if an expression node n has two sons n' and n'' which are expression nodes (see Fig. 144), then the generated code will

---

[4] The other sons are constants, names, or symbols, such as $+, -, (,)$, etc.

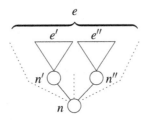

Fig. 144. Expression node n deriving expression e with two sons n' and n'' which are expression nodes and derive expressions e' and e''

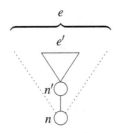

Fig. 145. Expression node n deriving expression e with one son n' which is an expression node and derives expression e'

1. evaluate as intermediate results the expressions e' and e'' derived from nodes n' and n'';
2. evaluate expression e derived from n and discard the intermediate results (i.e., the registers where they were stored can be overwritten).

If n has only one son n' which is an expression node (see Fig. 145), one proceeds in an analogous way. Expression nodes which have no expression nodes as sons generate variable names or constants.

We will always evaluate expressions in registers. We have 32 general-purpose registers. Register $gpr(0_5)$ is tied to zero, registers $gpr(28_5)$ through $gpr(31_5)$ are reserved for base pointer, stack pointer, heap pointer, and link address. For expressions with multiplications and divisions we additionally need registers $gpr(20_5)$ through $gpr(27_5)$ for the software multiplication and division algorithms from Sect. 9.1. Thus, registers $gpr(1_5)$ through $gpr(19_5)$ are free to be used for expression evaluation. We will see that for the translation of assignment statements, one of these registers will temporarily not be available. Thus, we will be left with 'only' 18 registers to be used for the evaluation of expressions. Fortunately, this allows one to evaluate *all* expressions whose derivation tree has at most

$$2^{18} = 262144$$

expression nodes — *if* we evaluate subexpressions in the right order. Such an order was identified in [SU70] and is very easily described:

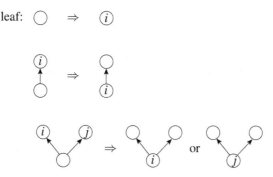

Fig. 146. Permitted moves of the pebble game

For brother nodes n with subtrees T, evaluate the subtree with the larger number of expression nodes first; keep only the result of this evaluation while evaluating the smaller subtree.

For the purpose of analyzing register consumption, one models expression evaluation by a *pebble game* on rooted trees T where every node n has at most two sons. It is played by placing and removing pebbles from the nodes of the tree according to certain rules. Pebbles have numbers i, and placing pebble i on node n means that the result of evaluating the subtree of T with root n is placed in register $gpr(i_5)$. One maintains a list of free pebbles. Initially, all pebbles are free and no node in T has a pebble. The game ends when a pebble is placed on the root of T. Three kinds of moves are allowed. They are illustrated in Fig. 146:

1. One is always allowed to place a pebble from the free list on a leaf of the tree. This removes the pebble from the free list.
2. If the single son of node n has pebble i then one can slide the pebble from the son of n onto n.
3. If the sons of node n have pebbles i and j then one can slide one of these pebbles onto n and remove the other pebble from the graph and return it to the free list.

We define the size $|T|$ of a tree as the number of nodes in the tree

$$|T| = \#T.V.$$

We define $P(n)$ as the smallest number of pebbles which suffices to play the game successfully on any tree of size n, i.e., with at most n nodes, and show the following by induction on n.

Lemma 91.

$$P(n) \leq max\{1, \lceil \log n \rceil\}.$$

Proof. by induction on n. For the start of the induction, consider Fig. 147. There is only a single tree consisting of one node. This node is at the same time leaf and root.

Fig. 147. The unique trees with one and two nodes

Fig. 148. The root has one son. The subtree T' has at most n nodes

Fig. 149. The root has two sons. The larger tree has at most $n-1$ nodes. The smaller tree has at most $n/2$ nodes

Placing a single pebble on this node by rule 1 ends the game, and we have $P(1) = 1$. Similarly, there also exists only one tree with $n = 2$ nodes. Except for a root and a leaf it has no other nodes. A single pebble is placed on the leaf by rule 1. We slide it to the root by rule 2. This ends the game and we have $P(2) = 1$.

For the induction step consider a tree T with $n > 2$ nodes, which requires $P(n)$ pebbles in order to place a pebble on its root. There are two cases:

- the root n of T has a single son n' which is the root of a subtree T' with $n-1$ nodes (see Fig. 148). Then one places a pebble on n' using at most $P(n-1)$ pebbles and then slides it onto the root. In this case, we get

$$P(n) \le P(n-1) = \lceil \log(n-1) \rceil \le \lceil \log n \rceil.$$

- the root n of T has two sons n' and n'' which are roots of subtrees T' and T'' (see Fig. 149). Without loss of generality, we assume

$$|T'| \le |T''|.$$

Then,

$$|T'| \le n/2 \quad \text{and} \quad |T''| \le n-1.$$

In this case, we i) place a pebble on the root n'' of T'' using at most $P(n-1)$ pebbles and leave it there. Then, we ii) place a pebble on the root n' of T' using at most $P(n/2)$ pebbles. Last, we iii) slide one of the pebbles left on the tree to n and return the other pebble to the free list. We conclude

$$P(n) \le \max\{P(n-1), P(n/2)+1\}.$$

If the first term is the maximum we proceed as above. If the second term is the maximum we conclude

$$
\begin{aligned}
P(n) &\le P(n/2)+1 \\
&\le \lceil \log(n/2) \rceil + 1 \\
&= \lceil \log(n/2) + 1 \rceil \\
&= \lceil \log n \rceil.
\end{aligned}
$$

Implementation Hints

Let e be an expression with derivation tree T, and let r be the root of this tree. If we wish to generate code for the evaluation of e we play the pebble game on the subtree T' of T which consists only of expression nodes (i.e., with labels in the set *en* as defined at the start of Sect. 11.4). Now, one can proceed in three phases:

1. For a node $n \in T.A$ we denote the number of descendants $n \circ u \in T.A$ which are expression nodes by

$$size(n) = \#\{u \mid n \circ u \in en\}.$$

   Clearly, in order to determine the moves of the above strategy, we need to know $size(n)$ for each expression node $n \in en$. One computes this in the following way: one augments records of type *DTE* (representing derivation tree elements) from Sect. 11.1.8 with integer components *.size* and *.pebble*. In a first phase one traverses tree T' and records the sizes of subtrees (counting only expression nodes) in the *.size* components.

2. Before generating code for expression e, we create an 'initial free list' as a linked list with all numbers of registers available for expression evaluation. We then traverse tree T recursively, starting at the root r. At any node n with sons n' and n'' in T', we traverse the subtree with larger *.size* component first. If sizes are equal, we traverse the left subtree first. We maintain the free list as required by the moves of the pebble game and for every derivation tree element we record in its *.pebble* component the number of the pebble placed on the node modeled by that *DTE*.

3. We also generate code by recursive tree traversal, again traversing larger subtrees first and — if sizes are equal — traversing the left subtree first. For operations performed at expression nodes n, we generate code by taking operands from the registers indicated by the pebbles of the sons of the node and storing the result in the register indicated by the pebble on n. Details for this are specified in later subsections. For expressions, the generated code will later be executed in the order in which it was generated. Thus it will use registers in the efficient order prescribed by the Sethi-Ullman strategy described above.

Observe that phases 2 and 3 can be merged into a single phase.

Fig. 150. Expressions in an assignment statement

12.2.2 The *R*-label

In Sect. 11.4 on expression evaluation we simultaneously defined expression type $etype(e, f)$, left value $lv(e, c)$, and (right) value $va(e, c)$ of expressions. In Sect. 11.5, we sometimes needed left values and right values of expressions in order to define the semantics of statement execution. To the expression nodes $i \in en$ of derivation trees we attach a binary label

$$R : en \to \mathbb{B}$$

which signals to the code generation algorithm whether code for a left value or for a right value should be generated. If $R(i) = 0$, we generate code for a left value for node i, and respectively, code for a right value if $R(i) = 1$. Label R is defined by a (recursive) case split on the expression nodes of the derivation trees. We start by considering expression nodes i that are sons of a statement node n and split cases on the statement $bw(n)$ that is derived from node n:

- assignments (see Fig. 150): $bw(n)$ has the form

$$e = e'.$$

Node n has sons $i, i' \in en$ and

$$bw(i) = e,$$
$$bw(i') = e'.$$

The semantics of statement execution uses for this statement the left value of e and the right value of e'. Thus we set

$$R(i) = 0,$$
$$R(i') = 1.$$

- while loops (see Fig. 151): $bw(n)$ is

$$\text{while } e \ \{ \ body \ \}.$$

Node n has a son $i \in en$ with

$$bw(i) = e.$$

The semantics of statement execution uses for this statement the right value of e. Thus, we set

$$R(i) = 1.$$

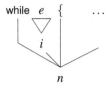

Fig. 151. Expression in a while loop

Fig. 152. Expression in a conditional statement

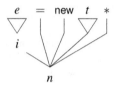

Fig. 153. Expression in a new statement

- conditional statements (see Fig. 152): $bw(n)$ has the form

$$\text{if } e \;\{\; \textit{if-part} \;\}$$

or

$$\text{if } e \;\{\; \textit{if-part} \;\} \text{ else } \{\; \textit{else-part} \;\}$$

Node n has a son $i \in en$ and

$$bw(i) = e.$$

The semantics of statement execution uses for this statement the right value of e. Thus we set

$$R(i) = 1.$$

- new statement (see Fig. 153): $bw(n)$ has the form

$$e = \text{new } t * .$$

Node n has a son $i \in en$ and

$$bw(i) = e.$$

The semantics of statement execution uses for this statement the left value of e. Thus we set

$$R(i) = 0.$$

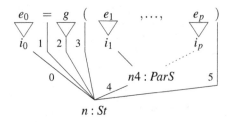

Fig. 154. Expressions in a function call statement

- function call (see Fig. 154): $bw(n)$ has the form

$$e_0 = g(e_1, \ldots, e_p) \quad \text{or} \quad e_0 = g().$$

Node n has a son $i_0 \in en$ and

$$bw(i_0) = e_0.$$

The semantics of statement execution uses for this statement the left value of e. Thus we set

$$R(i_0) = 0.$$

If function g has no parameters we are done. Otherwise, the parameter sequence is the border word of son 4 of n, which is labeled $ParS$, and for $j \in [1 : p]$ parameter e_j is derived from node $i_j \in en$ where

$$i_j = se(n4, j).$$

The semantics of statement execution uses for parameters the right value of e. Thus we set

$$R(i_j) = 1$$

for all j.
- return statement (see Fig. 155): $bw(n)$ has the form

$$\text{return } e.$$

Node n has a son $i \in en$ with

$$bw(i) = e.$$

The semantics of statement execution uses for this statement the right value of e. Thus we set

$$R(i) = 1.$$

Next, we define the R-label for expression nodes i that are not sons of a statement node. The border word $bw(i)$ of these nodes is a subexpression of the border word $bw(n)$ of their father. We split cases on the expression $bw(n)$:

Fig. 155. Expression in a return statement

Fig. 156. Subexpressions of an expression with a binary operator

- binary operators (see Fig. 156): $bw(n)$ has the form

$$e \circ e'$$

where $\circ$ is a binary operator including the comparison operators.
Node n has as sons expression nodes i and i' with

$$bw(i) = e,$$
$$bw(i') = e'.$$

Binary operators operate only on right values. Thus we set

$$R(i) = R(i') = 1.$$

- unary operators (see Fig. 157): $bw(n)$ has the form

$$\circ e$$

where $\circ \in \{-_1, !\}$ is a unary operator.
Node n has a son $i \in en$ with

$$bw(i) = e.$$

Unary operators also operate only on right values. Thus we set

$$R(i) = 1.$$

- constants: $bw(n)$ has the form

$$c$$

where $c \in [0:9]^+ \cup \{\text{true}, \text{false}\} \cup \{'_\sqcup', \ldots, '^\sim'\}$ is a numerical, Boolean, or character constant.

Fig. 157. Subexpression of an expression with a unary operator

Node n has a son $i \in en$ with $\ell(i) \in \{C, BC, CC\}$ and

$$bw(i) = c.$$

Constants only have right values. Thus we set

$$R(i) = 1.$$

- bracketed terms: $bw(n)$ has the form

$$(e).$$

Node n has a son $i \in en$ with

$$bw(i) = e.$$

The semantics prescribes determination of the right value of arithmetic or Boolean expression e. Thus we set

$$R(i) = 1.$$

- struct elements (see Fig. 158): $bw(n)$ has the form

$$e.x.$$

Node n has a son $i \in en$ with

$$bw(i) = e.$$

The semantics of expression evaluation defines left values and right values of e, but only left values are used subsequently. Thus we set

$$R(i) = 0.$$

- array elements (see Fig. 159): $bw(n)$ has the form

$$e[e'].$$

Node n has two sons $i, i' \in en$ with

$$bw(i) = e,$$
$$bw(i') = e'.$$

Fig. 158. Subexpression of an expression selecting a struct component

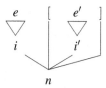

Fig. 159. Subexpressions of an expression selecting an array element

Fig. 160. Subexpression of an expression dereferencing a pointer

The semantics of expression evaluation defines left values and right values of e, but only left values are used subsequently. The semantics uses the right value of e'. Thus we set

$$R(i) = 0,$$
$$R(i') = 1.$$

- pointer dereferencing (see Fig. 160): $bw(n)$ has the form

$$e*.$$

Node n has a son $i \in en$ with
$$bw(i) = e.$$
The semantics of expression evaluation uses the right value of e. Thus we set

$$R(i) = 1.$$

- address of (see Fig. 161): $bw(n)$ has the form

$$e\&.$$

Node n has a son $i \in en$ with

Fig. 161. Subexpression of an 'address of' expression

Fig. 162. Expression node n with only one son i

$$bw(i) = e.$$

The semantics of expression evaluation uses the left value of e. Thus we set

$$R(i) = 0.$$

Note that expression $e\&$ itself only has a right value, i.e., $R(n) = 1$.
- node n has only one son i and $i \in en$, for instance if $\ell(n) = E$ and $\ell(i) = T$ (see Fig. 162). Here we use recursion and set

$$R(i) = R(n).$$

Alternatively, we could have constructed a simplified tree by eliminating such edges in the tree.

Implementation Hints

In order to compute values $R(n)$ for expression nodes n, one adds components .R to derivation tree elements of type *DTE* which store the values $R(n)$ of the R-labels. Computation of these labels can be done together with the computation of the *.size* components.

12.2.3 Composable MIPS Programs

Let d and d' be MIPS configurations, let $p \in (\mathbb{B}^{32})^*$ be a MIPS ISA program and let $q \in \mathbb{B}^{30} \circ \{00\}$ be an aligned byte address. Further, assume that program p is stored in the memory of configuration d starting at address q:

$$d.m_{4 \cdot |p|}(q) = p.$$

Moreover assume that program p is stored in the code region

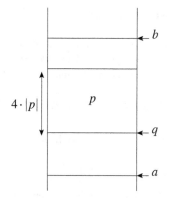

Fig. 163. Program p starts at address q in the code region $[a : b]$

$$[q : q +_{32} (4 \cdot |p| - 1)_{32}] \subseteq [a : b]$$

as illustrated in Fig. 163. We define that execution of program p stored in this way ends in configuration d' and write

$$d \to^*_{p,q} d'$$

if there is a MIPS computation $(d^0, \ldots, d^K)$ with the following properties:

1. the computation starts in configuration d and ends in configuration d':

$$d^0 = d \wedge d^K = d'.$$

2. the program counter is initially at the start of program p and is finally just behind the end of program p:

$$d.pc = q \wedge d'.pc = q +_{32} (4 \cdot |p|)_{32}.$$

3. for all j the program does not modify itself or any other part of the code region:

$$s(d^j) \;\to\; ea(d^j) \notin [a : b].$$

4. for $j < K$ the program counter of configuration d^j stays in the region occupied by program p:

$$j < K \;\to\; d^j.pc \in [q : q +_{32} (4 \cdot |p| - 1)_{32}].$$

If the above properties hold no matter where we place program p in the code region, we write $d \to^*_p d'$:

$$d \to^*_p d' \;\equiv\; \forall q \in [a : b -_{32} (4 \cdot |p| - 1)_{32}] \cap (\mathbb{B}^{30} \circ \{00\}) : d \to^*_{p,q} d'.$$

An easy proof shows that programs p and p' satisfying this property can be composed.

Lemma 92.

$$d \to^*_p d' \wedge d' \to^*_{p'} d'' \;\to\; d \to^*_{p;p'} d''.$$

12.2.4 Correctness of Code for Expressions

Expression translation is a part of statement translation which is supposed to maintain predicate $consis(c,d)$ at certain points of the two computations under consideration. Recall that predicate $consis$ is the conjunction of all compiler consistency predicates, i.e.,

$$consis(c,d) = \bigwedge_X X\text{-}consis(c,d)$$

with $X \in \{e, p, rds, bsp, code, pc, clr\}$. When we have a step

$$c' = \delta_C(c)$$

on the C0 level starting in configuration c, then the generated code $code(hd(c.prn))$ is executed on the MIPS level starting in configuration d and arriving after a number s of steps in a configuration d' with a MIPS computation

$$u[0:s] \quad \text{with} \quad u[0] = d \quad \text{and} \quad u[s] = d'.$$

While this code is executed, the program counter changes. Thus for the resulting MIPS configurations $u[i]$ with $i > 0$ consistency with configuration c is certainly not maintained. We define the weaker predicate

$$consis'(c,d) = \bigwedge_{X \neq pc} X\text{-}consis(c,d)$$

which states X-consistency only for components X other than the pc. After we have specified code generation, straightforward bookkeeping will show that predicate

$$consis'(c, u[i])$$

is maintained for a long time. Indeed it will be maintained both by

- code for expressions, and
- any portions of the code for the statement $hd(c.pr)$ before the code for the last expression in that statement.

This motivates a detailed recipe for generating code $code(i)$ for expression nodes i in the derivation tree for expressions $bw(i)$ in bodies of functions f (see Fig. 164):

- identify top-level expression nodes n, i.e., nodes whose fathers are statement nodes and which belong to the body of function f

$$n \in en \wedge father(n) \in stn \wedge fu(n) = f.$$

- run the Sethi-Ullman pebble strategy on the subtree with root n resulting in moves $k \in [1:K]$. We define the predicate

$$pebble(i, j, k) \leftrightarrow \text{node } i \text{ has pebble } j \text{ after move } k.$$

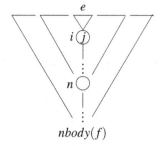

Fig. 164. Pebble j lies on expression node i in the subtree of top-level expression node n in the body of function f

- for each move k generate code $ecode(n,k)$ according to the rules of the following subsections and define $Ecode(n,k)$ as the code generated for the first k moves of the strategy

$$Ecode(n,k) = ecode(n,1) \circ \ldots \circ ecode(n,k).$$

Then set

$$code(n) = Ecode(n,K).$$

Code $ecode(n,k)$ for move k is generated such that the following lemma follows by induction over the moves k of the pebble strategy.

Lemma 93. *Assume consis' holds initially between configurations c and d. Let d^k be the configuration obtained by executing program $Ecode(n,k)$ from configuration d:*

$$consis'(c,d) \wedge d \rightarrow^*_{Ecode(n,k)} d^k.$$

Then

1. *consis' is maintained:*

$$consis'(c,d^k).$$

2. *Suppose node i has pebble j after move k, and the border word of i is expression e (see Fig. 164) which has in function f expression type t:*

$$pebble(i,j,k) \wedge bw(i) = e \wedge etype(e,f) = t.$$

Then, register $gpr(j_5)$ contains in configuration d^k the following data

$$d^k.gpr(j_5) = \begin{cases} ba(lv(e,c),c), & R(i) = 0, \\ ba(va(e,c),c), & R(i) = 1 \wedge pointer(t) \wedge va(e,c) \neq null, \\ enc(va(e,c),t), & R(i) = 1 \wedge t \in ET, \\ 0^{32}, & otherwise. \end{cases}$$

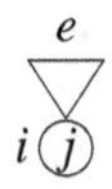

Fig. 165. Expression node i generating a constant e

Proof. by induction on the number of moves k. For $k = 0$ the lemma is trivially true, because initially there are no pebbles on the tree. In the induction step we have

$$d \rightarrow^*_{Ecode(n,k-1)} d^{k-1} \rightarrow^*_{ecode(n,k)} d^k$$

and we can use the lemma for $k - 1$ as induction hypothesis. Expression code $ecode(n,k)$ will only update registers with addresses lower than 28_5. This does not affect *consis'*. The second part of the lemma is the recipe for generating code: for each move we generate the code $ecode(n,k)$ for that move such that we can conclude the lemma.

In the sequel, we consider moves k placing pebble j on an expression node i in the subtree of a top-level expression node n and specify the generated code $ecode(n,k)$ for this move. Recall from Fig. 164 that any expression node i belongs to the derivation tree of the body of a function

$$f = fu(i).$$

This node derives an expression

$$e = bw(i)$$

and if evaluated in function f this expression has type

$$t = etype(e, f).$$

12.2.5 Constants

As illustrated in Fig. 165 node i generates a constant

$$e = bw(i)$$

if

$$\ell(i) \in \{C, CC, BC\}.$$

We generate code $ecode(n,k)$ which places the binary encoding $enc(va(e,c),t) \in \mathbb{B}^{32}$ of constant e into register i. Let $va(e,c) = v$, then the code generated for such a node is simply the macro

```
gpr(i)  =  enc(v,t)
```

$$nbody(f)$$

Fig. 166. Expression node i generating a variable name x in the body of a function f

from Sect. 9.1. Note that the compiler has to compute this encoding. For Boolean constants, character constants, and the null pointer computing the encoding is trivial. For integer constants or unsigned integer constants, however, the compiler has to convert the decimal number e into a binary or two's complement number.

Part 2 of Lemma 93 follows for node i by construction of the code. Pebbles on nodes i' other than i are not changed in move k and the corresponding register contents are not updated by $ecode(n,k)$:

$$i' \neq i \wedge pebble(i',j',k) \leftrightarrow pebble(i',j',k-1) \wedge d^k.gpr(j'_5) = d^{k-1}.gpr(j'_5).$$

Thus the lemma follows for these nodes i' directly from the induction hypothesis.

12.2.6 Variable Names

As illustrated in Fig. 166 node i generates a variable name

$$x = bw(i)$$

if

$$\ell(i) = id \wedge x \in L(Na).$$

By context conditions we expect x to be the name of a global variable or a parameter or local variable in function f:

$$x \in VN \cup ft(f).VN.$$

In general, compilers generate error messages if they detect the violation of context conditions. Here we omit the details of how this is done. There are two cases

- $x \in ft(f).VN$, i.e., x is the name of a component of a function frame for f and

$$lv(x,c) = top(c).x.$$

By invariants for statement execution we know that a frame for $f = fu(i)$ is the top frame of the $C0$ stack when expression $x = bw(i)$ is evaluated in $C0$ configuration c, i.e., we have $f = cf(c)$. By compiler consistency we know that for consistent MIPS configurations d the stack pointer

$$sptr(d) = d.gpr(spt_5)$$

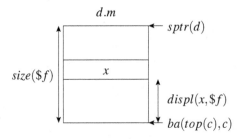

Fig. 167. Computing the base address of a local variable x of a top function frame belonging to function f

points in MIPS configuration d to the first free address after the top stack frame $top(c)$ (see Fig. 167). As the size of this stack frame is determined by the type $\$f$ of function f as $size(\$f)$, we get the base address of this frame as

$$ba(top(c),c) = sptr(d) -_{32} size(\$f)_{32}$$

and the base address of component $top(c).x$ as

$$ba(top(c).x,c) = ba(top(c),c) +_{32} displ(x,\$f)_{32}.$$

This base address is loaded into register j by the instruction

```
addi j spt displ(x,$f)-size($f)
```

If $R(i) = 0$ this is all the code that we generate. Otherwise we *dereference* register j, i.e., we load memory word $d.m_4(d.gpr(j_5))$ into register j.

```
lw    j j 0
```

We abbreviate the last instruction by

```
deref(j)
```

• $x \in VN \setminus ft(f).VN$, i.e., x is the name of a global variable and

$$lv(x,c) = gm.x.$$

By compiler consistency we know that for consistent MIPS configurations d the base pointer

$$bptr(d) = d.gpr(bpt_5)$$

points in MIPS configuration d to the base address of global memory as shown in Fig. 168:

$$ba(gm,c) = bptr(d).$$

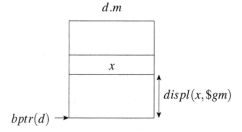

Fig. 168. Computing the base address of a global variable x

The base address of component $gm.x$ is

$$ba(gm.x, c) = ba(gm, c) +_{32} displ(x, \$gm)_{32}.$$

This base address is loaded into register j by the instruction

```
addi j bpt displ(x,$gm)
```

If $R(i) = 0$ we are done. Otherwise we *dereference* register j with

```
deref(j)
```

For node i and register j, part 2 of Lemma 93 follows immediately by construction if $R(i) = 0$, i.e., if we only have to compute the correct base address. If we also have to dereference, we need $e\text{-}consis(c, d^{k-1})$ or $p\text{-}consis(c, d^{k-1})$, which are both parts of $consis'(c, d^{k-1})$, to conclude the lemma. For nodes $i' \neq i$ one argues as in the case of constants.

12.2.7 Struct Components

Node i generates an access to a struct component if for some expression e and name n we have

$$bw(i) = e.n,$$
$$bw(i0) = e,$$
$$bw(i2) = n,$$

as shown in Fig. 169.
We have

$$t = etype(e, f) = \{t_1 \ n_1; \dots; t_s \ n_s\}$$

and context conditions require

$$n \in \{n_1, \dots, n_s\}.$$

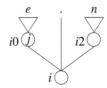

Fig. 169. Expression node i generating an access to component $.n$ of a struct

By the rules of the pebble game, move k slides pebble j from node $i0$ to node i. Thus we have $pebble(i0, j, k-1)$ and can use the induction hypothesis for node $i0$, pebble j, and move $k-1$. By the rules for generating R-labels we have $R(i0) = 0$. Thus we know

$$d^{k-1}.gpr(j_5) = ba(lv(e,c),c).$$

By the rules of expression evaluation and for computing base addresses we have

$$ba(lv(e.n,c),c) = ba(lv(e,c).n,c) = ba(lv(e,c),c) +_{32} displ(n,t)_{32}.$$

We compute the base address of $lv(e.n,c)$ by the code

```
addi j j displ(n,t)
```

If $R(i) = 0$ we are done, otherwise we dereference

```
deref(j)
```

Note that for subvariables of the form $e.n_1.n_2 \ldots .n_s$, where a number s of struct component selectors is applied consecutively, the compiler generates s `addi` instructions summing up the displacements of components n_i. An obvious optimization would be to replace these instructions by a single one, adding the sum of all displacements to the base address of e stored in register j.

12.2.8 Array Elements

Node i generates an array element if for some expressions e and e' we have

$$bw(i) = e[e'],$$
$$bw(i0) = e,$$
$$bw(i2) = e',$$

as shown in Fig. 170.
We have

$$etype(e,f) = t[n] \wedge etype(e',f) \in \{int, uint\}$$

for some $n \in \mathbb{N}$. Context conditions require

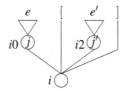

Fig. 170. Expression node i generating an access to an array component

$$va(e',c) \in [0:n-1]$$

but here we will not generate code which checks this. By the rules of the pebble game sons $i0$ and $i2$ of i have pebbles j and j' before move k. Move k slides pebble j:

$$pebble(i0,j,k-1) \wedge pebble(i2,j',k-1) \wedge pebble(i,j,k).$$

By the rules for generating R-labels we have

$$R(i0) = 0,$$
$$R(i2) = 1.$$

By induction hypothesis we have

$$d^{k-1}.gpr(j_5) = ba(lv(e,c),c) \wedge d^{k-1}.gpr(j'_5) = enc(va(e',c),etype(e',f)).$$

By the rules of expression evaluation and for computing base addresses we have

$$ba(lv(e[e'],c),c) = ba(lv(e,c)[va(e',c)],c) = ba(lv(e,c),c) +_{32} (va(e',c) \cdot size(t))_{32}.$$

We write the binary encoding of $size(t)$ into register 23 and then use the software multiplication routine from Sect. 9.2.[5] The code we generate is

```
gpr(23) = enc(size(t),uint)
mul(j',j',23)
add  j j j'
```

If $R(i) = 0$ we are done, otherwise we dereference

```
deref(j)
```

12.2.9 Dereferencing Pointers

Node i dereferences a pointer if for some expression e we have

$$bw(i) = e*,$$
$$bw(i0) = e,$$

Fig. 171. Expression node i where a pointer is dereferenced

as shown in Fig. 171.
For some type t we have

$$etype(e*, f) = t \wedge etype(e, f) = t*.$$

By the rules of the pebble game we slide pebble j in move k

$$pebble(i0, j, k-1) \wedge pebble(i, j, k).$$

By the rules of generating R-labels we have

$$R(i0) = 1.$$

By induction hypothesis and the rules of expression evaluation we get

$$d^{k-1}.gpr(j_5) = ba(va(e, c), c) = ba(lv(e*, c), c).$$

Therefore, if $R(i) = 0$ we generate no code at all. If $R(i) = 1$ we dereference register j.

```
deref(j)
```

12.2.10 'Address of'

Node i takes an address if for some expression e

$$bw(i) = e\&,$$
$$bw(i0) = e,$$

as shown in Fig. 172.
For some type t we have

$$etype(e\&, f) = t* \wedge etype(e, f) = t.$$

By the rules of the pebble game we slide pebble j in move k

$$pebble(i0, j, k-1) \wedge pebble(i, j, k).$$

---

[5] Register 23 is used only by software division, thus it is free to be used temporarily here.

Fig. 172. Expression node i where an address is taken

Fig. 173. Expression node i where a unary operator is evaluated

By the rules of generating R-labels we have

$$R(i) = 1,$$
$$R(i0) = 0.$$

By induction hypothesis and the rules for expression evaluation we get

$$d^{k-1}.gpr(j_5) = ba(lv(e,c),c) = ba(va(e\&,c),c).$$

Thus no code needs to be generated.

12.2.11 Unary Operators

Node i evaluates a unary operator if for some expression e and some operator $\circ \in \{-,!\}$

$$bw(i) = \circ e,$$
$$bw(i1) = e,$$

as shown in Fig. 173.
For some type $t \in \{int, uint, bool\}$ we have

$$etype(\circ e, f) = etype(e, f) = t.$$

By the rules of the pebble game we slide pebble j in move k

$$pebble(i0, j, k-1) \wedge pebble(i, j, k).$$

By the rules of generating R-labels we have

$$R(i) = R(i1) = 1.$$

By induction hypothesis and the rules for expression evaluation we get

$$d^{k-1}.gpr(j_5) = enc(va(e,c),t).$$

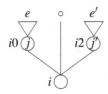

Fig. 174. Expression node i where a binary operator is evaluated

Unary Minus

- If $\circ = -$ and $t = int$ we generate the code

```
sub   j 0 j
```

- If $\circ = -$ and $t = uint$ we generate the code

```
subu j 0 j
```

The last case split is not essential for the correctness of the compiler. This will change when we introduce interrupts.

Negation

If $t = bool$ and $\circ = \,!$ we generate

```
subi j j 1
setlt j 0
```

12.2.12 Binary Arithmetic Operators

Node i evaluates a binary arithmetic operator if for some expressions e and e', and for some operator $\circ \in \{+,-,*,/\}$

$$bw(i) = e \circ e',$$
$$bw(i0) = e,$$
$$bw(i2) = e',$$

as shown in Fig. 174.
For some type $t \in \{int, uint\}$ we have

$$etype(e \circ e', f) = etype(e, f) = etype(e', f) = t.$$

By the rules of the pebble game, sons $i0$ and $i2$ of node i have pebbles j and j', and we slide pebble j in move k:

$$pebble(i0, j, k-1) \wedge pebble(i2, j', k-1) \wedge pebble(i, j, k).$$

By the rules of generating R-labels we have

$$R(i) = R(i0) = R(i2) = 1.$$

By induction hypothesis and the rules for expression evaluation we get

$$d^{k-1}.gpr(j_5) = enc(va(e,c),t) \wedge d^{k-1}.gpr(j'_5) = enc(va(e',c),t).$$

Addition

- If $\circ = +$ and $t = int$ we generate

```
add   j j j'
```

- If $\circ = +$ and $t = uint$ we generate

```
addu j j j'
```

Subtraction

- If $\circ = -$ and $t = int$ we generate

```
sub   j j j'
```

- If $\circ = -$ and $t = uint$ we generate

```
subu j j j'
```

Multiplication

If $\circ = *$ and $t \in \{int, uint\}$ we generate the program for software multiplication from Sect. 9.2.

```
mul(j,j,j')
```

Division

- If $\circ = /$ and $t = int$ we generate the program for integer division of two's complement numbers from Sect. 9.3.4.

```
divt(j,j,j')
```

- If $\circ = /$ and $t = uint$ we generate the program for integer division of binary numbers from Sect. 9.3.3.

```
divu(j,j,j')
```

12.2.13 Comparison

Node i evaluates a comparison if for some expressions e and e', and for some comparison operator $\circ \in \{==, !=, <=, <, >=, >\}$

$$bw(i) = e \circ e',$$
$$bw(i0) = e,$$
$$bw(i2) = e',$$

as shown before in Fig. 174. Note that we are treating the comparison operators like keywords, i.e., as single elements of the terminal alphabet of $C0$.

For some type $t \in \{int, uint\}$ we have

$$etype(e \circ e', f) = bool \wedge etype(e, f) = etype(e', f) = t.$$

By the rules of the pebble game sons $i0$ and $i2$ of node i have pebbles j and j', and we slide pebble j in move k:

$$pebble(i0, j, k-1) \wedge pebble(i2, j', k-1) \wedge pebble(i, j, k).$$

By the rules of generating R-labels we have

$$R(i) = R(i0) = R(i2) = 1.$$

By induction hypothesis and the rules for expression evaluation we get for elementary types

$$d^{k-1}.gpr(j_5) = enc(va(e, c), t) \wedge d^{k-1}.gpr(j_5') = enc(va(e', c), t),$$

whereas for pointer types we get

$$d^{k-1}.gpr(j_5) = ba(va(e, c), c) \wedge d^{k-1}.gpr(j_5') = ba(va(e', c), c).$$

We work out only some of the cases that arise. The remaining cases are simple exercises.

Less Than

- If $\circ \equiv <$ and $t = int$ we generate

```
slt   j j j'        // gpr(j) = (gpr(j)<gpr(j'))
```

- If $\circ \equiv >$ and $t = int$ we generate

```
slt   j j' j        // gpr(j) = (gpr(j)>gpr(j'))
```

Less Than or Equal

- If $o \equiv <=$ and $t = int$ we generate

```
slt  j j' j     // gpr(j) = !(gpr(j)<=gpr(j'))
xori j j 1      // gpr(j)[0] = !gpr(j)[0]
```

- If $o \equiv <=$ and $t = uint$ we generate

```
sltu j j' j
 xori j j 1
```

Equal

If $o \equiv ==$ we generate

```
0: bne  j j' 3    // if gpr(j)!=gpr(j') goto 3
1: addi j 0 1     // gpr(j) = 1
2: blez 0 2       // goto 4
3: add  j 0 0     // gpr(j) = 0
```

Observe that it is allowed to test expressions of the same pointer type for equality or inequality, or to test an expression of a pointer type with a null pointer constant. The code generated above tests only the equality of the base addresses. We have to use the context condition for $t = etype(e, f)$ and $t' = etype(e', f)$:

$$t = t' \vee pointer(t) \wedge e' = null.$$

If t and t' are both pointer types, we use Lemma 88 in order to show

$$va(e, c) \neq va(e', c) \leftrightarrow ba(va(e, c), c) \neq ba(va(e', c), c).$$

If we have

$$pointer(t) \wedge va(e, c) \neq null \wedge e' = null$$

we use

$$ba(va(e, c), c) \geq sbase > 0.$$

12.2.14 Translating Several Expressions and Maintaining the Results

In some situations it is necessary to evaluate several expressions e_i, obtaining the results of expression evaluation of *all* these expressions in some GPR registers j_i. This problem is solved in a very easy way. For a set of pebbles

$$J \subset [1 : 27]$$

and an expression node n we generate $code(n, J)$ as follows:

- in the pebble game, we initially place the pebbles $j \in J$ on the used list. They will stay there, because they are never removed during the pebble game for n.
- now play the pebble game for node n and generate code as described above. Because pebbles in $j \in J$ are not written by this code, registers $gpr(j_5)$ with $j \in J$ maintain their value.

Thus we have the following.

Lemma 94.

$$d \to^*_{code(n,J)} d' \wedge j \in J \ \to \ d.gpr(j_5) = d'.gpr(j_5).$$

12.3 Translation of Statements

We start with configurations c and d satisfying $consis(c,d)$. As usual we denote

$$c' = \delta_C(c).$$

Recall from the statement of Lemma 81 that we have to generate code such that we can show

$$\exists s : consis(c', \delta^s_M(d)).$$

Recall also that

$$consis(c,d) \equiv consis'(c,d) \wedge pc\text{-}consis(c,d).$$

Naively, one would proceed by a case split on the head $c.pr[1]$ of the program rest, but as we have explained before, this does not work — we split cases on the first program rest node

$$n = c.prn(1)$$

instead. Also we denote by

$$f = fu(n)$$

the function of statement node n. In each case, the proof of the following lemma will be an easy bookkeeping exercise.

Lemma 95.

$$consis(c,d) \wedge c' = \delta_C(c) \ \to \ \exists s : consis'(c', \delta^s_M(d)).$$

We will do the easy bookkeeping as we go and postpone the proof that $pc\text{-}consis$ is maintained to Sect. 12.4.5.

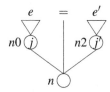

Fig. 175. Statement node n where an assignment statement is derived

12.3.1 Assignment

A statement node n generates an assignment if for some expressions e and e' we have

$$bw(n) \equiv e = e',$$
$$bw(n0) = e,$$
$$bw(n2) = e',$$

as shown in Fig. 175. The sons $n0$ and $n2$ of n are top-level expression nodes.
For some type t we have

$$t = etype(e', f).$$

By the rules for generating R-labels we have

$$R(n0) = 0,$$
$$R(n1) = 1.$$

We generate expression code

```
code(n0)
```

In the end we have a pebble j on node $n0$, i.e., the result of evaluation of expression e is finally stored in register $gpr(j_5)$. We prevent it from being overwritten by generating expression code

```
code(n2,{j})
```

which in the end places a pebble j' on node $n2$.
Let d' be the configuration reached from d by running the code generated so far:

$$d \to^*_{code(n0) \circ code(n2,\{j\})} d'.$$

By Lemmas 93 and 94 we conclude

$$d'.gpr(j_5) = ba(lv(e,c),c),$$
$$d'.gpr(j_5') = \begin{cases} ba(va(e',c),c), & pointer(t), \\ enc(va(e',c),t), & t \in ET. \end{cases}$$

We generate code for the obvious store operation

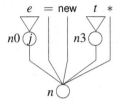

Fig. 176. Statement node n where a new statement is derived

```
sw    j' j 0
```

Let d'' be the configuration reached by executing this statement in configuration d'. One concludes by easy bookkeeping

$$consis'(c',d'') \wedge d''.pc = end(n) +_{32} 1_{32}.$$

12.3.2 New Statement

As shown in Fig. 176, a statement node n generates a new statement if for some expression e and type name t we have

$$bw(n) \equiv e = \text{new } t*,$$
$$bw(n0) = e,$$
$$bw(n3) = t.$$

Son $n0$ of n is a top-level expression node.

By the rules for generating R-labels we have

$$R(n0) = 0.$$

We generate expression code

```
code (n0)
```

In the end we have a pebble j on node $n0$. Let d' be the configuration reached by running the code generated so far:

$$d \rightarrow^*_{code(n0)} d'.$$

From Lemma 93 we get

$$d'.gpr(j_5) = ba(lv(e,c),c).$$

Let

$$c' = \delta_C(c).$$

From the semantics of statement execution and $bsp\text{-}consis(c, d')$ we get

$$d'.gpr(hpt_5) = ba(c'.nh - 1).$$

As this is the base address of the newly created heap variable, we generate the store operation

```
0: sw    hpt j 0
```

Next we increase the heap pointer by $size(t)$ and test whether it reaches the maximal allowed heap address *hmax*. If it does, the compiled program will execute a system call instruction with an appropriate parameter x and cause an interrupt. Details of interrupt handling will be explained in Chap. 14.

```
1: addi hpt hpt size(t)    // hptr = hptr + size(t)
2: subi 1 hpt hmax         // gpr(1) = hptr - hmax
3: bltz 1 4                // if hptr<hmax skip
                           // error handling
4-5: gpr(1) = x            // macro
6: sysc                    // system call
```

It remains to initialize the newly accessible space on the heap with zeros. Using a program from Sect. 9.1 this is achieved by

```
7: addi 1 hpt -size(t)  // gpr(1) = 'old heap pointer'
8: addi 2 0 size(t)/4   // gpr(2) = size(t)/4
9-12: zero(1,2)
```

Let d'' be the configuration reached from d' by running the assembly program with the lines 0-12. Then one concludes by easy bookkeeping

$$consis'(c', d'') \wedge d''.pc = end(n) +_{32} 1_{32}.$$

12.3.3 While Loop

As shown in Fig. 177 a statement node n derives a while loop if there is an expression e and a statement sequence *body* such that

$$bw(n) = \text{while } e \ \{body\},$$
$$bw(n1) = e,$$
$$bw(n3) = body.$$

Son $n1$ of n is a top-level expression node.

By the rules for generating R-labels we have

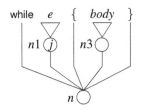

Fig. 177. Statement node n where a while loop is derived

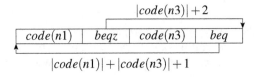

Fig. 178. Illustration of jump distances in the code of a while loop

$$R(n1) = 1.$$

We generate expression code

```
code (n1)
```

In the end, we have a pebble j on node $n1$.

Let d' be the configuration reached from configuration d by running the code generated so far:

$$d \rightarrow^*_{code(n1)} d'.$$

From Lemma 93 we get

$$d'.gpr(j_5) = enc(va(e,c),bool).$$

For the computation of jump distances, recall that we denote for any node n by $|code(n)|$ the length of $code(n)$ measured in words. We continue $code(n1)$ with code whose jump distances are illustrated in Fig. 178.

```
beqz j |code (n3) |+2              // if va (e, c) =0
                                   // jump out of loop
code (n3)                          // code of the body
blez 0 -(|code (n1) |+|code (n3) |+1) // jump back to
                                   // start of loop
```

Let d'' be the configuration reached from d' by executing just the *beqz* instruction. Easy bookkeeping shows

$$consis'(c',d'')$$

and

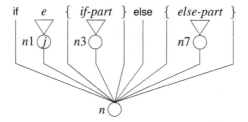

Fig. 179. Statement node n where an if-then-else statement is derived

$$d''.pc = \begin{cases} start(n3), & va(e,c) = 1, \\ end(n) +_{32} 1_{32}, & va(e,c) = 0. \end{cases}$$

12.3.4 If-Then-Else

As shown in Fig. 179, a statement node n derives an if-then-else statement if there is an expression e and statement sequences a (if-part) and b (else-part) such that

$$bw(n) = \text{if } e \ \{a\} \text{ else } \{b\},$$
$$bw(n1) = e,$$
$$bw(n3) = a,$$
$$bw(n7) = b.$$

Code generation is very similar to the case of while loops. Jump distances are illustrated in Fig. 180.

```
code(n1)                    // gpr(j) = enc(va(e,c),bool)
beqz j |code(n3)|+2         // if va(e,c)=0
                            // jump to else-part
code(n3)                    // code of if-part
blez 0 |code(n7)|+1         // jump over else-part
code(n7)                    // code of else-part
```

Let d' be the configuration after execution of $code(n1)$. Let d'' be the configuration reached from d' by executing just the $beqz$ instruction. Easy bookkeeping shows

$$consis(c',d'')$$

and

$$d''.pc = \begin{cases} start(n3), & va(e,c) = 1, \\ start(n7), & va(e,c) = 0. \end{cases}$$

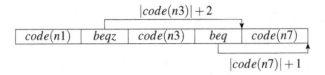

Fig. 180. Illustration of jump distances in the code of an if-then-else statement

12.3.5 If-Then

As shown in Fig. 181, a statement node n derives an if-then statement if there is an expression e and statement sequences a (if-part) such that

$$bw(n) = \text{if } e \{a\},$$
$$bw(n1) = e,$$
$$bw(n3) = a.$$

Jump distances for the code generated are illustrated in Fig. 182.

```
code(n1)              // gpr(j) = enc(va(e,c),bool)
beqz $j |code(n3)|+1  // if va(e,c)=0
                      // jump over if-part
code(n3)              // code of if-part
```

We have devoted an individual subsection to this case because, due to the absence of an else-part, the result of the easy bookkeeping is different from the result for if-then-else statements.

Let d' be the configuration after execution of $code(n1)$. Let d'' be the configuration reached from d' by executing just the $beqz$ instruction. As for while loops, easy bookkeeping shows

$$consis'(c', d'')$$

and

$$d''.pc = \begin{cases} start(n3), & va(e,c) = 1, \\ end(n) +_{32} 1_{32}, & va(e,c) = 0. \end{cases}$$

12.3.6 Function Call

Figure 183 shows a statement node n deriving a function call with p parameters. Note that the parameter declaration sequence is derived from node $n4$. Thus, for $y \in [1 : p]$, parameter y is derived from node

$$m_y = se(n4, y)$$

and the top-level expression node for parameter y is

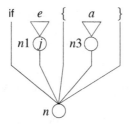

Fig. 181. Statement node n where an if-then statement is derived

$$|code(n3)| + 1$$

| $code(n1)$ | $beqz$ | $code(n3)$ |

Fig. 182. Illustration of jump distances (measured in bytes) in the code of an if-then statement

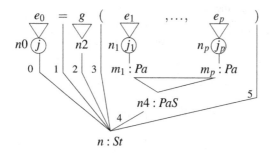

Fig. 183. Statement node n where a call of a function g with p parameters is derived

$$n_y = m_y 0.$$

Thus, node n derives a function call with p parameters if there are expressions $e, e_1, \ldots, e_p$ and a function name $g \in FN$ such that

$$bw(n) \equiv e = g(e_1, \ldots, e_p),$$
$$bw(n0) = e,$$
$$bw(n2) = g,$$

and for all $y \in [1 : p]$:

$$e_y = bw(n_y).$$

By the rules for generating R-labels we have

$$R(n0) = 0$$

and for all $y \in [1 : p]$:

$$R(n_y) = 1.$$

From the rules of statement execution and $bsp\text{-}consis(c,d)$ we conclude

$$d.gpr(spt_5) = ba(top(c'),c') -_{32} 8_{32}.$$

This is illustrated in Fig. 184.

Below, we generate code $c1$ to check whether there is enough room on the stack for the new frame, initialize the newly created function frame in the memory of the MIPS machine, and then jump to the code of the body of the called function. The MIPS configuration obtained by executing this code starting from configuration d is called d':

$$d \rightarrow^*_{c1} d'.$$

Testing Available Space on the Stack

The generated code is similar to the code which tests for new statements whether there is enough space on the heap

```
addi 1 spt size($g)    // gpr(1) = sptr + size($g)
                       //        = 'new stack pointer'
subi 1 1 smax          // gpr(1) = 'new stack pointer'
                       //          - smax
blez 1 4               // if 'new stack pointer'<=smax
                       // skip error handling
gpr(1) = x             // macro
sysc                   // system call
```

Adjusting the Stack Pointer

The stack pointer is increased by $size(\$g) + 8$ (see Fig. 184).

This ensures

$$spt\text{-}consis(c',d').$$

Increasing the stack pointer immediately guarantees that all accesses of the compiled code to the stack are below the stack pointer, even during function calls. This allows interrupt handlers to use the space above the stack pointer for its own stack.

Return Destination

```
code(n0)
```

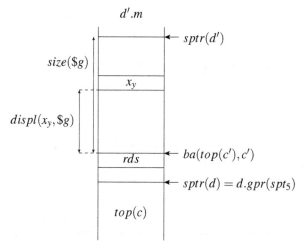

Fig. 184. The new top frame $top(c')$ created by the call of function g in the memory of the MIPS configuration d'

In the end we have a pebble j on node $n0$. From Lemma 93, we get for the configuration d_0 after execution of this code

$$d_0.gpr(j_5) = ba(lv(e,c),c).$$

We store this base address in the memory word $size(\$g) + 4$ bytes *below* the new stack pointer $sptr(d')$ (see Fig. 184).

```
sw    j spt  -(size($g)+4)
```

This ensures

$$rds\text{-}consis(c',d').$$

Parameters and Local Variables

Let

$$\$g = \{t_1\ x_1; \ldots; t_r\ x_r\}$$

be the type of the function frame of the called function g. For each index $y \in [1 : p]$ of a parameter, we generate code evaluating the parameter x_y.

```
code (ny)
```

In the end, we have a pebble j_y on node n_y.

From Lemma 93 we get for the configuration d_y after execution of this code

$$d_y.gpr(j_{y5}) = \begin{cases} ba(va(e_y,c),c), & pointer(t_y), \\ enc(va(e_y,c),t_y), & t_y \in ET. \end{cases}$$

We follow $code(n_y)$ immediately by a store instruction which initializes parameter x_y in the new function frame (see Fig. 184).

```
sw    jy spt -size($g)+displ(xy,$g)
```

All simple subvariables of local variables x_y with indices $y \in [p+1:r]$ are initialized with encodings of their default values, i.e., with 0^{32}. The displacement of the first local variable of function g is

$$displ(x_{p+1},\$g).$$

We denote the number of memory words occupied by all local variables of function g by

$$z = \left(\sum_{y>p} size(t_y)\right)/4.$$

Local variables are then initialized by

```
add    1 spt -size($g)+displ(xp+1,$g)
addi   2 0 z
zero(1,2)
```

This ensures

$$e\text{-}consis(c',d')$$

and

$$p\text{-}consis(c',d').$$

Jump and Link

Next — and almost finally — we jump to the start address of the code generated for the body of the called function g.

```
jal   0bstart(nbody(g))[27:2]
```

This completes the code generated for a function call. Two remarks are in order:

- We have not yet achieved $consis'(c',d')$ because the link address saved in register 31 is not yet stored in the function frame. This will be done by the first instruction of $code(nbody(g))$.
- Function bodies are translated in some order. If the body of the called function g is translated before the body of the calling function f we know address $start(nbody(g))$ and no problem arises. Otherwise, we translate in two passes: in the first phase the *iindex* field of the jump-and-link instruction is set to an

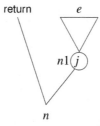

Fig. 185. Statement node n where a return statement is derived

arbitrary value. After all function bodies are translated, the second phase is performed: now, all desired start addresses are known and the *iindex* fields are filled in.

Bookkeeping gives

$$\bigwedge_{X\neq pc,clr} X\text{-}consis(c',d') \wedge d'.pc = start(nbody(g)).$$

12.3.7 Return

As shown in Fig. 185, a node n derives a return statement if there is an expression e such that

$$bw(n) = return\ e,$$
$$bw(n1) = e.$$

For some type t we have

$$t = etype(e,f).$$

By the rules for generating R-labels we have

$$R(n1) = 1.$$

We generate code for evaluating expression e

```
code (n1)
```

At the end there is a pebble j on node n.
Let

$$d \to^*_{code(n1)} d'.$$

From $consis(c,d)$ and Lemma 93 we conclude

$$d'.gpr(j_5) = \begin{cases} ba(va(e,c),c), & pointer(t) \wedge va(e,c) \neq null, \\ enc(va(e,c),t), & t \in ET, \\ 0^{32}, & \text{otherwise.} \end{cases}$$

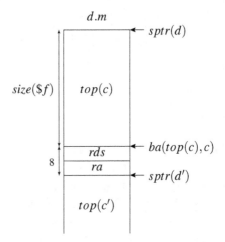

Fig. 186. The old top frame $top(c)$ is deleted by the return statement of function f

Now consider Fig. 186. Using $bsp\text{-}consis(c,d')$ and $rds\text{-}consis(c,d')$ we conclude

$$d'.m_4(sptr(d') -_{32} (size(\$f)+4)_{32}) = ba(c.rds(c.rd),c).$$

We store the result of evaluating e, which is presently in register j, at this address.

```
sw    j    spt  -(size($f)+4)
```

Next we store the address ra of the return jump in register 1, decrease the stack pointer and jump to the return address.

```
lw    1    spt  -(size($f)+8)    // gpr = ra
addi  spt  spt  -(size($f)+8)    // decrease spt
jr    1                          // goto ra
```

Let d'' be the resulting configuration. By bookkeeping and using $clr\text{-}consis(c,d')$ we conclude

$$consis'(c',d'') \wedge d''.pc = end(c.clr(c.rd)) +_{32} 1_{32}.$$

The return statement of function *main* necessarily receives special treatment. There is no C0 function to return to when this statement is reached. Instead we return to the operating system kernel with a system call *sysc*, an instruction which is discussed in detail in Chap. 14. Parameters x for the system call (here: informing the kernel that it is called because the running program has terminated) are usually passed via the general-purpose registers. In the application binary interface (ABI) discussed in Sect. 15.7.4 this parameter is passed via $gpr(1_5)$. Therefore in the generated code for the return statement of function *main* the system call instruction is preceded by instructions writing the desired parameter x into register 1. Thus the generated code is

```
gpr(1) = x        // specifying kind of syscall
sysc              // system call to kernel
```

More parameters can be passed in other registers.

12.3.8 Summary of Intermediate Results

For the configurations d'' we have shown so far:

1. for conditional statements and for while loops with $va(e,c) = 1$ we have

$$consis(c',d'')$$

   and we are done with the proof of Lemma 81, i.e., the induction step for compiler correctness.

2. after the call of a function g we have (with $d'' = d'$)

$$\bigwedge_{X \neq pc,clr} X\text{-}consis(c',d'') \wedge d''.pc = start(nbody(g)).$$

3. after a return statement we have

$$consis'(c',d'') \wedge d''.pc = end(c.clr(c.rd)) +_{32} 1_{32}.$$

4. in all other cases we have

$$consis'(c',d'') \wedge d''.pc = end(c.prn(1)) +_{32} 1_{32}.$$

In order to complete the proof of Lemma 81 we have to reach from d'' in zero or more steps a MIPS configuration q which satisfies $consis(c',q)$. In particular we have to establish $pc\text{-}consis(c',q)$, i.e.,

$$q.pc = start(c'.prn(1)).$$

For this purpose we specialize Lemma 90 as follows.

Lemma 96.

1. *If $c.prn(1)$ does not derive a return statement, then*

$$c.prn(2) = succ(c.prn(1)).$$

2. *If $c.prn(1)$ derives a return statement, then*

$$c.prn(2) = succ(c.clr(c.rd)).$$

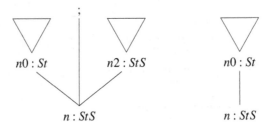

Fig. 187. Node *n* deriving a statement sequence

12.4 Translation of Programs

It remains to specify how the code for statement sequences, bodies, function declaration sequences, and programs is constructed from the code of the single statements. This is less obvious than it seems.

12.4.1 Statement Sequences

This is straightforward. As shown in Fig. 187, node *n* labeled with *StS* always has a son *n0* labeled *St* and possibly a son *n2* labeled *StS*. In case there is a single son we set

$$code(n) = code(n0).$$

In case son *n2* exists we recursively define

$$code(n) = code(n0) \circ code(n2).$$

In this situation, node *n0* always has a statement node as its direct successor:

$$dsucc(n0) = n20.$$

Moreover, the code for *n2* starts with the code for the successor *n20* of *n0*, hence

$$start(n2) = start(n20) = end(n0) +_{32} 1_{32}.$$

We return to the summary of intermediate results (Sect. 12.3.8) and finish cases 3 and 4 for situations where $c.clr(c.rd)$ or $c.prn(1)$ is not the last statement in a statement sequence and thus has a direct successor.

Let

$$s = \begin{cases} c.clr(c.rd), & \ell(c.prn(1)) = rSt, \\ c.prn(1), & \text{otherwise.} \end{cases}$$

We conclude from $\neg lastst(s)$ with the help of Lemma 96:

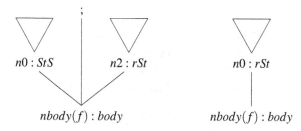

Fig. 188. Node $nbody(f)$ deriving the body of function f

$$d''.pc = end(s) +_{32} 1_{32}$$
$$= start(dsucc(s))$$
$$= start(c.prn(2))$$
$$= start(c'.prn(1)),$$

i.e., we have

$$consis(c'.d'').$$

12.4.2 Function Bodies

As shown in Fig. 188, a node $nbody(f)$ always has a son $n0$ labeled with StS and a son $n2$ labeled with rSt. Recalling that the code for function calls does not save the link address (which was saved in register 31 by the jump-and-link instruction), we start the code $code(nbody(f))$ for function bodies with a preamble performing that task. We continue with the code for $n0$ and $n2$.

```
sw    31 spt -(size($f)+8)
code(n0)
code(n2)
```

We return to case 2 of the intermediate results (Sect. 12.3.8). Let q be the configuration reached from d'' by execution of the store word instruction of the preamble. On one hand, the preamble establishes

$$clr\text{-}consis(c',q),$$

on the other, we have

$$q.pc = start(nbody(g)) +_{32} 4_{32}$$
$$= start(n0)$$
$$= start(c'.prn(1)).$$

Thus we have

$$consis(c',q)$$

and the induction step is completed for function calls.

We also can finish cases 3 and 4 for situations where $c.prn(1)$, resp. $c.clr(c.rd)$, are the last statements in the statement sequence of the body of a function f.

Let s be defined as in Sect. 12.4.1 and let it be the last statement node derived from $n0$. Then we have

$$n2 = succ(s)$$

and

$$start(n2) = end(n0) +_{32} 1_{32} = end(s) +_{32} 1_{32}.$$

We have $lastst(s) \wedge inbody(s)$ and conclude with the help of Lemma 96:

$$\begin{aligned} d''.pc &= end(s) +_{32} 1_{32} \\ &= start(succ(s)) \\ &= start(c.prn(2)) \\ &= start(c'.prn(1)), \end{aligned}$$

i.e., we have

$$consis(c',d'').$$

12.4.3 Function Declaration Sequences

Obviously for a node m labeled FuD we need to generate the code for the body of the function. Recall the definition of $nbody(f_i)$ in Sect. 11.2.3 where we defined the node which derives the body of function f_i depending on the outdegree of its function declaration node. Using that definition we set

$$code(m) = code(nbody(fu(m))).$$

As in the case of statement sequences, a node n labeled $FuDS$ always has a son $n0$ labeled FuD and possibly a son $n2$ labeled $FuDS$. In case there is a single son we set

$$code(n) = code(n0).$$

In case son $n2$ exists we recursively define

$$code(n) = code(n0) \circ code(n2).$$

12.4.4 Programs

The root ε of the derivation tree is labeled with $prog$ and has a son $n \in [0:2]$ labeled $FuDS$. The code for programs consists of a prologue $plog$ and the code $code(n)$ of the son labeled $FuDS$:

$$code(\varepsilon) = plog \circ code(n).$$

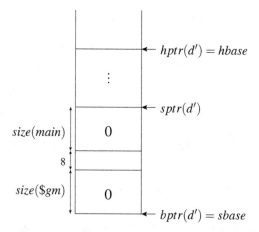

Fig. 189. The prologue of a translated program initializes global memory and the frame for function *main* with zeros and sets the base/stack/heap pointer registers appropriately

Let c^0 be the initial configuration of the $C0$ computation. The prologue *plog* starts in a configuration d^0 and produces a configuration d' satisfying $consis(c^0, d')$. As shown in Fig. 189 it has to initialize the base pointer to *sbase*, the heap pointer to *hbase*, and the stack pointer to $sbase +_{32} size(\$gm)_{32} +_{32} size(\$main)_{32} +_{32} 8_{32}$. If there are no global variables declared we set $size(\$gm) = 0$.

```
gpr(hpt) = hbase
gpr(bpt) = sbase
addiu spt bpt size($gm)+size($main)+8
```

The memory region between the base pointer and the stack pointer, i.e., the global memory and the function frame for function *main*, is initialized with zeros.

```
subu 1 spt bpt // gpr(1) = size($gm)+size($main)+8
srl  1 1 2     // gpr(1) = gpr(1)/4
zero(bpt,1)
```

Note that this area also contains return address and return value destination fields for function *main*. It does not matter what we write there, because the return statement of function *main* is compiled to a system call (see Sect. 12.3.7). Finally we jump to the start of the code of *nbody*(*main*).

```
j    0bstart(nbody(main))[27:2]
```

Easy bookkeeping establishes the base case of the compiler correctness result.

Lemma 97.
$$consis(c^0, d').$$

12.4.5 Jumping Out of Loops and Conditional Statements

We are ready to complete the proof of Lemma 81. We have to finish cases 3 and 4 when s is the last node in a loop body or in an if-part or an else-part, i.e., s satisfies predicate

$$Q(s) \equiv lastst(s) \wedge (inwhile(s) \vee incond(s)).$$

We have to argue about the MIPS instructions which jump out of the bodies of while loops and over the else-part of if-then-else statements (but not of if-then statements, which have no else-part). This is somewhat technical. We define predicates distinguishing between these cases:

$$lastwhile(s) \equiv lastst(s) \wedge inwhile(s),$$
$$lastif\text{-}in\text{-}ifte(s) \equiv lastst(s) \wedge Fa(s)5 \in T.A \wedge prefix(Fa(s)3,s),$$
$$no\text{-}jump\text{-}out(s) \equiv lastst(s) \wedge incond(s) \wedge \neg lastif\text{-}in\text{-}ifte(s).$$

Note that condition $Fa(s)5 \in T.A$ identifies the Father of s as deriving an if-then-else statement, because all other possible fathers of statement sequences have at most five sons. For any node s labeled with $\ell(s) = St$, we now define a very special nesting depth $depth(s)$ of s as

$$depth(s) = \begin{cases} 0, & \neg lastst(s) \vee inbody(s), \\ 1 + depth(Fa(s)), & \text{otherwise.} \end{cases}$$

We also define the jump distance $jdis(s)$ by

$$jdis(s) = \begin{cases} 0, & \neg lastst(s) \vee inbody(s), \\ 1, & lastwhile(s), \\ 1 + jdis(Fa(s)), & lastif\text{-}in\text{-}ifte(s), \\ jdis(Fa(s)), & no\text{-}jump\text{-}out(s). \end{cases}$$

We proceed to show the following.

Lemma 98. *For any MIPS configuration k and any statement node s with $\ell(s) = St$, the following holds: if the pc in k points directly behind the code of s, then after $jdis(s)$ steps it points to $start(succ(s))$:*

$$k.pc = end(code(s)) +_{32} 1_{32} \rightarrow \delta_M^{jdis(s)}(k).pc = start(succ(s)).$$

Proof. by induction on $r = depth(s)$. For $r = 0$ we have $\neg lastst(s) \vee inbody(s)$ and hence $jdis(s) = 0$. The claim

$$k.pc = start(succ(s))$$

follows immediately by the code generation for statement sequences and function bodies, and the definition of function $succ$.

For the induction step assume that the lemma is proven for all nodes s' with $depth(s') \leq r - 1$ and consider a node s with $depth(s) = r$. We always have

$$k.pc = end(code(s)) +_{32} 1_{32}.$$

There are three cases

1. $lastwhile(s)$: Let
$$h = \delta_M(k)$$

be the result of jumping out of the loop body at the last instruction in the $code(Fa(s))$. That instruction jumps to

$$h.pc = start(Fa(s)) = start(succ(s)).$$

2. $lastif\text{-}in\text{-}ifte(s)$: Let
$$h = \delta_M(k)$$

be the result of jumping out of the if-part behind the code of the if-then-else statement derived from $Fa(s)$

$$h.pc = end(Fa(s)) +_{32} 1_{32}.$$

Because
$$depth(Fa(s)) = r - 1$$

we can use the definition of $jdis$, the induction hypothesis, and the definition of the successor function to conclude

$$\begin{aligned}
\delta_M^{jdis(s)}(k).pc &= \delta_M^{jdis(Fa(s))+1}(k).pc \\
&= \delta_M^{jdis(Fa(s))}(h).pc \\
&= start(succ(Fa(s))) \\
&= start(succ(s)).
\end{aligned}$$

3. $no\text{-}jump\text{-}out(s)$: Statement s is derived from an if-statement or the else-part of an if-then-else statement. From the translation of conditional statements we have

$$k.pc = end(code(s)) +_{32} 1_{32} = end(Fa(s)) +_{32} 1_{32}.$$

Because
$$depth(Fa(s)) = r - 1$$

we can use the definition of $jdis$, the induction hypothesis, and the definition of the successor function to conclude

$$\begin{aligned}
\delta_M^{jdis(s)}(k).pc &= \delta_M^{jdis(Fa(s))}(k).pc \\
&= start(succ(Fa(s))) \\
&= start(succ(s)).
\end{aligned}$$

With the help of this lemma we immediately finish cases 3 and 4.

Let

$$q = \delta_M^{jdis(s)}(d''),$$

then

$$\begin{aligned} q.pc &= start(succ(s)) \\ &= start(c.prn(2)) \\ &= start(c'.prn(1)). \end{aligned}$$

We conclude

$$pc\text{-}consis(c', q)$$

and

$$consis(c', q).$$

12.5 Compiler Correctness Revisited

We started this chapter with the formulation of a compiler correctness theorem. This turned out to be a great help in the construction of the compiler: code was generated such that the induction step of the correctness proof would work. In the theorem the *existence* of a sequence $(s(i))$ of step numbers is stated. In the proof we showed more than the mere existence of these numbers: we *constructed* them although we did not bother to formalize this fact. Fixing this will be technically easy and not very interesting. In contrast it is very interesting to see why the statement of the compiler correctness theorem without this fix is flawed in an absolutely terrible way. We illustrate the problem with a subsection whose title might be surprising in a book about computer science. But the reader will see that it is absolutely to the point.

12.5.1 Christopher Lee and the Truth About Life After Death

We start by describing a scene from the Edgar Wallace movie 'The Devil's Daffodil' from 1961. In a scene Christopher Lee plays a Chinese detective who applies torture during the interrogation of a criminal. When the horrified criminal cries: 'Stop it! I will tell you everything,' the detective answers with a serene smile: 'Not everything! Only the truth.' Inspired by this we effortlessly state the truth about life after death:

- There is life after death.
- There is no life after death.

Clearly, the truth is there. But unfortunately in addition to the truth the contrary of the truth is also there, and no hint is given about which is which. It turns out that we can satisfy the present formulation of the compiler correctness theorem in the same catastrophically uninformative way. We don't even look at the *C*0 program and there is a single piece of code which translates every statement of every *C*0 program.

Let

$$N = \langle hmax \rangle - \langle sbase \rangle$$

be the number of memory cells used to store the encoded global memory, stack, and heap in the MIPS machine. We interpret this portion of the memory together with the stack pointer and the heap pointer as the binary encoding of a number

$$n \in \mathbb{B}^{N \cdot 8 + 2 \cdot 32}.$$

The code of every statement is a single infinite loop, and the body of the loop simply increases this number n modulo

$$M = 2^{N \cdot 8 + 2 \cdot 32}.$$

Programming this is an easy exercise and after M iterations we have enumerated *all* possible combinations of heap, stack, global memory, and pointers. We have pc-consistency at the start of each loop, and for *every* C0 configuration c which can be encoded in the given space we have $consis(c, d)$ for the MIPS configuration d after some iteration of the loop. We don't even have to worry about the initial configuration. After some number of iterations we are in a MIPS configuration d' satisfying $consis(c^0, d')$, and we have the basis of the induction. Suppose we have $consis(c, d)$ and $c' = \delta_C(c)$. Then after at most M iterations of the loop we are in a MIPS configuration d' satisfying $consis(c', d')$, and we have the induction step. Thus our compiler correctness theorem can also be shown for this completely useless 'translation' and the statement of the compiler correctness theorem in its present nonconstructive form is catastrophically flawed. Fortunately, the proof is not. All we have to extract from the proof is a more precise formulation of the step numbers $s(i)$.

12.5.2 Consistency Points

We say that MIPS configuration d is a *consistency point* if the program counter points to the start of the code generated for a statement node n:

$$cpoint_M(d) \equiv \exists n \in stn : d.pc = start(n).$$

Clearly compiler consistency cannot possibly hold outside of consistency points. In what follows we make use of an iterated MIPS transition function $\delta\text{-}iter_M$ defined by

$$\delta\text{-}iter_M(d, 0) = d,$$
$$\delta\text{-}iter_M(d, n+1) = \delta_M(\delta\text{-}iter_M(d, n)).$$

For MIPS configurations d we denote the number of instructions executed until the next consistency point is reached by

$$nsc_M(d) = \min\{n > 0 \mid cpoint_M(\delta\text{-}iter_M(d, n))\}.$$

The next consistency point reached after d is

$$nc_M(d) = \delta\text{-}iter_M(d, nsc_M(d)).$$

Now we can strengthen the induction step of the compiler correctness theorem: if we have $consis(c,d)$ and $c' = \delta_C(c)$, then consistency with c' is reached at the next consistency point reached after d.

Lemma 99.

$$consis(c,d) \wedge c' = \delta_C(c) \;\rightarrow\; consis(c', nc_M(d))$$

Proof. We simply observe, that between the MIPS configurations d, for which we showed consistency, the *pc* of the MIPS machine is never in a consistency point.

12.5.3 Compiler Correctness for Optimizing Compilers

Optimizing compilers often generate code for several high-level language statements together. This often results in a very substantial speed-up of the translated program. However the additional freedom to optimize across several statements comes at a price: consistency does not hold any more after every $C0$ step. A compiler correctness theorem for optimizing compilers as well as for non-optimizing compilers can now be stated in the following way:

- one identifies a set of addresses

$$CP_M \subset \mathbb{B}^{32}.$$

A MIPS configuration d is then at a consistency point if its *pc* points into this set

$$cpoint_M(d) \;\equiv\; d.pc \in CP_M.$$

In the case of our non-optimizing compiler we had

$$CP_M = \{start(n) \mid n \in stn\}.$$

- one also identifies a set of statement nodes

$$CP_C \subset stn.$$

A $C0$ configuration c is at a consistency point if the head $hd(c.prn)$ of the program rest nodes component lies in this set

$$cpoint_C(c) \;\equiv\; hd(c.prn) \in CP_C.$$

In the non-optimizing compiler one uses the full set of statement nodes

$$CP_C = stn.$$

Next we define consistency points for $C0$ configurations c in the obvious way:

$$\delta\text{-}iter_C(c,0) = c,$$
$$\delta\text{-}iter_C(c,n+1) = \delta_C(\delta\text{-}iter_C(c,n)),$$
$$nsc_C(c) = \min\{n > 0 \mid cpoint_C(\delta\text{-}iter_C(c,n))\},$$
$$nc_C(c) = \delta\text{-}iter_C(c, nsc_C(c)).$$

- one makes an observation about the connection between $C0$ and MIPS consistency points. Obviously, in a MIPS consistency point the pc should point to the start of the compiled code for a $C0$ consistency point. Thus MIPS consistency points are defined by the choice of $C0$ consistency points, i.e.,

$$CP_M \equiv \{start(n) \mid n \in CP_C\}.$$

- one assumes $consis(c,d)$ in the induction step. If c' is obtained by running the $C0$ computation until the next $C0$ consistency point, then a consistent MIPS configuration d is reached by running the MIPS computation until the next MIPS consistency point. In the induction step one would show the following.

Lemma 100.

$$consis(c,d) \wedge c' = ncc(c) \rightarrow consis(c', nc_M(d)).$$

12.6 Final Remarks

- First of all, remember that the construction and correctness proof of a compiler for a useful language does not take more than 70 pages, even in a textbook.
- The key technical definition is of course the simulation relation $consis(c,d)$, because it guides the compiler construction and the correctness proof. Do not try to memorize this definition literally. Proceed instead in the following way:
  - the construction of allocated base addresses using the size of types and displacements is reasonably obvious. Remember that the address ranges for two subvariables are disjoint or one is contained in the other.
  - in the end you are interested in the value of elementary subvariables x. This is captured in e-$consis(c,d)$, where one requires

    $$d.m_4(ba(x,c)) = enc(va(x,c), vtype(x,c)).$$

  - the counterpart for pointers is p-$consis(c,d)$. If x points to y, then the MIPS memory at address $ba(x,c)$ contains $ba(y,c)$:

    $$y = va(x,c) \wedge y \in SV(c) \rightarrow d.m_4(ba(x,c)) = ba(y,c).$$

  - memorize the layout for stack frames (you can choose a different layout, but then you must adjust a few definitions). From this you derive bsp-$consis(c,d)$ and rds-$consis(c,d)$.
  - $code$-$consis(c,d)$ is obvious; the code should not be self-modifying.
  - remember that code is generated for nodes of the derivation tree. Therefore the ghost components prn and clr were introduced into the configurations.
  - coupling pc and the program rest is done with pc-$consis(c,d)$:

    $$d.pc = start(hd(c.prn)).$$

- now for the only tricky stuff: remember the definition of the successor function *succ* and the reconstruction of the program rest using $hd(prn)$ (and *clr* for return statements).
- For expression translation remember:
  - the Sethi-Ullman algorithm: size of trees = number of their nodes. Recursively move the pebble to the root of the larger subtree first. The analysis is then fairly obvious.
  - there is an *R*-label distinguishing between right values and left values (addresses).
  - what has to hold if node *i* for expression *e* in the body of function *f* has pebble *j* after move *k*. Three cases: i) left value computation, ii) pointer value computation, iii) elementary value computation. If $etype(e,f) = t$, then

$$d^k.gpr(j_5) = \begin{cases} ba(lv(e,c),c), & R(i) = 0, \\ ba(va(e,c),c), & R(i) = 1 \wedge pointer(t) \wedge va(e,c) \neq null, \\ enc(va(e,c),t), & R(i) = 1 \wedge t \in ET, \\ 0^{32}, & \text{otherwise.} \end{cases}$$

  This definition guides the code generation in a fairly obvious way.
- For statement translation remember
  - preservation of *consis* guides the code generation in a fairly obvious way.
  - because of the mismatch between *pc* semantics and program rest semantics, one has to use the successor function *succ* to argue about jumps out of if-parts and out of the bodies of while loops. This is used in the proof of *pc-consis*.
  - there is a prelude to the code of function *main* setting up pointers and initializing global memory with zeros. It guarantees consistency in the beginning.
- Remember — also for parties — that the non-constructive formulation of the compiler correctness theorem is as useful as the truth about life after death in a computer science textbook. Fortunately the proof is constructive such that the points at which consistency holds can be identified.

The idea of presenting a correctness proof for a compiler in an introductory computer science lecture is not new. It goes back to [LMW89]; indeed Chaps. 11 and 12 of this text are strongly influenced by this old textbook. A formal verification of roughly the compiler presented here is reported in [LPP05]. The formal verification of an optimizing compiler is reported in [Ler09] and subsequent work. Very recently even the formal verification of an optimizing compiler for a multi-threaded *C* dialect (running on multi-core machines of the x86 type) has been achieved [SVZN+13].

12.7 Exercises

1. Consider the following declarations:

```
typedef struct {int a, int b} y;
typedef y[2] x;
x z;
int a;
```

Compute the base addresses $ba(z,c)$ and $ba(a,c)$. Note that base addresses in global memory do not depend on c.

2. Consider the following code:

```
int z;
int a;

int main()
{
   z=4;
   a=10;
   if z>1 {z = f1(z,z-1)};
   return 0
};

int f1(int a, int b)
{
   int z;
   if a>1 {z = (a-1)*f2(a-1,a)} else {z = 1};
   return z
};

int f2(int a, int b)
{
   int z;
   if a>1 {z = (a-1)*f1(a-1,a)} else {z = 1};
   return z
}
```

For each configuration below illustrate the $C0$ configuration and the memory map of the MIPS configuration. For configuration C^x specify $ba(z,C^x)$ and $ba(a,C^x)$.

a) C^0 before the first call of $f1$.
b) C^1 after the first call of $f1$.
c) C^2 after the call of $f1$, $f2$, $f1$.

3. Consider the graphs from Fig. 190. Apply the Sethi-Ullman algorithm to each of the graphs. Specify the moves with the following commands:
   - $set(j,n)$: place pebble j on node n.
   - $remove(j,n)$: remove pebble j from node n.
   - $slide(j,n,n')$: slide pebble j from node n to node n'.

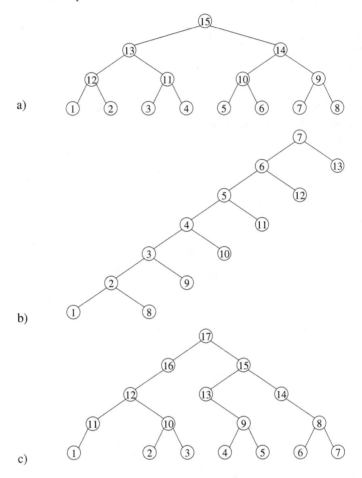

Fig. 190. Example trees for the Sethi-Ullman algorithm

4. We defined $P(n)$ as the smallest number of pebbles which suffices to apply the Sethi-Ullman algorithm successfully on any tree of size n. Show

$$P(n) \geq \log n.$$

Hint: consider a complete binary tree with depth $\log(n)$. Consider the last move before the point when every path from a leaf to the root has a pebble.

5. For the following expressions sketch the derivation tree and determine the number of registers needed for the evaluation of the expression:
   a) $((500+60)-(90*30))+9+10+11$.
   b) $-(((-x_1+(-x_2))*(-(-x_3-x_4)*(-x_5)))/(x_6*(-x_7+x_8)))$.
   c) $-((a+b)*(c+d))/((e-f)*(g-h)-x)$.
   d) $-((-(--a_1+(-b_3)))+((-(x_9+z)*(-a_3))/((m_4+n_3)*-b)))$.

6. Prove Lemma 92.

   Hint: draw the counterpart of Fig. 163 for program $p; p'$.

7. Consider the following declarations:

```
typedef int[3] vector;
vector c;
int a;
int* b;
```

In the body of function main on the right-hand side of the assignment we have:

```
a + b* - c[2]
```

What MIPS code is generated?

8. Assume that a, b, and c are simple local variables of function f and that nxt is a global pointer. What code is generated for the following statements in the body of function f?

   a) `a = a + 1`

   b) `nxt* = c`

   c) `if a==0 {a = b} else {a = -b}`

   d) `while a==1 {a = a}`

9. What MIPS code is generated for the following assignments (assume that the types involved permit the evaluation)?

   a) $e = e_1 < e_2 \;||\; e_1 == 0$

   b) $e = (e_1 - e_2) * e_3$

   c) $e = e_1 * \&$

   d) $e = e_1 * . n\&$

   e) $e = e_1[e_2 + e_3]$

10. Consider the following C0 code:

```
typedef int[3] vector;
typedef vector* vector_p;

int a;
int z;
vector c;
vector_p b;

int main()
{
    a = 3;
    z = 0;
    b = new vector*;
    while z<a {c[z] = z; z = z + 1}
}
```

What MIPS code is generated?

11. Let us change the layout of the stack in the MIPS memory: we switch the locations of result destination and return address for every stack frame. Thus, the return address is now in the word directly below the frame and the return destination is in the second word below. Indicate the changes (if any) in
    a) the base address computation for the function frames.
    b) the definition of compiler consistency.
    c) code generation.

12. Recall Horner's rule, which says that a polynomial $P(x) = \sum_{i=0}^{n} a_i x^i$ can be evaluated by computing $h^{n+1}(x)$:

$$h^0(x) = 0,$$
$$h^{i+1}(x) = a_{n-i} + x \cdot h^i(x).$$

    a) Prove that
    $$P(x) = h^{n+1}(x).$$

    b) Prove that you can use Horner's rule to easily compute the value of a decimal string.

       That is, write a *C0* function *Ecode_ds* that takes as input a pointer p of type *DTEp* from Sect. 11.1.8. Assume that p was generated by a parser for *C0* when parsing a valid *C0* program and $p*.label$ is $\langle DiS \rangle$, deriving a sequence of decimal digits. The function should return a *LELp* such that the labels of the elements of the list are instruction codes that load the binary value (modulo 2^{32}) of that decimal sequence into some register r.

13. Jump distances and code size. Code generation as described here puts limitations on the size of certain portions of programs.
    a) What is the greatest number of memory words over which we can jump with branch instructions?
    b) What is the greatest number of memory words over which we can jump with jump instructions?
    c) What is the largest possible size of the code of if-parts, else-parts, and while bodies, if our code generation is supposed to work?
    d) How far apart can the start addresses of the code of functions be?
    e) How could you overcome these limitations?
       Hint: use *jr* and *jalr* instructions.
    f) Using *jr* and *jalr* instructions everywhere slows down the compiled code. By how much?
       Hint: the register with the jump target must be loaded prior to the jump register instruction.
    g) Assume that you replace one jump instruction by a jump register instruction. Can this force you to do a second such replacement in a different part of the program? Why or why not?

14. We changed the type *DTE* from Sect. 11.1.8 by adding two new components: *.size* and *.pebble*.

- Write a function which takes as input a pointer p to a derivation tree entry which codes an expression node and computes in field *.size* the number of binary operators in the subtree with root $p*$.
  Hint: compute this recursively for all nodes i in the subtree with root $p*$.
- Now write a function which for each node i in the subtree with root $p*$ stores in the field *.pebble* of i the number of the pebble used in the Sethi-Ullman strategy for this node.

When you have solved the last three exercises you should have no difficulty in programming the entire code generation for the compiler.

13

Compiler Consistency Revisited

By now we have laid the foundations which permit us to deal with system programming — here the programming of a kernel — in a precise way. Kernels perform so-called process switches, where they save or restore processor registers in $C0$ variables, which are called process control blocks. In higher-level programming languages such as $C0$ one can only access C variables; the processor registers are not directly accessible. Thus in order to program 'process save' and 'restore' one has to be able to switch from $C0$ to assembly code and back. We call the resulting language '$C0$ + assembly' or for short 'C+A'. A semantics for this language is developed in this chapter.

We begin in Sect. 13.1 by studying a seemingly academic question: given an ISA configuration d, what are all possible well-formed $C0$ configurations c satisfying $consis(c, d)$? We call every such configuration a possible C *abstraction* of d. It turns out that the construction of such C abstractions is unique except for the heap. During the reconstruction of possible C abstractions Lemma 90 from Sect. 12.1.6 is reused for the reconstruction of the program rest. Reachable portions of the heap are found essentially by a graph search starting from pointer variables in the global memory and on the stack. In general, heaps cannot be uniquely reconstructed, but an easy argument shows that different reconstructed heaps are basically isomorphic in the graph theoretic sense. We call $C0$ configurations c and c' equivalent if their heaps are isomorphic and their other components are identical. Then we show: i) computations starting in equivalent configurations continue in equivalent configurations, and ii) expression evaluation in equivalent computations leads to the same result, unless the expression has pointer type.

This result opens the door to define in Sect. 13.3 the semantics of C+A in a quite intuitive way. As long as abstracted C code is running, we run the ISA computation and C abstractions in lockstep. If the computation switches to inline assembly or jumps completely outside of the C program (this happens if a kernel starts a user process), we simply drop the C abstraction and continue with ISA semantics. When we return to translated C code we reconstruct a C abstraction at the earliest possible ISA step. Which C abstraction we choose does not matter: the value of expressions which don't have pointer type does not depend on the (nondeterministic) choice.

© Springer International Publishing Switzerland 2016
W.J. Paul et al., *System Architecture*, DOI 10.1007/978-3-319-43065-2_13

Pointer chasing on the heap is also known from garbage collection. For interested readers we have therefore included in Sect. 13.2 a short treatment of garbage collection, although we will not make use of it later.

Finally, there is a situation in which a user process, whose program lies outside of the kernel, returns control to some instruction in the middle of the kernel, and the kernel performs the aforementioned 'process save'. This will turn out to be very similar to a 'goto', a construct absent in the original definition of the structured programming language $C0$. In preparation for the later treatment of this situation, in the exercises we augment $C0$ with labels and gotos. Compilation is trivial. Providing small-step semantics for it is not, but the problem is again easily solved with Lemma 90.

In the classroom we only treat Sects. 13.1 and 13.3. The lemmas stating that computations starting in equivalent configurations continue in equivalent configurations and that expression evaluation in equivalent computations leads to the same result (unless the expression has pointer type) are stated but not proven in the classroom, because their proofs are of the bookkeeping type.

13.1 Reconstructing a Well-Formed $C0$ Configuration

We begin this chapter with a technical result: given a MIPS configuration d, we characterize the set of well-formed $C0$ configurations c satisfying $consis(c,d)$, i.e., which can be abstracted from the MIPS configuration as a possible original $C0$ configuration under the translation process, if such configurations exist. Later this result will be crucial for defining the semantics of $C0$ programs with inline assembly portions.

Recall that well-formed c configurations without ghost components were defined in Sect. 11.3.5 by predicate inv-$conf(c)$ which comprises the conditions:

$$
\begin{aligned}
inv\text{-}conf(c) \equiv\ & tc(c) \land \\
& p\text{-}targets(c) \land \\
& inv\text{-}pr(c) \land \\
& inv\text{-}rds(c).
\end{aligned}
$$

In Chap. 12 we augmented configurations by ghost components $c.prn$ and $c.clr$. In Sect. 12.1.5 the ghost program rest component $c.pr$ was coupled with the program rest nodes component $c.prn$ using the very simple invariant inv-prn-$pr(c)$ and invariant inv-$pc(c)$ translated into inv-$prn(c)$. For the caller stack we noted invariant inv-$clr(c)$. In Sect. 12.1.6 we related program rest nodes, caller stack, and derivation tree by the crucial Lemma 90 and collected the statement of this lemma into invariant inv-prn-$clr(c)$. We call c configurations with ghost components well-formed if they obey all the invariants identified above:

$$wfc(c) \equiv inv\text{-}conf(c) \wedge$$
$$inv\text{-}prn\text{-}pr(c) \wedge$$
$$inv\text{-}prn(c) \wedge$$
$$inv\text{-}clr(c) \wedge$$
$$inv\text{-}prn\text{-}clr(c).$$

In a nutshell we have just collected all lemmas which we have shown to hold for all configurations of any C0 computation into the well-formedness predicate. Thus, intuitively, a single configuration c is well-formed if it could have been produced from an initial configuration c^0 by a C0 computation.

13.1.1 Associating Code Addresses with Statements and Functions

We are given a MIPS configuration d. We know the derivation tree $T = (A, \ell)$ of the translated program and the parameters like a, b, $sbase$, etc., of the translation. We are looking for well-formed C0 configurations c satisfying $consis(c,d)$, and we assume that such a configuration exists.

Recall that the set of statement nodes in T was defined as the set of nodes labeled St or rSt.

$$stn = \{i \in A \mid \ell(i) \in \{St, rSt\}\}.$$

The subset of the statement nodes generating function calls is denoted by $calln$ and the set of nodes generating conditional statements is denoted by $condn$. Formal definitions are left as an exercise. For any statement node $n \in stn$ we generated code $code(n)$ occupying in the MIPS memory $d.m$ addresses from $start(n)$ to $end(n)$:

$$d.m_{4 \cdot |code(n)|}(start(n)) = code(n),$$

and we call the set of addresses occupied by $code(n)$ the code region of node n:

$$coder(n) = [start(n) : end(n)].$$

In particular we know by $code\text{-}consis(c,d)$ that the code $code(\varepsilon)$ of the translated program occupies address range

$$[a : b] = [start(\varepsilon) : end(\varepsilon)]$$

of the MIPS memory $d.m$, s.t.

$$d.m_{4 \cdot |code(\varepsilon)|}(start(\varepsilon)) = code(\varepsilon).$$

Code regions of different nodes are nested or disjoint.

Lemma 101.

$$n \neq n' \rightarrow (coder(n) \subseteq coder(n') \vee$$
$$coder(n) \supseteq coder(n') \vee$$
$$coder(n) \cap coder(n') = \emptyset).$$

Nesting occurs in two situations:

- the code of while loops includes the code of the statements of the bodies.
- the code of conditional statements includes the code for the if-part and (if present) the else-part.

Now one easily concludes the following.

Lemma 102.

1. *The code of different statement nodes has different start addresses.*

$$n \neq n' \rightarrow start(n) \neq start(n').$$

2. *If the code regions of statements share the same end address, one of them is a conditional statement.*

$$end(n) = end(n') \rightarrow (n \in condn \vee n' \in condn).$$

3. *The code of different call statements has different end addresses.*

$$n, n' \in calln \wedge n \neq n' \rightarrow end(n) \neq end(n').$$

The proof of the lemmas is left as an exercise. We define the set of start addresses of code regions and the set of end addresses of code regions of function calls as

$$START = \{start(n) \mid n \in stn\},$$
$$ENDC = \{end(n) \mid n \in calln\}.$$

The last lemma permits one for any address $x \in START$ to identify the statement node $stn(x)$ from which the code starting at address x was translated:

$$stn(x) = \varepsilon\{n \in stn \mid start(n) = x\}.$$

This is illustrated in Fig. 191.

Also we can identify for any address $y \in ENDC$ the node $calln(y)$ generating the function call whose code ends at address y

$$calln(y) = \varepsilon\{n \in calln \mid end(n) = y\}.$$

Every statement node n belongs to the body of a unique function f, i.e., it satisfies $n \in f$ for a unique function f which was denoted in Sect. 11.3.3 as $fu(n)$. Therefore we can determine for each start address $x \in START$ the name $fun(x)$ of the function from whose body the ISA instruction at address x was translated:

$$fun(x) = fu(stn(x)).$$

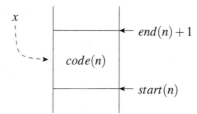

Fig. 191. Reconstructing the node *n* in the derivation tree from which the code at an address *x* was generated

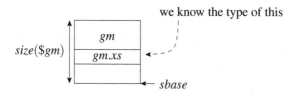

Fig. 192. Reconstructing the global memory by means of *sbase* and $gm

13.1.2 Reconstructing Everything Except Heap and Pointers

Values of Elementary Subvariables of gm

By *bsp-consis*(*c*, *d*) and the definition of base addresses we have for the base pointer general-purpose register *bpt*:

$$ba(gm, c) = sbase = d.gpr(bpt_5).$$

The type $gm and the size *size*($gm) of the global memory are known from the *C*0 program. By this, we know the memory region in *d* where the global memory resides (see Fig. 192). By *tc*(*c*) we know the simple subvariables *gm.xs* of the global memory as well as their base addresses and types (which depend only on *ba*(*gm*, *c*) and $gm).

$$ba = ba(gm.xs, c),$$
$$t = vtype(gm.xs, c).$$

If *t* is elementary we use *e-consis*(*c*, *d*) and reconstruct the values in *C*0 by inverting function *enc*:

$$va(gm.xs, c) = c.m(gm.xs) = enc^{-1}(d.m_4(ba), t).$$

The reconstruction of pointer values is done later together with the reconstruction of the heap.

The First Program Rest Node prn(1)

By *pc-consis*(*c*, *d*) we have

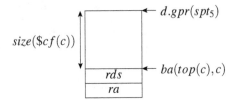

Fig. 193. Reconstructing the top-most stack frame by means of spt and the current function $cf(c)$

$$start(c.prn(1)) = d.pc.$$

Thus $d.pc$ is the start address of the code generated by the first program rest node, which can be identified as

$$c.prn(1) = stn(d.pc).$$

Function Name and Base Address of the Top Stack Frame

We do not yet know the recursion depth $c.rd$, but by $inv\text{-}prn(c)$ we know that the function $fu(c.prn(1))$ generating program rest node $c.prn(1)$ is the function of the top stack frame, i.e., the function $cf(c)$ currently executed:

$$c.st(c.rd) = cf(c) = fu(c.prn(1)) = fun(d.pc).$$

From the type table we get type $\$cf(c)$ and its size $size(\$cf(c))$. We find the base address of the top frame using $bsp\text{-}consis(c,d)$ for the stack pointer register spt:

$$ba(top(c),c) = d.gpr(spt_5) -_{32} size(\$cf(c))_{32},$$

as illustrated in Fig. 193.

Function Name and Base Address of Lower Stack Frames and the Caller Stack

Assume we know for some index $i \in [1 : c.rd]$ the function name $c.st(i)$ and base address $ba(ST(i,c),c)$ of the stack frame $ST(i,c)$ at recursion depth i (although we may not know index i yet). As illustrated in Fig. 194 two words below the base address, i.e., at

$$ra = d.m_4(ba(ST(i,c),c) -_{32} 8_{32}),$$

we find the address of the next MIPS instruction executed after the return from the function. From $clr\text{-}consis(c,d)$ we get

$$ra = end(c.clr(i)) +_{32} 1_{32},$$

thus

$$ra -_{32} 1_{32} = end(c.clr(i)).$$

From this address we find the statement node which generated the call statement as

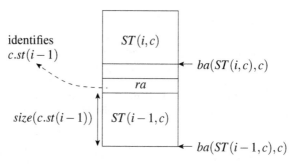

Fig. 194. Using the return address and base address of a frame to find the base address and function of the frame below

$$c.clr(i) = calln(ra -_{32} 1_{32}).$$

By invariant $inv\text{-}clr(c)$ this is a node of the body of function $c.st(i-1)$, thus we can determine the function name of the stack frame below $ST(i,c)$ as

$$c.st(i-1) = fu(c.clr(i)).$$

Using $clr\text{-}consis(c,d)$ we determine the base address of the newly found frame as

$$ba(c.st(i-1),c) = ba(c.st(i),c) -_{32} (size(\$c.st(i)) + 8)_{32}.$$

We do this reconstruction from top to bottom for $j = 0, 1, \ldots$ and indices

$$i = c.rd - j.$$

We stop the reconstruction when the base address $ba(c.st(i),c)$ of a frame we have found is two words above the end of the global memory

$$ba(c.st(i),c) = sbase +_{32} (size(\$gm) + 8)_{32},$$

because then $i = 0$ and we have

$$c.rd = j.$$

Elementary Local Subvariables and Parameter Values

Since we know the base addresses and types of $C0$ stack frames, we can compute the base addresses and types of all local variables and parameters. Using $tc(c)$ and $e\text{-}consis(c,d)$, we reconstruct the values of elementary local subvariables and parameters just as we reconstructed the elementary subvariables of the global memory.

Program Rest Nodes c.prn and Program Rest c.pr

We have already reconstructed the first program rest node $c.prn(1)$ as well as the entire caller stack $c.clr$. Reconstruction of the program rest follows directly from the nontrivial Lemma 90, resp. invariant $inv\text{-}prn\text{-}clr(c)$:

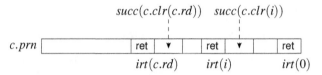

Fig. 195. Illustrating the $c.rd + 1$ return statements in the program rest nodes of c and the correspondence of the nodes $c.clr(i)$ on the caller stack to the nodes right after the returns

$$c.prn(j+1) = \begin{cases} succ(c.prn(j)), & \ell(c.prn(j)) \neq rSt, \\ succ(c.clr(i)), & j = irt(i,c), \end{cases}$$

as illustrated in Fig. 195. Reconstruction of the instructions

$$c.pr[i] = bw(c.prn(i))$$

follows directly from the trivial invariant *inv-prn-pr(c)*.

13.1.3 Reachable Subvariables

We have defined the set $SV(c)$ of subvariables of a C0 configuration c. During a computation, variables or subvariables on the heap may become unreachable for future computation. As a simple example consider the program from Sect. 11.6.4 where we constructed a linked list with pointers *first* and *last* to the first and last list elements. If we assign to these pointers the value *null*, then the list elements remain on the heap, but in the absence of address arithmetic we cannot access these heap variables any more.

We could also declare a pointer variable p of type $v = u*$ where $u = LEL*$, i.e., p is a pointer to a list element pointer, store in it a pointer to the next component of the first list element and then set pointer *first* to *null*.

```
v p;
p = first*.next&;
first = null;
```

Now the *content* component of the first list element is not reachable any more, but the *next* component stays reachable via pointer p.

We proceed to define the set *reach(c)* of reachable subvariables of a C0 configuration c as the union of sets *reach(i,c)*, s.t. subvariables in *reach(i,c)* can be reached from subvariables in the global memory or on the stack by chasing pointers at most i times:

- global memory and stack frames are reachable

$$reach'(0,c) = \{gm\} \cup \{ST(j,c) \mid j \in [0:c.rd]\}.$$

- subvariables of reachable variables are reachable

$$reach(i,c) = \{y \mid \exists x \in reach'(i,c) : subvar(y,x)\}.$$

- the reachable pointer variables in $reach(i,c)$ which are not *null* are

$$rpointer(i,c) = \{y \in reach(i,c) \mid pointer(y) \wedge c.m(y) \neq null\}.$$

- chasing such pointer variables gives new reachable subvariables

$$reach'(i+1,c) = \{c.m(y) \mid y \in rpointer(i,c)\} \cup reach(i,c).$$

- no other subvariables are reachable

$$reach(c) = \bigcup_{i=0}^{\infty} reach(i,c).$$

An easy argument shows that only finitely many sets $reach(i,c)$ have to be considered.

Lemma 103.
$$\exists R \in \mathbb{N} : reach(c) = \bigcup_{i=0}^{R} reach(i,c).$$

Proof. If new subvariables are found in step i, i.e., if

$$reach(i+1,c) \setminus reach(i,c) \neq \emptyset,$$

then a pointer p which was not yet reachable in step $i-1$ was followed in step i:

$$p \in rpointer(i,c) \setminus reach(i-1,c).$$

As a *C*0 configuration c contains only finitely many simple subvariables this can only happen finitely often. Thus there exists R with

$$reach(R+1,c) \setminus reach(R,c) = \emptyset$$

and no more new pointers are found.

For technical reasons we also collect the return destinations in the set

$$rdsreach(c) = \{c.rds(i) \mid i \in [1 : c.rd]\}.$$

These subvariables all have simple type, but they might not be reachable. Imagine a situation where function $f = ST(i,c)$ was called using global pointer variable g specifying a result destination on the heap, i.e., $bw(c.clr(i)) \equiv g* = f(\ldots)$ and $onheap(c.rds(i),c)$. By changing the value of g, heap variable $c.rds(i)$ may become unreachable subsequently.

A very easy induction on the structure of expressions shows that left values obtained in expression evaluations are reachable.

Lemma 104. *If $lv(e,c)$ exists, then*

$$lv(e,c) \in reach(e,c).$$

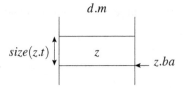

Fig. 196. Implementation subvariable $z = (z.ba, z.t)$ in the memory of MIPS configuration d

13.1.4 Implementation Subvariables

In what follows we aim at the reconstruction of the reachable portion of the heap. Again we will proceed by pointer chasing. In the course of this process we will identify in the MIPS configuration d what we call *implementation subvariables*. Implementation subvariables z have two components:

- a base address $z.ba \in \mathbb{B}^{32}$,
- a type $z.t \in ET \cup TN$.

This is illustrated in Fig. 196. An implementation subvariable z is an *implementation pointer* if its type is a pointer type. It is simple, resp. elementary, if its type is simple, resp. elementary:

$$ipointer(z) \equiv pointer(z.t),$$
$$isimple(z) \equiv simple(z.t),$$
$$iET(z) \equiv z.t \in ET.$$

Every C0 subvariable $x \in SV(c)$ is implemented by an implementation subvariable $is_c(x)$ defined by

$$is_c(x).ba = ba(x, c),$$
$$is_c(x).t = vtype(x, c).$$

A very simple argument shows that this mapping is injective.

Lemma 105.
$$x \neq y \rightarrow is_c(x) \neq is_c(y).$$

Proof. If the base addresses of x and y differ we immediately get

$$is_c(x).ba = ba(x, c) \neq ba(y, c) = is_c(y).ba$$

and we are done. Otherwise we apply Lemma 88 to conclude

$$is_c(x).t = vtype(x, c) \neq vtype(y, c) = is_c(y).t.$$

Let

$$IS(c) = is_c(SV(c)) = \{is_c(x) \mid x \in SV(c)\}$$

be the set of implementation subvariables obtained from the subvariables of c in this way. Then

$$is_c : SV(c) \rightarrow IS(c)$$

is bijective, and we can define the inverse function

$$sv_c = is_c^{-1}$$

associating with implementation subvariable $is \in IS(c)$ the $C0$ subvariable $x = sv_c(is)$ which is implemented by is:

$$sv_c(is_c(x)) = x.$$

From now on we only consider implementation subvariables in $IS(c)$. Subvariables zs of implementation subvariables z (overloading notation we denote this by $subvar(zs, z)$) are defined in the obvious way:

- every implementation subvariable is a subvariable of itself

$$subvar(z, z).$$

- if z has an array type, then array components $z[i]$ are subvariables

$$z.t = t[n] \rightarrow \forall i \in [0 : n-1] : subvar(z[i], z),$$

where

$$z[i].ba = z.ba +_{32} displ(i, z.t)_{32},$$
$$z[i].t = t.$$

- if z has a struct type, then struct components $z.n_i$ are subvariables

$$z.t = \{t_1\ n_1; \ldots; t_s\ n_s\} \rightarrow \forall i \in [1 : s] : subvar(z.n_i, z),$$

where

$$z.n_i.ba = z.ba +_{32} displ(n_i, z.t)_{32},$$
$$z.n_i.t = t_i.$$

- this can be applied recursively

$$subvar(zs', zs) \wedge subvar(zs, z) \rightarrow subvar(zs', z).$$

Trivial induction gives the following for selector sequences s.

Lemma 106.

$$zs.ba = z.ba +_{32} displ(s, z.t)_{32},$$
$$zs.t = ctype(s, z.t).$$

Application of Lemma 83 immediately allows us to conclude for selector sequences s: if z implements x, then zs implements xs

$$is_c(x) = z \;\rightarrow\; is_c(xs) = zs.$$

Lemma 107.

$$is_c(xs) = is_c(x)s.$$

Given implementation subvariables

$$z = is_c(x)$$

that we can observe in MIPS configuration d, we try to reconstruct the corresponding $C0$ variables

$$x = sv_c(z)$$

and the values $c.m(xs)$ of their simple subvariables. For this purpose we consider subvariables zs of z implementing subvariables xs of x and satisfying

$$is_c(xs) = zs.$$

Now suppose subvariable xs is simple and let

$$ba = ba(xs,c) = zs.ba,$$
$$t = vtype(xs,c) = zs.t.$$

If xs is elementary we use tc and $e\text{-}consis(c,d)$, and we conclude — as in the reconstruction of gm and the stack frames —

$$d.m_4(ba) = enc(c.m(xs),t).$$

Thus we reconstruct the current $C0$ value of xs as

$$c.m(xs) = enc^{-1}(d.m_4(ba),t). \tag{10}$$

Now consider Fig. 197. If xs is a pointer with a type

$$t = t'*$$

pointing to $C0$ variable

$$y = c.m(xs)$$

which has by $p\text{-}targets(c)$ type

$$vtype(y,c) = t',$$

then by $tc(c)$ implementation subvariable

$$zs = (ba,t)$$

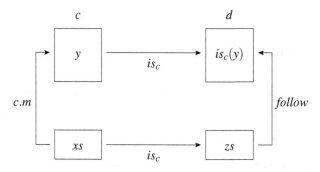

Fig. 197. Implementations of subvariables xs and y, where xs points to y

is an implementation pointer, and application of $p\text{-}consis(c,d)$ gives

$$d.m_4(ba) = ba(y,c).$$

We define the implementation subvariable $follow(zs)$ obtained by following implementation pointer zs by

$$follow(zs).ba = d.m_4(ba),$$
$$follow(zs).t = t'$$

and immediately conclude as follows.

Lemma 108. *If xs points to y, then following the implementation of xs gives the implementation of y*

$$zs = is_c(xs) \wedge c.m(xs) = y \ \rightarrow \ follow(zs) = is_c(y).$$

With a very small extra argument we can show the converse of this lemma.

Lemma 109. *If following the implementation of xs gives the implementation of y, then xs points to y*

$$zs = is_c(xs) \wedge follow(zs) = is_c(y) \ \rightarrow \ c.m(xs) = y.$$

Proof. Let xs point to y', i.e.,

$$c.m(xs) = y'.$$

Then by the previous Lemma 108 and the hypothesis we have

$$is_c(y') = follow(zs) = is_c(y).$$

We conclude

$$y = y'$$

by the injectivity of function is_c which was shown in Lemma 105.

Now the road is free: not for the reconstruction of $reach(c)$ but at least for its implementation $is_c(reach(c))$.

- We start with the implementations of global memory gm and the stack frames $ST(i,c)$:

$$ireach'(0) = \{is_c(gm)\} \cup \{is_c(ST(i,c)) \mid i \in [0:c.rd]\}.$$

- Subvariables of reachable implementation subvariables are reachable:

$$ireach(i) = \{y \mid \exists z \in ireach'(i) : subvar(y,z)\}.$$

- Reachable implementation pointers in $ireach(i)$ which are not *null* are

$$ripointer(i) = \{z \in ireach(i) \mid ipointer(z) \wedge d.m_4(z.ba) \neq 0_{32}\}.$$

- Following such implementation pointers gives new reachable implementation subvariables:

$$ireach'(i+1) = \{follow(z) \mid z \in ripointer(i)\} \cup ireach(i).$$

- No other implementation subvariables are reachable:

$$ireach = \bigcup_{i=0}^{\infty} ireach(i).$$

Using Lemma 108 and straightforward induction we get the following.

Lemma 110.
$$\forall i : ireach(i) = is_c(reach(i,c)).$$

We conclude
$$ireach = \bigcup_{i=0}^{R} ireach(i).$$

Thus we have reconstructed the set *ireach* of implementations of the reachable subvariables of $C0$ configuration c as well as the values $c.m(xs)$ of elementary subvariables. Moreover we have identified the reachable implementation pointers. For technical reasons we also have to identify the implementations of the return destinations:

$$irdsreach = is_c(rdsreach(c)).$$

Reconsider Fig. 194. We use $rds\text{-}consis(c,d)$ to conclude

$$is_c(c.rds(i)).ba = d.m_4(ba(ST(i,c),c) -_{32} 4_{32}).$$

By context condition 28 we get the type of the return result from the type component of the function table

$$is_c(c.rds(i)).t = ft(c.st(i)).t.$$

Thus we have the following.

Lemma 111.
$$is_c(c.rds(i)) = (d.m_4(ba(ST(i,c),c) -_{32} 4_{32}), ft(c.st(i)).t).$$

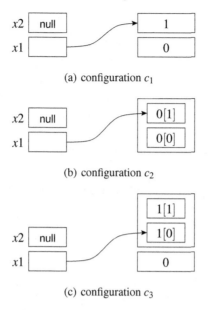

(a) configuration c_1

(b) configuration c_2

(c) configuration c_3

Fig. 198. $C0$ configurations resulting in the same reachable portion of the heap

13.1.5 Heap Reconstruction Is not Unique

We illustrate with a simple example that the reachable portion $reach(c)$ of the heap cannot be reconstructed uniquely. In Sect. 13.1.6 we will then show that reconstruction of $reach(c)$ is only possible up to what we call *heap isomorphisms*.

Consider a program with the following declarations:

```
typedef int* intp;

typedef int[2] array;
typedef array* arrayp;

intp x1;
arrayp x2;
```

Then each of the following three statement sequences produces on the heap a single implementation subvariable

$$z = (z.ba, z.t) = (hbase +_{32} 4_{32}, int)$$

which is reached via pointer variable $x1$:

```
x1 = new int*;
x1 = new int*;
```

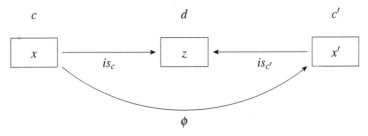

Fig. 199. Translating subvariables x of configuration c into subvariables x' of configuration c'

- the resulting configuration c_1 is illustrated in Fig. 198(a), and we have

$$va(x1, c_1) = 1.$$

```
x2 = new array*;
x1 = x2*[1]&;
x2 = null;
```

- the resulting configuration c_2 is illustrated in Fig. 198(b), and we have

$$va(x1, c_2) = 0[1].$$

```
x1 = new int*;
x2 = new array*;
x1 = x2*[0]&;
x2 = null;
```

- the resulting configuration c_3 is illustrated in Fig. 198(c), and we have

$$va(x1, c_3) = 1[0].$$

In all three configurations c_i we have

$$va(x2, c_i) = null.$$

13.1.6 Heap Isomorphisms and Equivalence of $C0$ Configurations

Let c and c' be two well-formed $C0$ configurations for the same program which are both consistent with *MIPS* configuration d, i.e.,

$$consis(c, d) \wedge consis(c', d).$$

Then, as shown in Fig. 199, we translate subvariables $x \in reach(c) \cup rdsreach(c)$ directly into their counterparts $x' \in reach(c') \cup rdsreach(c')$ which share the same implementation z.

We proceed with studying the properties of the bijective mapping

$$\phi : reach(c) \cup rdsreach(c) \to reach(c') \cup rdsreach(c'),$$

which is defined by

$$\phi(x) = is_{c'}^{-1}(is_c(x)).$$

On the stack and global memory the mapping is the identity, because there the reconstruction of subvariables was unique. Also the mapping trivially preserves type. Hence the following is trivial.

Lemma 112.

$$ingm(x,c) \vee onstack(x,c) \to \phi(x) = x$$

and

$$vtype(\phi(x),c') = vtype(x,c).$$

The following lemma states that mapping ϕ is consistent with the formation of subvariables.

Lemma 113.

$$\phi(xs) = \phi(x)s.$$

Proof. Let

$$\phi(x) = is_{c'}^{-1}(is_c(x)) = x'.$$

Applying $is_{c'}$ gives

$$is_c(x) = is_{c'}(x').$$

Applying Lemma 107 twice gives

$$is_c(xs) = is_c(x)s$$
$$= is_{c'}(x')s$$
$$= is_{c'}(x's).$$

Applying $is_{c'}^{-1}$ now proves the lemma.

Because reconstruction of values of elementary subvariables is unique we can conclude that elementary subvariables x and their translations $\phi(x)$ have the same value.

Lemma 114.

$$vtype(x,c) \in ET \to c'.m(\phi(x)) = c.m(x).$$

Proof. Let

$$vtype(x,c) = t = vtype(\phi(x),c')$$

and

$$is_c(x) = z = is_{c'}(\phi(x)).$$

Applying Equation 10 twice gives

$$c.m(x) = enc^{-1}(d.m_4(z.ba),t) = c'.m(\phi(x)).$$

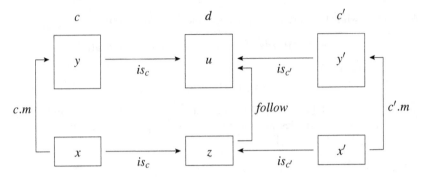

Fig. 200. Chasing a pointer x in configuration c and its translation x' in configuration c'

If a pointer x points in configuration c to subvariable y, then its translation $\phi(x)$ points in configuration c' to the translation $\phi(y)$ of y.

Lemma 115.

$$pointer(x,c) \wedge c.m(x) = y \rightarrow c.m(\phi(x)) = \phi(y).$$

Proof. Consider Fig. 200. Let

$$is_c(x) = z = is_{c'}(x')$$

and

$$is_c(y) = u = is_{c'}(y').$$

We conclude from Lemma 108 that following implementation pointer z gives implementation variable u

$$follow(z) = u.$$

Then we conclude with Lemma 109 that x' points to y'

$$c'.m(x') = y'.$$

As $x' = \phi(x)$ and $y' = \phi(y)$ this is the desired result.

Finally, using Lemma 111 we conclude the following for the return destinations.

Lemma 116.

$$\phi(c.rds(i)) = c'.rds(i).$$

Proof. Lemma 111 gives

$$is_c(c.rds(i)).t = ft(c.st(i)).t = ft(c'.st(i)).t = is_{c'}(c'.rds(i)).t$$

and

$$is_c(c.rds(i)).ba = d.m_4(ba(ST(i,c),c) -_{32} 4_{32})$$
$$= d.m_4(ba(ST(i,c'),c') -_{32} 4_{32})$$
$$= is_{c'}(c'.rds(i)).ba.$$

Thus

$$is_c(c.rds(i)) = is_{c'}(c'.rds(i)).$$

Application of $is_{c'}^{-1}$ gives the lemma.

We call a bijective mapping ϕ which satisfies Lemmas 112–116 a *heap isomorphism* between c and c'. An easy exercise shows the following.

Lemma 117.

- *The identity mapping is a heap isomorphism.*
- *If ϕ is a heap isomorphism, then so is ϕ^{-1}.*
- *If ϕ and ϕ' are heap isomorphisms, then so is $\phi' \circ \phi$.*

We call configurations c and c' equivalent and write $c \sim c'$ if the following conditions hold:

- Components X of the configurations that can be uniquely reconstructed are identical
$$X \in \{rd, st, pr, prn, clr\} \quad \rightarrow \quad c.X = c'.X.$$

- There is a heap isomorphism between c and c'.

From Lemma 117 we immediately conclude the following.

Lemma 118. $\sim$ *is an equivalence relation.*

The results obtained in this section so far are summarized as follows.

Lemma 119. *If c and c' are both well-formed $C0$ configurations for the same program and both are consistent with MIPS configuration d, then they are equivalent, i.e.,*

$$c \sim c'.$$

So reconstruction of consistent $C0$ configurations from MIPS configurations d is unique only up to heap isomorphisms.

13.1.7 Computations Starting in Equivalent Configurations

We proceed to show:

- in equivalent configurations c and c' expressions of elementary type have the same value.
- computations starting in equivalent $C0$ configurations continue in equivalent configurations.

This means: as long as we consider pointer values only as intermediate results it does not matter which one of the equivalent $C0$ configurations c we choose for the reconstruction: continuing the computation from c always gives the same results of elementary type. Technically we prove two lemmas.

Lemma 120. *Assume $c \sim c'$ related by heap isomorphism ϕ. Let $e \in f$ be an expression with simple type, i.e.,*

$$simple(etype(e,f)).$$

Then

- *if $lv(e,c)$ is defined, then*

$$lv(e,c') = \phi(lv(e,c)).$$

-

$$va(e,c') = \begin{cases} va(e,c), & etype(e,f) \in ET, \\ \phi(va(e,c)), & pointer(etype(e,f)). \end{cases}$$

Proof. The lemma is proven by an obvious induction on the structure of expressions. The bases cases are:

- constant evaluation, which is independent of c. Thus the lemma trivially holds.
- variable binding. Names x are bound to subvariables on the stack or in global memory:

$$onstack(lv(x,c)) \vee ingm(lv(x,c)).$$

By the properties formulated in Lemma 112, ϕ is the identity mapping for such subvariables:

$$lv(x,c') = lv(x,c) = \phi(lv(x,c)).$$

The interesting cases of the induction step are:

- computing the value $va(e,c)$ from a left value $x = lv(e,c)$. By induction hypothesis we have

$$lv(e,c) = x \wedge lv(e,c') = \phi(x).$$

If $etype(e,f)$ is elementary, we get from the properties formulated in Lemma 114

$$\begin{aligned} va(e,c') &= c'.m(\phi(x)) \\ &= c.m(x) \\ &= va(e,c). \end{aligned}$$

If $etype(e,f)$ is a pointer type, we get from the properties formulated in Lemma 115

$$\begin{aligned} va(e,c') &= c'.m(\phi(x)) \\ &= \phi(c.m(x)) \\ &= \phi(va(e,c)). \end{aligned}$$

- left value of struct components. Let $e = e'.n_i$. By induction hypothesis we have

$$lv(e',c') = \phi(lv(e',c)).$$

We get from the properties formulated in Lemma 113

$$\begin{aligned}
lv(e,c') &= lv(e',c').n_i \\
&= \phi(lv(e',c)).n_i \\
&= \phi(lv(e',c).n_i) \\
&= \phi(lv(e,c)).
\end{aligned}$$

- left value of an array element. Let $e = e'[e'']$. By induction hypothesis we get

$$lv(e',c') = \phi(lv(e',c)) \wedge va(e'',c') = va(e'',c).$$

Again we get from the properties formulated in Lemma 113

$$\begin{aligned}
lv(e,c') &= lv(e',c')[va(e'',c')] \\
&= \phi(lv(e',c))[va(e'',c)] \\
&= \phi(lv(e',c)[va(e'',c)]) \\
&= \phi(lv(e,c)).
\end{aligned}$$

- comparison of pointers. Suppose expression e is $e' == e''$, where e' and e'' have by the context condition the same pointer type:

$$etype(e',f) = etype(e'',f) = t' *.$$

We use the injectivity of ϕ to conclude the third line below and get

$$\begin{aligned}
va(e,c') &\leftrightarrow va(e',c') = va(e'',c') \\
&\leftrightarrow \phi(va(e',c)) = \phi(va(e'',c)) \\
&\leftrightarrow va(e',c) = va(e'',c) \\
&\leftrightarrow va(e,c).
\end{aligned}$$

We omit the remaining cases, which are much simpler.

Lemma 121.

$$c \sim c' \ \rightarrow \ \delta_C(c) \sim \delta_C(c').$$

Proof. The lemma is proven by the obvious case split on the first instruction of the program rest, which belongs to some function f. For the configurations after execution of this statement we use the abbreviations

$$u = \delta_C(c) \quad \text{and} \quad u' = \delta_C(c')$$

as illustrated in Fig. 201.

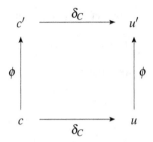

Fig. 201. Configurations c and c' and their successor configurations u and u' related by heap isomorphism ϕ. If configuration u has more variables then c, then ϕ has to be extended; if it has fewer variables, then ϕ is restricted for the new configurations

- assignment $e = e'$: in configurations c and c' we update subvariables $lv(e,c)$ and $lv(e,c')$ with values $va(e',c)$ and $va(e',c')$, resp. If $etype(e',f)$ is elementary we get by Lemma 120 for the configurations u and u':

$$va(e,u') = va(e',c')$$
$$= va(e',c)$$
$$= va(e,u).$$

This preserves the property stated in Lemma 114. If $etype(e,f)$ is a pointer type we get

$$va(e,u') = va(e',c')$$
$$= \phi(va(e',c))$$
$$= \phi(va(e,u)).$$

This preserves the property stated in Lemma 115.
- conditional statements or while statements: let expression e be the branch condition involved. It has Boolean type, and by Lemma 120 we conclude

$$va(e,c) = va(e,c').$$

This implies that program rest nodes (and hence the program rests) in u and u' are identical:

$$u'.prn = u.prn.$$

- new statement $e = new\ t*$: this creates in configurations u resp. u' the new heap variable $c.nt$, resp. $c'.nt$. One extends the heap isomorphism ϕ in the obvious way by

$$\phi(c.nt) = c'.nt$$

and then treats the following assignment like an ordinary assignment.

- function call $e = f(e_1, \ldots, e_p)$: the same stack frame is added in both configurations. On the new subvariables the extended heap isomorphism ϕ must be the identity mapping. Parameter passing is treated like a sequence of assignments. For the return destination we know from Lemma 120:

$$lv(e, c') = \phi(lv(e, c)).$$

Let

$$c.rd + 1 = i = c'.rd + 1$$

be the index of the new top frame. For the new item on the *rds*-stacks we know:

$$\begin{aligned} u'.rds(i) &= lv(e, c') \\ &= \phi(lv(e, c)) \\ &= \phi(u.rds(i)). \end{aligned}$$

This preserves the property stated in Lemma 116.

- return statement *return e*: In configuration c this is an assignment of $va(e, c)$ to subvariable $c.rds(c.rd)$; in configuration c' this is an assignment of value $va(e, c')$ to subvariable $c'.rds(c'.rd)$. By the property stated in Lemma 116 we have:

$$\begin{aligned} c'.rds(c'.rd) &= c'.rds(c.rd) \\ &= \phi(c.rds(c.rd)). \end{aligned}$$

We can finish this case like an assignment statement (plus the obvious restriction of the heap isomorphism ϕ).

13.2 Garbage Collection

If variables or subvariables on the heap become unreachable, then space is wasted in the implementation, because the machine memory $d.m$ contains locations that will never again be accessed. Garbage collection is a process where a C0 configuration c is replaced by an equivalent configuration c' in which all subvariables on the heap are reachable. If we implement c' with the base address function from Sect. 12.1, then no space is wasted on the heap. Garbage collection might be invoked when the execution of a 'new' statement fails in an error situation (the *sysc* statement generated in Sect. 12.3.2), because insufficient storage space is available on the heap. The garbage collection algorithm[1] proceeds in three stages:

- We start with configurations d and c satisfying

$$consis(c, d).$$

By pointer chasing from MIPS configuration d we identify the subvariables in $reach(c)$ up to heap isomorphism.

---

[1] The application of this algorithm to compiled C0 code running under an operating system kernel is treated as an exercise in Chap. 15.

$$d \quad \underline{\quad consis \quad} \quad c \quad \sim \quad c' \quad \underline{\quad consis \quad} \quad d'$$

garbage collection

Fig. 202. Structure of garbage collection

- We choose a particular heap isomorphism ϕ and a new $C0$ configuration

$$c' \sim c .$$

In the new configuration all heap variables are reachable, however it will in general not obey $consis(c', d)$ because implementation variables on the heap are now allocated to the wrong addresses. With the help of the usual base address function ba we then reallocate the heap variables, obtaining a new MIPS configuration d' which is consistent with c',

$$consis(c', d').$$

This is illustrated in Fig. 202. Because only space for reachable subvariables is allocated, the heap of d' will not be larger than the heap of d.

- Using properties of compiler consistency and heap isomorphism we show in a bookkeeping exercise how to calculate d' from d. In a nutshell we will i) copy reachable subvariables of the heap to their new locations on the heap, and ii) adjust pointers such that they point to the new locations.

From $c \sim c'$ we conclude with Lemmas 120 and 121 that continuing the $C0$ computation from c' instead of c will have no observable consequences for the elementary values computed. From $consis(c', d')$ and compiler correctness we conclude that these results are produced by the compiled program if we resume the MIPS computation at d' instead of d. Filling in the details will require only very moderate effort.

13.2.1 Pointer Chasing

For sets of subvariables or implementation subvariables $Y \subseteq SV(c) \cup IS(c)$ we define the set $os(Y)$ of *outermost* subvariables in Y. A subvariable in Y is an outermost subvariable if it is not contained in any subvariable in Y except itself.

$$os(Y) = \{y \in Y \mid \forall z \in Y : subvar(y, z) \to z = y\}.$$

Inspecting the definitions of $reach(i, c)$ and $ireach(i)$ we see that it obviously suffices to determine the sets $os(ireach(i))$ and then to determine from them the sets $os(reach(i, c))$.

The address range $ar(is)$ of an implementation subvariable is defined as

$$ar(is) = [is.ba : is.ba +_{32} (size(is.t) - 1)_{32}]$$

and from Lemma 87 we conclude directly that either the address ranges of implementation subvariables are disjoint or one is a subvariable of the other.

Lemma 122.

$$subvar(is, is') \leftrightarrow ar(is) \subseteq ar(is'),$$
$$\neg subvar(is, is') \wedge \neg subvar(is', is) \leftrightarrow ar(is) \cap ar(is') = \emptyset.$$

Now we compute the sets $os(ireach(i))$ in a straightforward way:

- the global memory and the stack frames are outermost subvariables. Thus

$$os(ireach(0)) = ireach'(0).$$

- suppose we have to follow k pointers

$$ripointer(i) = \{z_1, \ldots, z_k\}.$$

Following these pointers successively for $j = 0$ to k we determine the sets

$$or(ireach(i+1), 0) = os(ireach(i)),$$
$$or(ireach(i+1), j) = os(or(ireach(i+1), j-1) \cup \{follow(z_j)\}),$$
$$os(ireach(i+1)) = \bigcup_{j=0}^{k} or(ireach(i+1), j).$$

Implementation of the second line is very simple. Suppose we have found a new implementation subvariable

$$z = follow(z_j).$$

We abbreviate the outermost reachable subvariables found so far by

$$F = or(ireach(i+1), j-1).$$

Then because of Lemma 122 only three cases are possible, as illustrated in Fig. 203:

- the newly found subvariable is contained in a subvariable found before:

$$\exists y \in F : ar(z) \subseteq ar(y).$$

Then we found nothing new and we discard z:

$$or(ireach(i+1), j) = F.$$

- z is disjoint from all outermost subvariables found before:

$$\forall y \in F : ar(y) \cap ar(z) = \emptyset.$$

Then z is (so far) an outermost subvariable:

$$or(ireach(i+1), j) = F \cup \{z\}.$$

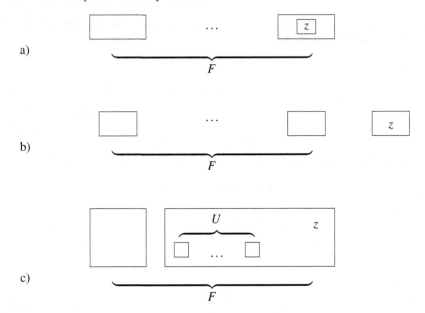

Fig. 203. F is the set of previously found outermost subvariables. The newly found implementation subvariable is either a) subsumed in a previously found subvariable in F, b) disjoint from all previously found subvariables in F, or c) itself subsumes a set $U \subseteq F$ of previously found subvariables

- the address range of z contains certain subvariables found before:

$$U = \{y \in or(ireach(i+1), j-1) \mid ar(y) \subseteq ar(z)\}.$$

In this case we include z and cancel the subvariables in U:

$$or(ireach(i+1), j) = F \setminus U \cup \{z\}.$$

$$os(ireach) \; = \; os\left(\bigcup_{i=0}^{R} ireach(i)\right) \; = \; \bigcup_{i=0}^{R} os(ireach(i)).$$

Successively chasing the pointers in $irdsreach$ as characterized in Lemma 111 we obtain the set

$$OrIS(c) = os(ireach \cup irdsreach) = os(ireach) \cup os(idsreach).$$

Note that an implementation subvariable $z \in OrIS(c)$ is on the heap if its base address is at least $hbase$.

$$onheap(z) \leftrightarrow z.ba \geq_{32} hbase.$$

Let T be the number of variables in $OrIS(c)$. We choose an ordering of subvariables $z_i \in OrIS(c)$

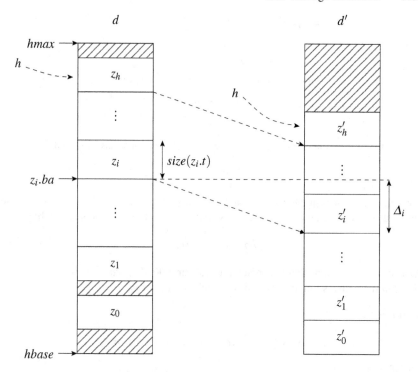

Fig. 204. In MIPS configuration d the outermost reachable implementation subvariables z_i on the heap might be separated by non-reachable portions of the heap. After garbage collection, in configuration d', their counterparts z_i' (we have to adjust pointers) will be stored consecutively at the bottom of the heap

$$OrIS(c) = \{z_0, \ldots, z_{h-1}, z_h, \ldots, z_{T-1}\}$$

such that $z_0, \ldots, z_{h-1}$ are heap variables and $z_i.ba \le z_{i+1}.ba$ for $i < h - 1$. This is illustrated in the left part of Fig. 204.

Every implementation subvariable $z_i \in OrIS(c)$ that we have found implements some reachable subvariable $y_i \in SV(c)$, i.e., we have

$$z_i = is_c(y_i).$$

While we do not know the y_i, we *do* know from their implementations the type and base address of all their subvariables $y_i s$.

$$vtype(y_is, c) = is_c(y_is).t$$
$$= is_c(y_i)s.t$$
$$= z_is.t,$$
$$ba(y_is, c) = ba(y_i, c) +_{32} displ(s, vtype(y_i, c))_{32}$$
$$= z_i.ba +_{32} displ(s, z_i.t)_{32}$$
$$= z_is.ba.$$

Notice that every reachable subvariable y' of configuration c has the form

$$y' = y_is$$

for some selector sequence s and reachable variable y_i. By the above equations the base address of y' then lies in the address range of implementation subvariable z_i:

$$ba(y_is, c) \in ar(z_i).$$

Conversely, we claim that any reachable subvariable $y' \in reach(c)$ with a base address in the address range of z_i must be a subvariable of y_i.

Lemma 123.

$$z_i = is_c(y_i) \wedge ba(y', c) \in ar(z_i) \rightarrow \exists s : y' = y_is.$$

Proof. Let

$$y' = y_xs \quad \text{and} \quad z_x = is_c(y_x).$$

Then as we have already shown

$$ba(y', c) \in ar(z_x).$$

Thus

$$ba(y', c) \in ar(z_x) \cap ar(z_i).$$

Because the address ranges overlap we know by Lemma 122

$$subvar(z_x, z_i) \vee subvar(z_i, z_x).$$

Since both z_x and z_i are in $OrIS(c)$ we conclude in both cases

$$i = x.$$

13.2.2 Garbage-Collected Equivalent Configurations

We transform c into a new well-formed C0 configuration c' which is equivalent to c. In configuration c' there are exactly h heap variables

$$c'.nh = h$$

and each heap variable $i \in [0 : h-1]$ has type $z_i.t$

$$c'.ht(i) = z_i.t.$$

Because for heap isomorphisms ϕ we have

$$\phi(xs) = \phi(x)s,$$

it suffices to define heap isomorphisms on outermost subvariables on the heap. Here, not surprisingly, we rename the subvariables y_i on the heap, i.e., with $i < h$, with i. On the other subvariables (in global memory and on the stack) there is no choice: we are not allowed to rename them. Abbreviating

$$x_i = \phi(y_i)$$

we set

$$x_i = \begin{cases} i, & i < h, \\ y_i, & \text{otherwise.} \end{cases}$$

Note that the x_i are actually variables of configuration c'.

By the rules formulated in Lemmas 112–116 this also determines $c'.X$ for components $X \in \{m, rds\}$. Equivalent configurations have identical components $X \in \{rd, st, pr, prn, clr\}$. Thus we set

$$X \in \{rd, sr, pr, prn, clr\} \rightarrow c.X = c'.X$$

and can conclude the following.

Lemma 124.

$$c \sim c'.$$

Among the C0 configurations equivalent to c, the chosen configuration c' is the representative with a non-redundant heap and where the order of heap variables of c' is consistent with the order of heap variables in c. We call it the *garbage-collected version* of c. With the help of the usual base address function ba we construct from c' a MIPS configuration d', which we call the *garbage-collected version* of d. We leave the program counter and the register file unchanged:

$$X \in \{pc, gpr\} \rightarrow d'.X = d.X.$$

For the base addresses of the heap variables we find:

$$ba(x_i, c') = \begin{cases} hbase, & i = 0, \\ ba(x_{i-1}, c') +_{32} size(z_{i-1}.t)_{32}, & i \in [1 : h-1], \\ ba(y_i, c), & ba(y_i, c) < hbase. \end{cases}$$

This is illustrated on the right side of Fig. 204.

The rules for $e\text{-}consis(c', d')$, $p\text{-}consis(c', d')$, and $rds\text{-}consis(c', d')$ determine the new memory content $d'.m$. The other components $X\text{-}consis(c', d')$ concern components which were not changed and follow from $consis(c, d)$. Thus we get the following.

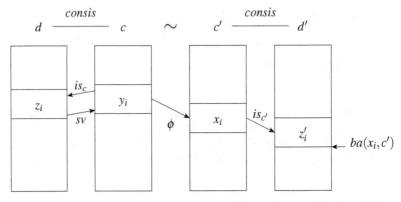

Fig. 205. For the bookkeeping we have to keep in mind: i) implementation subvariables z_j of d implement subvariables y_i of c via is_c, ii) outermost subvariables y_i are transformed into variables x_i of c' via heap isomorphism ϕ, and iii) for proving $consis(c', d')$ we mainly consider the new base addresses $ba(x_i, c')$

Lemma 125.

$$consis(c', d').$$

13.2.3 Construction of a Garbage-Collected MIPS Configuration

Let

$$\Delta_i = ba(y_i, c) -_{32} ba(x_i, c')$$
$$= z_i.ba -_{32} ba(x_i, c').$$

This distance applies also to subvariables $x_i s$ and $y_i s$ since we have

$$ba(y_i s, c) -_{32} ba(x_i s, c') = (ba(y_i, c) +_{32} displ(s, z_i.t)_{32}) -_{32}$$
$$(ba(x_i, c') +_{32} displ(s, z_i.t)_{32})$$
$$= \Delta_i.$$

Thus we obtain the following.

Lemma 126.

$$ba(x_i s, c') = ba(y_i s, c) -_{32} \Delta_i.$$

This means that all y_i and their subvariables in configuration d are shifted down in configuration d' by $\langle \Delta_i \rangle$ byte addresses. From now on the construction of $d'.m$ from $d.m$ is a bookkeeping exercise. For illustration also see Fig. 205.

- Elementary subvariables: assume implementation subvariable $z_i s$ has an elementary type, i.e.,

$$z_i s.t \in ET,$$

and let

$$x_i s = \phi(y_i)s = \phi(y_i s).$$

By the rules for heap isomorphisms

$$c'.m(x_i s) = c'.m(\phi(y_i s))$$
$$= c.m(y_i s).$$

From $e\text{-}consis(c,d)$ we know

$$d.m_4(ba(y_i s, c)) = c.m(y_i s).$$

For ensuring $e\text{-}consis(c',d')$ we have to set

$$d'.m_4(ba(x_i s, c')) = c'.m(x_i s).$$

Using also Lemma 126 we see that we have to set

$$d'.m_4(ba(x_i s, c')) = d.m_4(ba(y_i s, c))$$
$$= d.m_4(ba(x_i s, c') +_{32} \Delta_i).$$

This is exactly what we intuitively expect. Elementary subvariables $z_i s$ of implementation subvariables z_i are moved down by $\langle \Delta_i \rangle$ addresses. Because for $i \geq h$ we have $\Delta_i = 0_{32}$, the elementary subvariables in the global memory and the stack stay exactly where they are.

- Pointers: assume implementation subvariable $z_i s$ has a pointer type, i.e.,

$$pointer(z_i s.t).$$

Again, let

$$x_i s = \phi(y_i)s = \phi(y_i s).$$

Assume moreover that in configuration c pointer $y_i s$ points to subvariable u:

$$c.m(y_i s) = u.$$

Clearly, u is reachable and has the form

$$u = y_j r$$

for some index j and selector sequence r. Then $x_i s$ is a pointer of configuration c'. By the rules for heap isomorphisms it points to some subvariable u' of c':

$$u' = c'.m(x_i s)$$
$$= c'.m(\phi(y_i s))$$
$$= \phi(c.m(y_i s))$$
$$= \phi(u)$$
$$= x_j r.$$

From p-$consis(c,d)$ we conclude

$$ba(u,c) = d.m_4(ba(y_is,c))$$
$$= d.m_4(z_is.ba).$$

By Lemma 123 we determine j from the address ranges of the implementation subvariables as

$$j = \in\{x \mid ba(u,c) \in ar(z_x)\}.$$

Aiming at p-$consis(c',d')$ we have to set

$$d'.m_4(ba(x_is,c')) = ba(u',c').$$

We instantiate Lemma 126 twice. First we get as above

$$ba(x_is,c') = ba(y_is,c) -_{32} \Delta_i$$
$$= z_is.ba -_{32} \Delta_i,$$

i.e., the memory word $\langle\Delta_i\rangle$ bytes below the original pointer is updated. Then we conclude for heap variables u'

$$ba(u',c') = ba(x_jr,c')$$
$$= ba(y_jr,c) -_{32} \Delta_j$$
$$= ba(u,c) -_{32} \Delta_j$$
$$= d.m_4(z_is.ba) -_{32} \Delta_j.$$

This means that the pointer value is decreased exactly by the amount $\langle\Delta_j\rangle$ by which its target was moved down, which is also intuitively quite reassuring.

- Return destinations: in the stack region of configuration d' we also have to update the implementations of the return destinations. For $i > 0$ we know that $c.rds(i)$ is a reachable subvariable of the form

$$c.rds(i) = u = y_jr.$$

Rules for heap isomorphism require

$$u' = c'.rds(i)$$
$$= \phi(c.rds(i))$$
$$= \phi(y_jr)$$
$$= \phi(u)$$
$$= x_jr.$$

From rds-$consis(c,d)$ we know

$$ba(u,c) = ba(c.rds(i),c)$$
$$= d.m_4(ba(ST(i,c),c) -_{32} 4_{32}).$$

As above we determine

$$j = \in\{x \mid ba(u,c) \in ar(z_x)\}$$

and do not worry about the exact form of selector sequence r.

Aiming at rds-$consis(c',d')$ we have to update the location $[ba(ST(i,c),c) -_{32} 4_{32}]$ for $c'.rds(i)$, which is the same in configurations d and d', by

$$d'.m_4(ba(ST(i,c),c) -_{32} 4_{32}) = ba(c'.rds(i),c')$$
$$= ba(u',c').$$

As above we invoke Lemma 126 to conclude

$$ba(u',c') = ba(x_j r,c')$$
$$= ba(y_j r,c) -_{32} \Delta_j$$
$$= ba(u,c) -_{32} \Delta_j$$
$$= d.m_4(ba(ST(i,c),c) -_{32} 4_{32}) -_{32} \Delta_j.$$

Thus, as in the case of pointers, the return destination pointer is decremented by exactly the amount Δ_j by which its target was moved down.

13.3 C0 + Assembly

In order to implement operating system kernels, one needs to access data that is not stored in the variables of high-level programming languages, namely

- register contents, and
- memory addresses outside the address range of C0 variables.

Therefore one usually programs kernels in a mixture of high-level language code and inline assembly code (here C0 + inline assembly). Before we proceed in this way we had better make sure that the programs we intend to write in a mixed programming language have well-defined semantics. In this short section we treat syntax, semantics, and compilation of C0 + assembly. Most of the time we consider situations in which the *reset* signal is inactive; thus we can write the next configuration in a MIPS computation as

$$d' = \delta_M(d).$$

13.3.1 Syntax

We extend the C0 syntax by three new statements: one statement with plain inline assembly code and two statements supporting the exchange of data between C0 variables and registers.

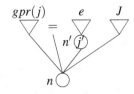

Fig. 206. Derivation tree for an assignment $gpr(j) = e \quad J$

1. A statement that allows us to inline blocks of assembly code:

$$\text{asm}(MIPS\text{-}Asm)$$

   where *MIPS-Asm* is a sequence of MIPS Assembler instructions.
2. A statement that allows us to assign a $C0$ expression to a general-purpose register, while keeping the register values of registers with numbers in a specified set $J \subset [0:31]$ unchanged:

$$\text{gpr}(j) = e \quad J$$

   where $j \in [0:31]$, $J \subset [0:27]$, and e is a $C0$ expression of a simple type. The set J of registers, whose value is not allowed to change, is enumerated in the form

$$J = \{j_1, \ldots, j_k\}.$$

   If $J = \emptyset$ we simply write

$$\text{gpr}(j) = e.$$

3. A statement that allows us to assign a general-purpose register to a $C0$ expression, while keeping the register values of registers with numbers in a specified set J unchanged:

$$e = \text{gpr}(j) \quad J$$

   where $j \in [0:31]$, $J \subset [0:31]$, and e is an identifier of a simple $C0$ variable. If $J = \emptyset$ we simply write

$$e = \text{gpr}(j).$$

We classify the first kind of statement as plain inline assembly; the two mixed language constructs are considered as statements of extended $C0$.

Interested readers are encouraged to go through the easy exercise of specifying the corresponding extension of the grammar and the context conditions.

13.3.2 Compilation

Code generation for the newly introduced statements is straightforward.

- asm(*MIPS-Asm*).

  The generated code is obviously just

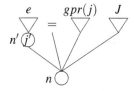

Fig. 207. Derivation tree for an assignment $e = gpr(j)$ J

MIPS-Asm

- $gpr(j) = e$ J.

  A derivation tree with a statement node n and an expression node n' for this statement is shown in Fig. 206. We label node n' with $R(n') = 1$ and run a pebble strategy which does not use pebbles with indices in J, i.e., we generate

  $code(n', J)$

  as introduced in Sect. 12.2.14. If the pebble game leaves pebble j' on node n' we complete the code for the assignment with

  ```
  addi j j' 0
  ```

- $e = gpr(j)$ J.

  A derivation tree with a statement node n and an expression node n' for this statement is shown in Fig. 207. We label node n' with $R(n') = 0$ and run a pebble strategy which does not use pebbles with indices in J and does not change $gpr(j_5)$, i.e., we generate

  $code(n', J \cup j)$

  If the pebble game leaves pebble j' on node n' we complete the code for the assignment with

  ```
  sw   j j' 0
  ```

Note that the pebble strategies use only pebbles with numbers at most 27. Registers with higher numbers always keep their value during expression evaluation. Thus it suffices to specify $J \subset [0:27]$.

13.3.3 Semantics and Compiler Correctness

We define the semantics of programs p written in a mixed programming language like *C0* + assembly via two computations:

- the low-level computation of the translated program; this is a computation in MIPS ISA and has well-defined semantics.

- a high-level $C0$ computation. As long as we are running translated $C0$ code this has the known semantics. But when we switch to assembly portions of the program we lose the high-level abstraction, and we are left with the low-level computation only. The main technical problem we have to solve is to regain the high-level abstraction after we leave the assembly portion.

Configurations

Formally we define that a $C0$ + assembly (or for short C+A) configuration y is either a MIPS configuration

$$y = d \tag{i}$$

or a pair

$$y = (d, c), \tag{ii}$$

where d is a MIPS configuration and c is a $C0$ configuration satisfying

$$wfc(c) \wedge consis(c, d).$$

In the first case we say that y has no C abstraction. In the second case we say that y has C abstraction c.

Depending on where a MIPS configuration d runs — inside or outside of the code region of the program — we abbreviate:

$$inside(d) \equiv d.pc \in [start(\varepsilon) : end(\varepsilon)],$$
$$outside(d) \equiv \neg inside(d).$$

A MIPS configurations d runs inline assembly if the program counter $d.pc$ of the configuration addresses code generated by a node n with border word $asm(x)$, i.e.,

$$inline(d) \equiv \exists n, x : bw(n) = \mathsf{asm}(x) \wedge d.pc \in [start(n) : end(n)].$$

Configuration d runs translated (possibly extended) $C0$ code if the program counter lies in the code region $[start(\varepsilon) : end(\varepsilon)]$ of the program but d does not run inline assembly:

$$transc(d) \equiv inside(d) \wedge \neg inline(d).$$

Note that we have not formulated any restrictions on the inline assembly portion of programs. Thus the assembly instructions may contain ISA instructions which jump completely out of the code region of the program. This will for instance be the case, when an operating system kernel which is written in $C0$ + assembly starts a user process.

Next State

We *always* refer to the next MIPS configuration by

$$d' = \delta_M(d).$$

For *C0* + assembly configurations $y = d$ or $y = (d,c)$ we will not always be able to define a unique next *C0* + assembly configuration y' as a function $\delta_{C+A}(y)$ of y. Therefore we use the notation

$$y \rightarrow_{C+A} y'$$

to indicate, that y' is *a* possible successor configuration after y. Computations are then sequences (y^i) satisfying

$$y^i \rightarrow_{C+A} y^{i+1}$$

for all i. Using notation from Sect. 12.5 we proceed to specify the possible successor configurations y' of y by an obvious case split:

- running in inline assembly and staying in inline assembly or running outside of the code region:

$$y = d \wedge (inline(d) \wedge inline(d') \vee outside(d')).$$

  We simply make a step of the MIPS computation:

$$y' = d'.$$

  Note that this case also covers jumps out of the code region in inline assembly.
- running translated not extended *C0* code:

$$y = (d,c) \wedge transc(d) \wedge hd(c.pr) \in L(St) \cup L(rSt).$$

  We step the MIPS configuration until the next consistency point and we step the *C* computation:

$$y' = (d'',c')$$

  with

$$d'' = nc_M(d),$$
$$c' = \delta_C(c).$$

  In an optimizing compiler we would also step the *C0* computation until the next consistency point

$$c' = nc_C(c).$$

  That y' is a C+A configuration, i.e., that it satisfies

$$wfc(c') \wedge consis(c',d'),$$

  follows from the compiler correctness result of Lemma 99. For optimizing compilers one would have to use Lemma 100.
- switching to inline assembly:

$$y = (d,c) \wedge inline(d).$$

  We simply drop the high-level language abstraction and proceed only with the low-level computation:

$$y' = d'.$$

So far the next configuration y' was unique and could have been defined by an ordinary transition function δ_{C+A}. This changes for the next case.

- returning to translated C:

$$y = d \wedge transc(d').$$

We continue the low-level computation until we can reconstruct a C abstraction.[2] Then we continue the high-level portion of the computation with *some* consistent $C0$ configuration c':

$$y' \in \begin{cases} \{d'\}, & \neg \exists c' : wfc(c') \wedge consis(c',d'), \\ \{(d',c') \mid wfc(c') \wedge consis(c',d')\}, & \text{otherwise.} \end{cases}$$

This is an instance of so-called *nondeterministic* semantics. The next configuration is not a function of the previous configuration any more. The user only knows that the new configuration lies in a specified set of configurations. The concrete choice of the configuration is not under his control. Therefore he has to write his programs such that they work for *any* of the possible nondeterministic choices. At the first sight this sounds scary. Fortunately we know by Lemma 119 that all possible choices c' are equivalent and thus by the results from Sect. 13.1.7 we will compute exactly the same values of elementary expressions for any choice we make. Note that the semantics is deterministic if the $C0$ program does not use a heap. We will discuss nondeterminism in more detail in Sect. 14.3.3.

- extended $C0$ code; assignment of expression to register in function f:

$$y = (c,d) \wedge transc(d) \wedge hd(c.pr) \equiv gpr(j) = e \quad J.$$

Let

$$t = etype(e,f).$$

We step d to the next consistency point and define y' and in particular $c' = \delta_C(c)$.

$$y' = (d'',c')$$

with

$$d'' = nc_M(d),$$
$$c'.prn = tail(c.prn),$$
$$c'.pr = tail(c.pr).$$

From the way we generate code for expression evaluation, we conclude as follows.

Lemma 127.

$$d''.gpr(j_5) = \begin{cases} enc(va(e,c),t), & t \in ET, \\ ba(va(e,c),c), & pointer(t), \end{cases}$$

---

[2] This occurs for instance in the following scenarios: i) we return from a jump out of the compiled code region, ii) behind inline assembly code we return to translated $C0$ code, but we must perform some jumps before we get to a consistency point.

and

$$i \in J \cup [28:31] \rightarrow d''.gpr(i_5) = d.gpr(i_5).$$

- extended C0 code; assignment of register to expression in function f:

$$y = (c,d) \wedge transc(d) \wedge hd(c.pr) \equiv e = gpr(j) \quad J.$$

Let

$$t = etype(e,f).$$

Again, we step d to the next consistency point.

$$y' = (d'',c')$$

with

$$d'' = nc_M(d),$$
$$c' = \delta_C(c).$$

The next C0 configuration c' is obtained as in an assignment, where the value of GPR register j is interpreted as a value of type t and assigned to the subvariable $lv(e,c)$:

$$c'.m(lv(e,c)) \in \begin{cases} \{enc^{-1}(d.gpr(j_5),t)\}, & t \in ET, \\ \{x \in SV(c) \mid vtype(x,c) = t' \wedge ba(x,c) = d.gpr(j_5)\}, & t = t' *. \end{cases}$$

Note that with this statement we can perform type casting and destroy type correctness. In this case the next configuration y' is undefined and the computation ends with an error.

From the way we generate code we conclude that the values of register j, of registers specified by J, and of registers not used for expression evaluation stay unchanged.

Lemma 128.

$$i \in J \cup \{j\} \cup [28:31] \rightarrow d''.gpr(i_5) = d.gpr(i_5).$$

Along the usual lines of a compiler correctness proof one establishes for the last two cases that well-formedness and compiler consistency are preserved.

Lemma 129.

$$wfc(c) \wedge consis(c,d) \rightarrow wfc(c') \wedge consis(c',d'').$$

The above definition can be adapted to optimizing compilers if one forbids optimization across the statements of extended C0. One requires compiler consistency to hold immediately before and after assignments between registers and expressions:

$$cpoint_C(c) \wedge cpoint_C(c')$$

in which case one gets

$$c' = nc_C(c).$$

Note that the above definitions work only while the reset signal is inactive and if the compiled code does not signal error conditions via syscalls. We handle these situations in Sect. 14.3.8.

13.4 Final Remarks

At first sight, this chapter might appear quite involved, but the main technical points are quite easily remembered:

- reconstruction of C abstractions
    - $d.pc = start(hd(c.prn))$. From this you get i) the function belonging to the top frame, ii) the size of the top frame, iii) the non-pointer values in the top frame, and iv) the program rest nodes until the first return statement by the successor function.
    - for the frame below the top frame you can proceed in the same way starting with the return address instead of the pc. Now iterate until you arrive at $sbase + size(\$gm)$ at which point you also know the recursion depth.
    - starting from the pointers you have found so far (and the result destinations) do pointer chasing which performs a graph search on the heap. You know the types of the pointers and thus the size of the implementation subvariables you will find. Any two implementation subvariables you find are disjoint or one is contained in the other (you have kept this in mind as a key property of the base address computation).
    - if c is a C abstraction of d and x is a reachable subvariable on the heap of c, then the function $is_c(x)$ assigning to x its implementation subvariable is bijective (property of ba-computation). This allows one to construct a heap isomorphism ϕ between two C abstractions c and c' by

$$\phi(x) = is_{c'}^{-1}(is_c(x)).$$

    - for relating expression evaluation via isomorphism ϕ you only have to remember that if $lv(e,c)$ is defined, then

$$lv(e,c') = \phi(lv(e,c)).$$

For values you then obviously want to have

$$va(e,c') = \begin{cases} va(e,c), & etype(e,f) \in ET, \\ \phi(va(e,c)), & pointer(etype(e,f)). \end{cases}$$

- for garbage collection (optional to remember)
    - identify reachable subvariables on the heap of d via pointer chasing.
    - compress the heap and adjust pointers in d'.
    - construct c' whose heap variables are just the reachable subvariables of c on the heap (in the same order). Then

$$c \sim c' \wedge consis(c',d').$$

- for 'C+A' semantics
    - configurations are MIPS ISA configurations d or pairs (d,c) of MIPS ISA configurations with a C abstraction.

- as long as translated C code is running, the C+A computation proceeds in C steps.
- if translated C code is left, the C abstraction is dropped.
- if translated C code is entered, the computation proceeds with ISA steps until a configuration d with a C abstraction is reached.
- extended C instructions

$$gpr(j) = e \quad J$$

and

$$e = gpr(j) \quad J,$$

where J specifies the set of registers which stay unchanged, help to couple C portions and ISA portions of the computation.

The (almost) unique reconstruction of C abstractions from ISA configurations is a typical folklore theorem. Everybody believes it — after all everybody *does* run compiled code in order to find out the value of elementary subvariables of C computations — but a proof might be hard to find. Special cases of the C+A semantics presented here have been used in formal correctness proofs of a kernel [AHL$^+$09] and a hypervisor [PSS12].

13.5 Exercises

1. Prove
   a) Lemma 127.
   b) Lemma 129.
2. Consider the following program

```
1:  int a;
2:  int b;
3:
4:  int main()
5:  {
6:      a=256;
7:      gpr(1) = a;
8:      gpr(2) = b& {1};
9:
10:     asm(
11:         srl   1 1 4
12:         sw    1 2 0
13:     );
14:
15:     return 1
16: }
```

Let (d, c) be the configuration after the assembly portion.

a) Show
$$va(b,c) = 16.$$

b) Change the assignment at line 8 to

```
gpr (2)  = b&
```

Can you still show (a)?

3. We extend $C0$ with labels and gotos in a restricted way.

Informally, labeled statements have the form

$$na : s$$

where na is a name and s is a statement or return statement.

A goto statement has the form

goto na

We require as context conditions:
  i) in every function body every label can occur at most once,
  ii) if statement goto na occurs in the body of function f, then label na must occur in front of a statement or return statement in the body of function f.

After execution of statement goto na in function f, the computation continues with the unique statement in the body of f with label na.

  a) Extend the $C0$ grammar such that it includes goto statements and labeled statements.
  b) Extend the definition of flattened statement sequences.
  c) Generate ISA code for goto statements in the body of f.
     Hint: jump to $start(n)$ for an appropriate statement node in the body of f.
  d) Extend the successor function for nodes generating goto statements.
  e) Specify the semantics of goto statement execution.
     Hint: use the successor function to construct a new program rest.
  f) Now argue that well-formedness and consistency are maintained when a translated goto statement is executed.

4. Now we allow much more general gotos with the context conditions:
  i) labels are distinct in the entire program,
  ii) the target label of a goto needs to be defined in the current function or in any function of a frame on the stack.

Jumps now go to the topmost occurrence of a function on the stack with the target label.

  a) Specify the new semantics.
     Hint: you have to destroy a few stack frames.
  b) Specify code generation.

     Attention: generating code for the destruction of frames is quite easy, but the compiled program needs to maintain a data structure with the labels for each frame.

Operating System Support for MIPS Processors and $C0$

The MIPS ISA that we have specified so far is not the full instruction set architecture. Except for the system call instruction, which will be introduced here, it is the portion of the ISA that can be used when the processor is running in *user mode*. We proceed to specify the full ISA, which we also call ISA-sp (for system programmers). The restricted ISA together with the system call instruction will be called ISA-u (for users). Three fundamental new mechanisms become visible: i) interrupts (Sect. 14.1), ii) address translation (Sect. 14.2), and iii) devices (Sect. 14.3); in the main text we only treat disks which do not produce interrupts. In the course of the chapter we develop and justify a model for programming in 'C + assembly + disk + interrupts' (Sect. 14.3.8). In this model we program a disk driver and prove its correctness (Sect. 14.3.9).

Interrupts and address translation can be treated very quickly in the classroom: after a clean specification the hardware constructions are basically trivial. Introducing disk models is easy too, but ISA for 'processor + disk' interleaves processor steps and disk steps (where disk accesses end) in a nondeterministic way. We take our time explaining how such a model can be justified (you have to construct the interleaving from the hardware computation) and to develop the formalism which allows us to handle this. Reordering disk steps behind processor steps which ignore the disk is explained easily; the formal proof justifying the reordering of two steps is of the bookkeeping type. Reordering infinite sequences requires a diagonal argument, which is both fun and instructive. So we present it in the classroom. The semantics of 'C + A + disk + interrupts' uses reordered computations. Introducing this — with the concepts presented before — takes a few minutes. Because translated C code does not access I/O-ports (because no variables are allocated there) disk steps are reordered out of translated C code and can be justifiably ignored during C portions of the computations. Thus — far from being abstract nonsense — the reorder theorem gives a very quick justification for a very practical programming model. Programming a disk driver and proving that it works in this model is so simple that one can leave it as reading material. But then, it is also quite tempting to show off in the classroom with this ... because it is so effortless.

© Springer International Publishing Switzerland 2016
W.J. Paul et al., *System Architecture*, DOI 10.1007/978-3-319-43065-2_14

There is a tricky point: the disk is not always live in reordered computations. If a disk access is started and then the processor never accesses the disk, the disk step ending the access is reordered behind all subsequent processor steps. We mention this in the classroom and point to the material in this text which handles this (Sect. 14.3.7), but we do not consider the present proof instructive enough for an introductory lecture.

14.1 Interrupts

14.1.1 Types of Interrupts

Interrupts are triggered by interrupt event signals; they change the control flow of programs. Here we consider the 23 event signals $ev[22:0]$ from Table 13. We classify interrupts in three ways:

- internal or external. The first 16 interrupts are generated outside of the CPU and memory system. We collect their event signals into the vector of external interrupts
$$eev[15:0] = ev[15:0].$$

  The other interrupts $ev[j]$ are generated within the CPU and memory system. Their activation depends only on the current ISA configuration d. We collect them into the vector of internal interrupts

$$iev(d)[22:16] = ev[22:16].$$

  More internal interrupt signals must be introduced if one adds floating-point units [MP00]. In multi-core processors one additionally tends to implement inter-processor interrupts.
- maskability. We make reset, illegal instructions, misalignment, and page faults non-maskable. All other interrupts can be masked. When an interrupt is masked, the processor will not react to it when the corresponding event signal is raised.
- resume type. After the interrupt is handled, there are three possibilities: i) The program can be aborted; this is the case for reset, illegal instruction, and misalignment. ii) The interrupted instruction is repeated; this is the case for page faults and external interrupts other than reset. iii) Program execution continues after the interrupted instruction; this is the case for all other interrupts.

14.1.2 Special Purpose Registers and New Instructions

A special purpose register file

$$d.spr : \mathbb{B}^5 \to \mathbb{B}^{32}$$

is added as a new component to MIPS configurations d. Here we will only use eight special purpose registers $spr(a)$. For these registers, we use the synonyms from Table

index j	synonym for $ev[j]$	name	maskable	resume
0	*reset*	reset	no	abort
$j \in [1:15]$	*eev[j]*	external event signal from a device	yes	repeat
16	*malf*	misaligned fetch	no	abort
17	*pff*	page fault during fetch	no	repeat
18	*ill*	illegal instruction	no	abort
19	*malls*	misaligned load/store	no	abort
20	*pfls*	page fault during load/store	no	repeat
21	*sysc*	system call	yes	continue
22	*ovf*	arithmetic overflow	yes	continue

Table 13. Interrupts handled by our design

$\langle a \rangle$	synonym for $spr(a)$	name
0	*sr*	status register
1	*esr*	exception status register
2	*eca*	exception cause
3	*epc*	exception *pc*
4	*edata*	exception data
5	*pto*	page table origin
6	*ptl*	page table length
7	*mode*	mode register

Table 14. Special purpose registers

opcode	fun	rs	Mnemonic	Assembler-Syntax	Effect
System Call					
000000	001 100		sysc	sysc	System Call
Coprocessor Instructions					
010000	011 000	10000	eret	eret	Exception Return
010000		00100	movg2s	movg2s *rd rt*	spr[rd] := gpr[rt]
010000		00000	movs2g	movs2g *rd rt*	gpr[rt] := spr[rd]

Table 15. Instructions for operating system support

14. The exact purpose of these registers will be explained shortly. The mode register distinguishes between system mode, where $d.mode[0] = 0$, and user mode, where $d.mode[0] = 1$. We abbreviate

$$mode(d) = d.mode[0].$$

Four R-type instructions are provided for operating system support. We have mentioned them already in the tables of Sect. 8.1, and we repeat the relevant portion of the tables in Table 15.

We can now provide semantics for these instructions. We start with the move instructions:

$$move(d) \equiv movg2s(d) \vee movs2g(d).$$

Instruction $movg2s$ moves data from the general-purpose registers to the special purpose registers, and instruction $movs2g$ moves data from the special purpose registers to the general-purpose registers. The general-purpose registers are addressed with field rt and the special purpose registers are addressed with rd. Formally, the update of register files is defined by

$$movg2s(d) \;\rightarrow\; \begin{array}{l} d'.spr(x) = \begin{cases} d.gpr(rt(d)), & x = rd(d), \\ d.spr(x), & \text{otherwise,} \end{cases} \\ d'.gpr = d.gpr \end{array}$$

and

$$movs2g(d) \;\rightarrow\; \begin{array}{l} d'.gpr(x) = \begin{cases} d.spr(rd(d)), & x = rt(d) \wedge x \neq 0_5, \\ d.gpr(x), & \text{otherwise,} \end{cases} \\ d'.spr = d.spr. \end{array}$$

The memory does not change, and the pc is incremented for these instructions:

$$move(d) \;\rightarrow\; \begin{array}{l} d'.m = d.m, \\ d'.pc = d.pc +_{32} 4_{32}. \end{array}$$

The exact effect of the instructions $eret$ and $sysc$ will be explained shortly.

14.1.3 MIPS ISA with Interrupts

Recall that the transition function δ_M for MIPS ISA as it has been defined so far computes a new MIPS configuration d' from an old configuration d and the value of the reset signal:

$$d' = \delta_M(d, reset).$$

We rename this transition function to δ_M^{old} and define the new transition function which takes as inputs the old configuration d and the vector of external event signals:

$$d' = \delta_M(d, eev).$$

We proceed to define a predicate $jisr(d, eev)$ which indicates that a jump to the interrupt service routine is to be performed. If this signal is inactive, the machine should behave as it did before with an inactive reset signal:

$$\neg jisr(d, eev) \;\rightarrow\; d' = \delta_M(d, eev) = \delta_M^{old}(d, 0).$$

For the computation of the $jisr$ predicate we unify the notation for external and internal interrupts and define a vector $ca(d, eev)[22 : 0]$ of cause signals by

$$ca(d, eev)[i] = \begin{cases} eev[i], & i \leq 15, \\ iev(d)[i], & i \geq 16. \end{cases}$$

For maskable interrupts with index i we use bit i of the status register sr as the mask bit for this interrupt. Thus we define the vector $mca(d, eev)[22:0]$ of masked cause signals as

$$mca(d, eev)[i] = \begin{cases} ca(d, eev)[i] \wedge d.sr[i], & i \in [1:15] \cup \{20, 21\}, \\ ca(d, eev)[i], & \text{otherwise.} \end{cases}$$

We jump to the interrupt service routine if any of the masked cause bits is on:

$$jisr(d, eev) = \bigvee_i mca(d, eev)[i].$$

In case this signal is active we define the interrupt level $il(d, eev)$ as the smallest index of an active masked cause bit:

$$il(d, eev) = \min\{i \mid mca(d, eev)[i]\}.$$

We handle interrupts with small indices with higher priority than interrupts with high indices. Thus the interrupt level gives the index of the interrupt that will receive service. With an active $jisr(d, eev)$ bit many things happen at the transition to the next state d':

- we jump to the start address $sisr$ of the interrupt service routine. We fix this start address to 0_{32}, i.e., to the first address in the ROM:

$$d'.pc = sisr = 0_{32}.$$

- all maskable interrupts are masked:

$$d'.sr = 0_{32}.$$

The purpose of this mechanism is to make the interrupt service routine — which will take some instructions — not interruptible by external interrupts. Of course, the interrupt service routine should also be programmed in such a way that executing it does not produce internal interrupts.

- the old value of the status register is saved into the exception status register esr. Thus it can be restored later:

$$d'.esr = d.sr.$$

- we store in the exception pc register epc something similar to the link address of a function call. It is the address where program execution will resume after the handling of the interrupt. We must obviously distinguish between repeat and continue interrupts[1]. We classify external interrupts and page faults as repeat interrupts with the predicate

---

[1] If the interrupt is aborting it does not matter what we save.

$$repeat(d,eev) \equiv il(d,eev) \in [1:15] \cup \{17,20\},$$

and set

$$d'.epc = \begin{cases} d.pc, & repeat(d,eev), \\ \delta_M^{old}(d,0).pc, & \text{otherwise.} \end{cases}$$

- in the exception data register *edata*, we save the effective address $ea(d)$. After a page fault on load/store this provides the effective address which generated the interrupt to the page fault handler:

$$d'.edata = ea(d).$$

- we save the masked cause bits into the exception cause register so that the interrupt level of the interrupt can be computed by software:

$$d'.eca = 0^{10} \circ mca(d,eev).$$

- in case of continue interrupts we need to finish the interrupted instruction. Concerning the general-purpose register file this is only relevant for arithmetic overflow exceptions. System calls do not update the GPR:

$$d'.gpr = \begin{cases} \delta_M^{old}(d,0).gpr, & il(d,eev) = 22, \\ d.gpr, & \text{otherwise.} \end{cases}$$

The memory is unchanged because load and store instructions cannot be interrupted with resume type *continue*:

$$d'.m = d.m.$$

- finally, we switch to system mode:

$$d'.mode = 0^{32}.$$

This completes the definition of what happens on activation of the *jisr* signal.

During a return from exception, i.e., if predicate $eret(d)$ is active, several other things happen simultaneously

- the pc is restored from the exception pc:

$$d'.pc = d.epc.$$

- the status register is restored from the exception status register:

$$d'.sr = d.esr.$$

- the mode is set to user mode:

$$d'.mode = 0^{31}1.$$

14.1.4 Specification of Most Internal Interrupt Event Signals

Except for the page fault event signals, which obviously depend on the not yet defined mechanism of address translation, we can specify when internal event signals are to be activated:

- illegal instruction. By inspection of the tables in Sect. 8.1 we define a predicate *undefined*(*c*) which is true if the current instruction $I(d)$ is not defined in the tables. Moreover we forbid in user mode i) the access of the special purpose registers by move instructions, as well as ii) the execution of the *eret* instruction:

$$ill(d) \equiv undefined(d) \vee mode(d) \wedge (move(d) \vee eret(d)).$$

- misalignment. The restricted instruction set considered here permits only word accesses. Misalignment occurs during fetch if the low order bits of the *pc* are not both zero. It occurs during during load/store if the low order bits of the effective address *ea* are not both zero:

$$misaf(d) \equiv d.pc[1:0] \neq 00,$$
$$misals(d) \equiv ls(d) \wedge ea(d)[1:0] \neq 00.$$

- system call. This event signal is simply identical with the predicate decoding a syscall instruction:

$$sysc(d) \equiv opc(d) = 0^6 \wedge fun(d) = 001100.$$

- arithmetic overflow. This signal is taken from the overflow output of the ALU specification if an ALU operation is performed:

$$ovf(d) \equiv alu(d) \wedge ovfalu(lop(d), rop(d), af(d), itype(d)).$$

- page faults. For the time being we tie page fault signals to zero.

Except for the generation of the page fault signals this already completes the formal specification of the interrupt mechanism. Thus, this mechanism clearly is not overly hard to specify. We will see in the next subsection that its implementation — with the building blocks we already have — is also not so hard.

14.1.5 Hardware

The translation of the internal event signals $iev[i]$ and the new predicates *sysc*, *eret*, *movg2s*, and *movs2g* from the specification into hardware signals is obvious. The translation of the *jisr* signal is not completely trivial. Recall that so far we are simulating ISA instructions in two phases: fetch and execute. Figure 208 shows the computation of the new execute signal; we have just replaced the *reset* signal in the old construction by *jisr*. If we discover a non-masked external interrupt during the fetch phase there is no need to perform the execute phase. Thus we will allow the hardware

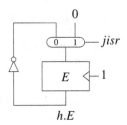

Fig. 208. Computation of the execute bit with interrupts

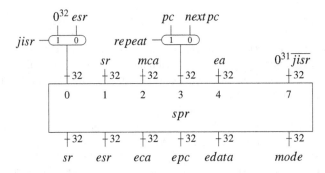

Fig. 209. Wiring special data inputs $sprdin[i]$ of the special purpose register file

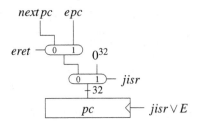

Fig. 210. The new pc environment supporting interrupts and return from exception

version $jisr(h, eev)$ of the $jisr$ signal to become active in the fetch phase, and if that happens we will skip the execute phase and immediately jump to the interrupt service routine. However, *during fetch we should ignore internal interrupts*: they depend on the instruction register, which contains leftover data from the previous instruction during the fetch phase or random data after power up. Therefore, during fetch we mask in the hardware all internal interrupts except *malf* and define the cause signals for the hardware by

$$ca(h, eev)[i] = \begin{cases} eev[i], & i \leq 16, \\ iev(h)[i] \wedge h.E, & i \geq 17. \end{cases}$$

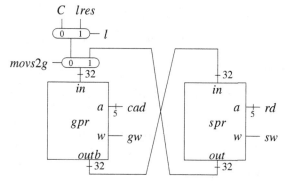

Fig. 211. Data paths between general-purpose registers and special purpose registers for the support of move instructions

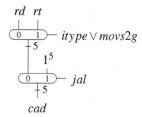

Fig. 212. Modified C-address computation supporting *movs2g* instructions

The main new component of the MIPS processor construction is obviously a $(5,32)$-SPR-RAM $h.spr$ as constructed in Sect. 6.5. Figure 209 shows for $i \in \{0,1,2,3,4,7\}$ how the special data inputs $sprdin[i]$ are computed. Mode and status register are clocked at $jisr$ and during the execute phase at $eret$. The other special purpose registers listed here are only clocked at $jisr$:

$$spr(i_5)ce(h,eev) = \begin{cases} jisr(h,eev) \vee h.E \wedge eret(h), & i \in \{0,7\}, \\ jisr(h,eev), & i \in [1:4]. \end{cases}$$

Figure 210 shows the obvious modification of the pc environment for the implementation of arbitrary jumps to interrupt service routines (as opposed to only resets) and for the implementation of exception returns. For the support of move instructions the ordinary data input $sprin$ and data output $sprout$ of the SPR-RAM are connected with the general-purpose register file as shown in Fig. 211. Moreover the address computations for the C-address of the general-purpose register file has to be modified in an obvious way as shown in Fig. 212.

For an efficient computation of the *repeat* signals we refer the reader to the exercises. The write signal sw of the special purpose register file should obviously be activated during the execute phase of *movg2s* instructions. However, *no* write should be performed for instructions that are interrupted by repeat interrupts. Thus

$$sw(h,eev) = h.E \wedge movg2s(h) \wedge \neg(jisr(h,eev) \wedge repeat(h,eev)).$$

In the same spirit the write signal to the memory has to be modified to

$$mw(h,eev) = h.E \wedge s(h) \wedge \neg(jisr(h,eev) \wedge repeat(h,eev)).$$

Finally, if $gprw^{old}$ is the previously defined write signal for the general-purpose register file, then the new signal is defined as

$$gw(h,eev) = h.E \wedge (gprw^{old}(h) \vee movs2g(h)) \wedge \neg(jisr(h,eev) \wedge repeat(h,eev)).$$

14.1.6 Hardware Correctness

At first glance the above modifications to the hardware are obviously correct. However, the correctness proof requires a few small adjustments. Clearly, we start with an initial hardware configuration h^0 after reset. We find a MIPS configuration d^0 satisfying $sim(d^0,h^0)$, where the old simulation relation is extended in the obvious way to take care of the special purpose register file:

$$sim(d,h) \equiv d \sim h \wedge d.spr = h.spr.$$

Next we construct the hardware computation (h^t) with the help of the sequence (eev_h^t) of external interrupt event signals *of the hardware*. We observe that simulation of ISA instructions takes one or two cycles. We define the sequence $(t(i))$ of hardware cycles in which simulation of ISA instruction $I(d^i)$ begins, by

$$t(0) = 0,$$

$$t(i+1) = t(i) + \begin{cases} 1, & jisr(h^{t(i)}, eev_h^{t(i)}), \\ 2, & \text{otherwise.} \end{cases}$$

In general, after t hardware cycles fewer than t ISA instructions have been executed. We define the sequence (eev_{isa}^i) of external interrupt event signals *seen by the ISA* by

$$eev_{isa}^i = \bigvee_{t=t(i)}^{t(i+1)-1} eev_h^t$$

and can prove the following effortlessly by induction on t.

Lemma 130.
$$sim(d^i, h^{t(i)})$$

14.2 Address Translation

One of the main tasks of operating system kernels is to provide on a single physical processor running *ISA-sp* the simulation of many virtual processors which run

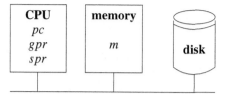

Fig. 213. Physical machines have a central processing unit (CPU) with the registers, as well as a memory and devices

$$
\begin{array}{c}
\quad 31 \qquad\qquad 11 \qquad 0 \\
va \quad \boxed{\begin{array}{c|c} va.px & va.bx \end{array}}
\end{array}
$$

Fig. 214. Virtual addresses *va* are partitioned into a page index *va.px* and a byte index *va.bx*

ISA-u. The physical processor is often also called the *host* and the simulated virtual processors are called the *guests*. The host has to store the memory contents of the guests in regions of its own memory and possibly in external memory located, e.g., on disks (see Fig. 213). This involves a translation of *virtual addresses* of the guests into *physical addresses* of the host. This translation is supported by the *memory management units* of the hardware. We will specify and implement here only a very simple address translation mechanism.

14.2.1 Specification

Memories of the host and of the guests are partitioned into pages each comprising

$$4K = 2^{12}$$

byte addresses. As illustrated in Fig. 214 we split virtual addresses $va \in \mathbb{B}^{32}$ into page index *va.px* and byte index *va.bx*:

$$va.px = va[31:12],$$
$$va.bx = va[11:0].$$

In any MIPS configuration d, two special purpose registers determine the region of memory $d.m$ whose content is interpreted as a single *page table*, namely

- the page table origin register $d.pto$. It stores the byte address where the page table starts. This address must be aligned to word boundaries, i.e., we require

$$d.pto[1:0] = 00.$$

- the page table length register $d.ptl$. It specifies the number of page table entries in the page table. Page table entries will contain four bytes, thus $d.ptl$ specifies the length of the page table measured in words.

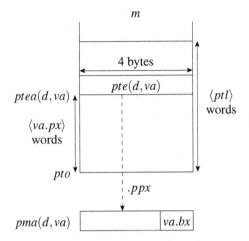

Fig. 215. Translation of a virtual address *va*. The page table starting at address *d.pto* is accessed with an offset of $\langle va.px \rangle$ words. The *ppx* field of the page table entry *pte(d, va)* at this address is concatenated with the byte index *va.bx* to form the physical memory address *pma(d, va)*

	31	12 11 10	0
pte	*ppx*	*v*	

Fig. 216. Every page table entry *pte* has a physical page index field *pte.ppx* and a valid bit *pte.v*

For virtual addresses *va* and MIPS configurations *d* the page table entry address

$$ptea(d, va) = d.pto +_{32} 0^{10} \circ va.px \circ 00$$

specifies the byte address where the page table entry for *va* can be found in memory in configuration *d*. As illustrated in Fig. 215, it is found $\langle va.px \rangle$ words resp. $\langle va.px \circ 00 \rangle$ bytes behind the page table origin. The corresponding page table entry in memory is

$$pte(d, va) = c.m_4(ptea(d, va)).$$

As illustrated in Fig. 216, we interpret certain fields in a page table *pte* entry as physical page index *ppx* and valid bit *v(d, va)*:

$$pte.ppx = pte[31 : 12],$$
$$pte.v = pte[11].$$

For the page table entry *pte(d, va)*, we get

$$ppx(d, va) = pte(d, va).ppx$$
$$= pte(d, va)[31 : 12]$$

and

$$v(d,va) = pte(d,va).v$$
$$= pte(d,va)[11].$$

If the page table entry cannot be found in the page table or if the valid bit of this page table entry is zero, we have a page fault:

$$pf(d,va) \equiv \langle va.px \rangle \geq \langle d.ptl \rangle \vee \neg v(d,va).$$

For later use we define predicates for page table length exceptions and for page faults due to invalid page table entry by

$$ptle(d,va) \equiv \langle va.px \rangle \geq \langle d.ptl \rangle,$$
$$ipf(d,va) \equiv \neg ptle(d,va) \wedge \neg v(d,va).$$

If there is no page fault, we can translate the virtual address into a physical memory address by appending the byte index $va.bx$ to the physical page index of the page table entry (see again Fig. 215):

$$\neg pf(d,va) \rightarrow pma(d,va) = ppx(d,va) \circ va.bx.$$

We now specify the page fault event signals and the instruction set with address translation.

In system mode there are neither address translation nor page faults. In user mode we have a page fault on fetch if address translation fails with the program counter as virtual address, and we have a page fault on load or store if address translation of the effective address fails:

$$pff(d) = mode(d) \wedge pf(d.pc,d),$$
$$pfls(d) = mode(d) \wedge ls(d) \wedge pf(ea(d),d).$$

Note that in any MIPS configuration d, at most one internal interrupt event signal can be active.

In the specification of the instruction set, we have to account for address translation in three cases, provided no page faults arise. In user mode the instruction $I(d)$ is determined by the translated pc:

$$I(d) = \begin{cases} d.m_4(pma(d,d.pc)), & mode(d) \wedge \neg pff(d), \\ d.m_4(d.pc), & \neg mode(d). \end{cases}$$

In load word instructions, we load in user mode from the translated effective address:

$$l(d) \rightarrow d'.gpr(rt(d)) = \begin{cases} d.m_4(pma(d,ea(d))), & mode(d) \wedge \neg pfls(d), \\ d.m_4(ea(d)), & \neg mode(d). \end{cases}$$

In store word instructions, we update in user mode the memory word starting at the translated effective address:

$$s(d) \rightarrow d.gpr(rt(d)) = \begin{cases} d'.m_4(pma(d,ea(d))), & mode(d) \wedge \neg pfls(d), \\ d'.m_4(ea(d)), & \neg mode(d). \end{cases}$$

14.2.2 Hardware

We construct a simple memory management unit and modify the control. For each instruction we may need up to four accesses to memory: i) translation for fetch, ii) instruction fetch, iii) translation for load/store, and iv) load/store. Thus, we split in user mode the fetch and execute phases into two subphases with the help of a new register $h.phase$ of the hardware. For an instruction in user mode, the bits $(E, phase)$ will go through the pattern:

$$(0,0), (0,1), (1,0), (1,1).$$

In system mode the phase bit is frozen to 0, and the execute bit E alternates through the usual pattern:

$$(0,0), (1,0).$$

Thus execution of an ISA instruction will take up to two cycles in system mode and up to four cycles in user mode. As in the processor construction of Sect. 14.1, we jump to the interrupt service routine immediately after discovery of an interrupt. This forces bits $(E, phase)$ of the control logic to $(0,0)$, and we define the inputs for these bits as

$$Ein(h, eev) = \neg jisr(h, eev) \wedge (\neg h.mode[0] \wedge \neg h.E \vee$$
$$h.mode[0] \wedge (h.E \oplus h.phase)),$$

$$phasein(h, eev) = \neg jisr(h, eev) \wedge h.mode[0] \wedge \neg h.phase.$$

In system mode, instruction execution is performed if $h.E = 1$. In user mode, instruction execution is performed if $h.E \wedge h.phase = 1$. We define the predicate signaling instruction execution in the current cycle by

$$ex(h) = h.E \wedge (\neg h.mode[0] \vee h.mode[0] \wedge h.phase).$$

With the help of this signal, the clock enable signal for the pc as well as the write signals for memory and general-purpose register file are modified in the obvious way:

$$pcce(h, eev) = ex(h) \vee jisr(h, eev),$$
$$mw(h, eev) = ex(h) \wedge s(h) \wedge \neg(jisr(h, eev) \wedge repeat(h, eev)),$$
$$gw(h, eev) = ex(h) \wedge (gprw^{old}(h) \vee movs2g(h)) \wedge \neg(jisr(h, eev) \wedge repeat(h, eev)).$$

Write signals and special clock enables for the special purpose register file need no modification, because, in user mode, *move* and *eret* instructions trigger an illegal interrupt before the register file is updated.

The data paths of the memory management unit (MMU) from Fig. 217 implement the address translation in a fairly obvious way. Assume

$$sim(d, h^t).$$

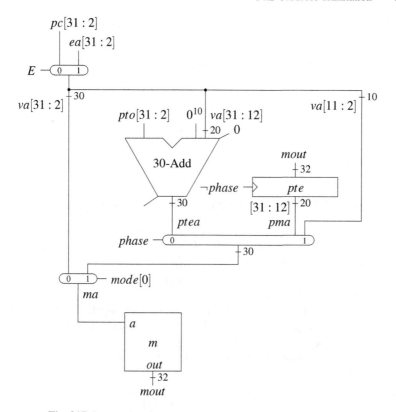

Fig. 217. Data paths of a simple memory management unit (MMU)

In system mode ($\neg h.mode[0]$), the memory address ma is computed as before and we have

$$ma(h^t) = d.pc[31:2],$$
$$ls(d) \rightarrow ma(h^{t+1}) = ea(d)[31:2].$$

In user mode ($h.mode[0]$), the adder computes page table entry addresses, and the page table entry register pte caches page table entries. Physical memory addresses are computed by concatenating the ppx field of this register with the bx field (except the last two bits) of the virtual address va. In the absence of interrupts, we have

$$ma(h^t) = ptea(d,d.pc)[31:2],$$
$$h^{t+1}.pte = pte(d,d.pc),$$
$$ma(h^{t+1}) = pma(d,d.pc)[31:2].$$

If a load or store operation (which did not cause a misalignment interrupt) was fetched, we have

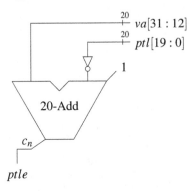

Fig. 218. Computation of the page table length exception signal *ptle*

$$ma(h^{t+2}) = ptea(d,ea(d))[31:2],$$
$$h^{t+2}.pte = pte(d,ea(d)),$$
$$ma(h^{t+3}) = pma(d,ea(d))[31:2].$$

We augment this unit by the circuit from Fig. 218, which computes a signal *ptle* signaling a page table length exception for the current translation. For the correctness, observe that it computes the negative bit of $S = \langle va(h).px[31:12]\rangle - \langle h.ptl[19:0]\rangle$ along the lines of Lemma 39. Here we distinguish the cases for the current virtual address $va(h)$ in the obvious way:

$$va(h) = \begin{cases} h.pc, & h.E = 0, \\ ea(h), & \text{otherwise.} \end{cases}$$

In the construction of Sect. 14.1, we masked the cause signals $ca(h,eev)[i]$ with execute cycle information. Now we use phase information immediately in the computation of the internal interrupt event signals:

- *malf*. In both modes a misaligned fetch is discovered by testing the last bits of the *pc* during the fetch phase:

$$malf \equiv h.pc[1:0] \neq 00 \wedge \neg h.E.$$

- *pff*. Page table length exceptions are discovered when $(E, phase) = 00$ whereas valid bits in the page table entry are tested when $(E, phase) = 01$:

$$pff(h) = h.mode[1] \wedge \neg h.E \wedge (ptle(h) \wedge \neg h.phase \vee \neg h.pte[11] \wedge h.phase).$$

- $X \in \{ill, malls, sysc, ovf\}$. These signals can all be computed once the instruction is fetched, which is the case in both modes as soon as $E = 1$:

$$X(h) = h.E \wedge X^{old}(h).$$

- *pfls*. Page table length exceptions are discovered when $(E, phase) = 10$ whereas valid bits in the page table entry are tested when $(E, phase) = 11$:

$$pfls(h) = h.mode[1] \wedge h.E \wedge ls(h) \wedge (ptle(h) \wedge \neg h.phase \vee \neg h.pte[11] \wedge h.phase).$$

Hardware correctness is shown as in Sect. 14.1. Only the definition of the cycle counts $t(i)$ has to be adjusted. We define the regular length $rle(i)$ of execution of instruction $I(d^i)$ by

$$rle(i) = \begin{cases} 2, & \neg mode(d^i), \\ 4, & mode(d^i). \end{cases}$$

We define the predicate $intd(i)$ stating that the execution of instruction $I(d^i)$ was interrupted by

$$intd(i) \equiv \exists j < rle(i) : jisr(h^{t(i)+j}, eev_h^{t(i)+j}).$$

Then the length $le(i)$ of the execution of the instruction is the number of cycles until the first interrupt in case there is one and the regular length of execution otherwise:

$$le(i) = \begin{cases} 1 + \min\{j \mid jisr(h^{t(i)+j}, eev_h^{t(i)+j})\}, & intd(i), \\ rle(i), & \neg intd(i). \end{cases}$$

Now we can define

$$t(i+1) = t(i) + le(i).$$

As before, interrupts are accumulated during execution of an instruction:

$$eev_{isa}^i = \bigvee_{t=t(i)}^{t(i+1)-1} eev_h^t.$$

14.3 Disks

We study here a single kind of device: a hard disk which uses an ATA-like protocol [AA94]. For simplicity we only treat devices which do not generate interrupts. Devices which generate interrupts are covered in the exercises. We model a disk both as a hardware component and in the ISA model. For this purpose we extend both hardware configuration h and ISA configuration d with two new components: i) a memory $h.ports$, resp. $d.ports$, of two pages whose locations are called I/O-ports, and ii) a page addressable swap memory of one terabyte. As a terabyte has $K^4 = 2^{40}$ bytes and a page has $4K = 2^{12}$ bytes, we have 2^{28} page addresses for the swap memory, both in the hardware model and in the ISA model:

$$h.sm, d.sm : \mathbb{B}^{28} \rightarrow \mathbb{B}^{8 \cdot 4K}.$$

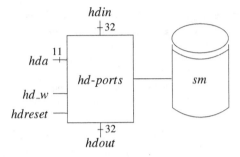

Fig. 219. Hardware disk (HD) with ports and swap memory. The ports have input and output signals like a RAM as well as a *reset* input

We represent the ports of the device as a two-page memory and the hardware memory is addressed by words while the ISA memory is addressed by bytes. Thus, formally, the disk ports in hardware are modeled as

$$h.ports : \mathbb{B}^{11} \to \mathbb{B}^{32}$$

while the disk ports in ISA are modeled as

$$d.ports : \mathbb{B}^{13} \to \mathbb{B}^{8}.$$

14.3.1 Hardware Model of a Disk

As a hardware component

$$h.ports : \mathbb{B}^{11} \to \mathbb{B}^{32}$$

the ports are word addressable. As shown in Fig. 219 the disk ports have

- data input $hdin \in \mathbb{B}^{32}$,
- data output $hdout \in \mathbb{B}^{32}$,
- address input $hda \in \mathbb{B}^{11}$, and
- write signal $hd\_w \in \mathbb{B}$

like an ordinary RAM. Moreover they have a reset signal $hdreset \in \mathbb{B}$.
Processor hardware communicates with the swap memory only via the ports. The different regions of the ports shown in Fig. 220 serve different purposes:

- Data written to swap memory and data read from memory must pass through the disk ports. This is done one page at a time. The region used for this is called the *buffer*. It consists of the first K words of the disk ports:

$$buf(h) = h.ports_K(0_{11}).$$

- The first word of the second page is used for addressing pages in the swap memory and is called the *swap page address register*. We define

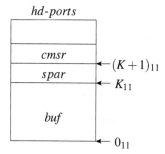

Fig. 220. Ports serve as buffer *buf*, swap page address register *spar*, and command and status register *cmsr*

$$spar(h) = h.ports(K_{11}).$$

We need only the last 28 bits to address all pages of the swap memory, thus only the last 28 bits of *spar* are used. They constitute the *swap page address*:

$$spa(h) = spar(h)[27 : 0].$$

- The second word of the second page is used to write commands to the disk and to poll its status. It is called the *command and status register*:

$$cmsr(h) = h.ports((K+1)_{11}).$$

Only the last two bits are used. They constitute the *command and status* of the disk:

$$cms(h) = cmsr(h)[1 : 0].$$

Bit 0 of this register is called the read bit. Setting it starts a read operation from swap memory to the buffer:

$$hdr(h) = cms(h)[0].$$

Bit 1 is called the write bit. Setting it starts a write operation from the buffer to swap memory:

$$hdw(h) = cms(h)[1].$$

The disk signals the completion of such an operation by clearing the command and status register. Thus we define

$$hdidle(h) \equiv cms(h) = 00.$$

A disk that is not idle is called *busy*:

$$hdbusy(h) \equiv \neg hdidle(h).$$

After reset (in cycle -1) the disk is idle:

$$hdidle(h^0).$$

Ports with other addresses $x \in \mathbb{B}^{11}$ are not used and are tied to zero:

$$\langle x \rangle > K+1 \rightarrow h.ports(x) = 0^{32}.$$

Two operating conditions should be obeyed:

- one shall not start simultaneous read and write operations

$$cms(h) \in \{00, 01, 10\}.$$

- one shall only write to the ports of an idle disk

$$hdbusy(h) \rightarrow \neg hd\_w(h).$$

In idle state, the ports behave like memory and the swap memory does not change:

$$hdout(h) = h.ports(hda(h)),$$

$$hdidle(h) \rightarrow \quad h'.ports(x) = \begin{cases} hdin(h), & hd\_w(h) \wedge x = hda(h) \wedge \langle x \rangle \leq K+1, \\ h.ports(x), & \text{otherwise}, \end{cases}$$

$$h'.sm = h.sm.$$

If the disk is busy it eventually becomes idle again in some later cycle:

$$hdbusy(h^t) \rightarrow \exists y > t : hdidle(h^y).$$

We say that a disk access ends in cycle t if the disk is busy in cycle t and idle in cycle $t+1$:

$$hdend(t) \equiv hdbusy(h^t) \wedge hdidle(h^{t+1}).$$

A hard disk access interval or *hd-interval* is an interval $[t : t']$ satisfying

$$hd\text{-}int(t,t') \equiv hdidle^{t-1} \wedge hdidle^{t'+1} \wedge \forall y \in [t : t'] : hdbusy^y.$$

Disk Writes

While the disk is busy writing, ports and swap memory don't change *in the hardware model*:

$$hdw(h^t) \wedge hd\text{-}int(t,t') \wedge y \in [t : t'] \wedge X \in \{ports, sm\} \rightarrow h^y.X = h^t.X.$$

Note that during this period, the buffer is written to some page of the swap memory. Thus on the more detailed physical level that page in swap memory *is* changing over time. But if we obey the operating conditions we have to wait until the write is completed and this gradual change is not observable on the hardware level. For this period we can define the content of this changing page any way we wish. The simplest definition is that it does not change.

After the write, the only port that changes is the command and status register, which is cleared. The swap memory page addressed by $spa(h^t)$ is updated with $buf(h^t)$:

$$h^{t'+1}.ports(x) = \begin{cases} 0^{32}, & \langle x \rangle = K+1, \\ h^t.ports(x), & \text{otherwise,} \end{cases}$$

$$h^{t'+1}.sm(x) = \begin{cases} buf(h^t), & x = spa(h^t), \\ h^t.sm(x), & \text{otherwise.} \end{cases}$$

Disk Reads

While the disk is busy reading, swap memory and ports other than the buffer are not affected.

$$hdr(h^t) \wedge hd\text{-}int(t,t') \wedge y \in [t:t'] \wedge \langle x \rangle \geq K \; \rightarrow \; \begin{matrix} h^y.ports(x) = h^t.ports(x), \\ h^y.sm = h^t.sm. \end{matrix}$$

While the read is progressing, the buffer gradually changes, and this change is observable on the detailed physical level. At the hardware level we prefer to leave the buffer undefined during this period. When the read is completed, the swap memory page addressed by $spa(h^t)$ is copied to the buffer. The command and status register is cleared.

$$buf(h^{t'+1}) = h^t.sm(spa(h^t)),$$

$$\langle x \rangle \geq K \rightarrow h^{t'+1}.ports(x) = \begin{cases} 0^{32}, & \langle x \rangle = K+1, \\ h^t.ports(x), & \text{otherwise,} \end{cases}$$

$$h^{t'+1}.sm = h^t.sm.$$

14.3.2 Accessing a Device with Memory-Mapped I/O

With an ordinary $(30,32)$-SRAM as hardware memory $h.m$, a $(10,32)$-ROM, denoted by $h.rom$, and the ports $h.ports$ we construct a memory system ms. In processor constructions, the memory system replaces the previous hardware memory $h.m$. Input and output signals are $msin$, $msout$, msa, and msw with the obvious intended meanings and of course the *reset* signal for the hard disk. We use the memory map from Fig. 221 and place the ROM at the first page, the ports of the disk at the two pages starting at address $hdbase$, and ordinary memory everywhere else. We assume that address $hdbase$ is two-page aligned, i.e.,

$$hdbase[12:0] = 0^{13}.$$

With the building blocks used in this text, we realize the memory system by an address decoder as shown in Fig. 222 and data paths shown in Fig. 223. For each component

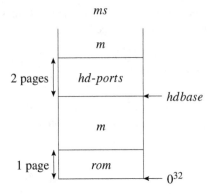

Fig. 221. Memory map of a memory system comprising one page of ROM, memory, and two pages of ports for a hard disk

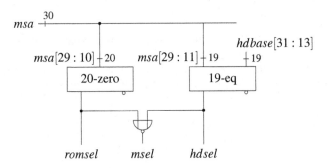

Fig. 222. Address decoder for a memory system comprising ROM, memory, and hard disk

$$X \in \{m, rom, hd\}$$

the address decoder computes signals $X sel$ indicating that component X is addressed by the memory system address msa according to the memory map of Fig. 221. Thus we have

$$romsel(h) \equiv msa(h)[29:10] = 0^{20},$$
$$hdsel(h) \equiv msa(h)[29:11] = hdbase[31:13],$$
$$msel(h) \equiv \neg(romsel(h) \vee hdsel(h)).$$

Component write signals are obviously computed as

$$hd\_w = msw \wedge hdsel,$$
$$mw = msw \wedge msel.$$

A processor with this memory system can now perform I/O-operations to and from the disk by read and write operations addressing the I/O-ports. This is called *memory-mapped I/O*.

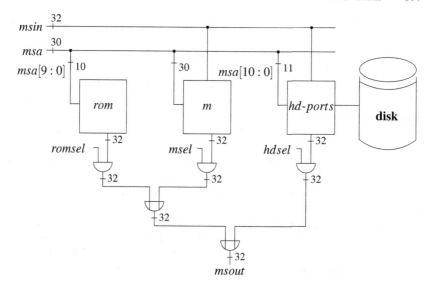

Fig. 223. Data paths of a memory system comprising ROM, memory and hard disk

With the hardware modified in this way we can not yet guarantee that the operating conditions for busy disks are met. The reason for this is that writes to memories (including ports) are performed under the control of explicit write signals, whereas reads are performed implicitly whenever write signals are not active. In more advanced memory systems explicit read signals are used in order to prevent cache misses or read side effects (for certain devices) due to implicit reads. But even here an implicit read to a device port other than the command register of an active disk would violate the operating conditions. In our simple memory system we fix this problem by forcing the effective address to 0^{32} unless a load or store operation is performed:

$$ea(h) = ea^{old}(h) \wedge ls(h).$$

This takes care of the problem when the instruction register contains meaningful data, i.e., after the fetch of the first instruction. Fortunately the disk is idle after reset and it can only become busy if a write instruction to the command and status register is performed, and this cannot be the case before an instruction is fetched. Thus, before the first instruction is fetched and executed the disk is clearly idle and implicit reads from the ports do not violate the operating conditions.

14.3.3 Nondeterminism Revisited

Formulating an extension of the MIPS ISA for a processor with the memory system constructed above and proving it correct looks at first glance like an easy exercise. We will proceed in two steps.

First we will formulate the ISA version of the memory system and formulate semantics in case the disk is idle. This step is indeed straightforward and involves basically just the usual translation between the byte addressable memory of ISA and the word addressable memories of hardware.

In a second step we will deal with busy disks and their return to idle state after a read or write operation. During each ISA instruction of the processor the disk might complete its work and become idle again or it might stay busy. In order to keep the ISA model simple we will hide this information from the user. If the disk is busy in configuration d then there will be *two* possible next configurations d': one where the disk is still busy and one where the disk has completed its work and is idle again. For single configurations d we will hide from the programmer which case occurs and only guarantee that in some later configuration the disk is idle again. This is called *nondeterminism*.

In a nondeterministic model one has to program such that the desired result is achieved for all possible nondeterministic choices. Note that we have encountered nondeterminism already in Sect. 13.3.3. Before we proceed we quickly review the ways in which nondeterministic computational models are formalized in the literature.

Ordinary deterministic computations are always modeled by a transition function

$$\delta : K \times E \to K,$$

where K is a set of states or configurations and E is the set of possible inputs. If a machine reads input $e \in E$ in configuration $k \in K$, then it makes a transition into configuration

$$k' = \delta(k, e).$$

So far we have seen hardware configurations, ISA configurations, and C0 configurations. Inputs were the reset signal ($reset \in \mathbb{B}$) and the external interrupt signals $eev \in \mathbb{B}^{32}$. We have also used notation like $k \to^e k'$ to indicate that k' is the *unique* next configuration if we read input e in configuration k:

$$k \to^e k' \leftrightarrow k' = \delta(k, e).$$

In nondeterministic computations a machine which reads input e in state k has several possible next states k'. One maintains the notation $k \to^e k'$, but it now indicates that k' is *a possible* next configuration if input e is read in configuration k. The three standard ways to formalize this are:

1. Transition relations: we make δ a *transition relation*

$$\delta \subseteq K \times E \times K.$$

The tuple (k, e, k') is included in the relation δ if a machine which reads e in state k can go to state k'. If nondeterminism is modeled this way, one defines formally

$$k \to^e k' \leftrightarrow (k, e, k') \in \delta.$$

2. Transition functions which specify the set of possible next states: in this case the range of the transition is the power set, i.e., the set 2^K of all subsets of K:

$$\delta : K \times E \to 2^K.$$

One includes in $\delta(k,e)$ all states k' to which the machine can go if it reads e in state k, and defines formally

$$k \to^e k' \leftrightarrow k' \in \delta(k,e).$$

In Sect. 13.3.3 we formalized the nondeterministic portion of computations in this style: we did a nondeterministic reconstruction of a consistent $C0$ state c' from an ISA state d' by specifying the set $\{(d',c') \mid consis(c',d')\}$ of all possible next $C0$ states.

3. Oracle inputs: the transition function gets an extra *oracle* input $o \in O$:

$$\delta : K \times E \times O \to K$$

specifying which of the possible nondeterministic choices is made. A machine reading input e in state k can go to state k' if the transition function allows this for *some* oracle input o:

$$k \to^e k' \leftrightarrow \exists o \in O : \delta(k,e,o) = k'.$$

The third alternative has turned out to be convenient in proofs where one abstracts sequential nondeterministic ISA computations from parallel hardware computations [KMP14]. Because this is what we plan, we will use both the second and third alternatives in this section.

In all formalizations a nondeterministic ISA computation for input sequence (e_i) is of course a sequence (k^i) of configurations satisfying for all i

$$k^i \to^{e_i} k^{i+1}.$$

In the third formalization one often collects the sequence of oracle inputs o_i permitting the transition from k^i to k^{i+1}, i.e.,

$$k^{i+1} = \delta(k^i, e_i, o_i),$$

into a stepping function

$$s : \mathbb{N} \to O$$

which for every step i provides the oracle input:

$$s(i) = o_i.$$

14.3.4 Nondeterministic ISA for Processor + Disk

In ISA memory, ROM and ports are byte addressable, thus ISA configurations d now have a new component $d.hd$ with

$$d.hd.ports : \mathbb{B}^{13} \to \mathbb{B}^8.$$

Definitions of buffer, swap page address, and command and status register are adapted in the obvious way:

$$buf(d) = d.hd.ports_{4K}(0_{13}),$$
$$spar(d) = d.hd.ports_4((4K)_{13}),$$
$$cmsr(d) = d.hd.ports_4((4K+4)_{13}),$$

$$spa(d) = spar(d)[27:0],$$
$$cms(d) = cmsr(d)[1:0].$$

Read, write, and idle predicates are

$$hdr(d) = cms(d)[0],$$
$$hdw(d) = cms(d)[1],$$
$$hdidle(d) \equiv cms(d) = 00,$$
$$hdbusy(d) \equiv \neg hdidle(d).$$

Unused ports are tied to zero:

$$\langle x \rangle \geq 4K+8 \to d.ports(x) = 0^8.$$

Initially the disk is idle:
$$hdidle(d^0).$$

We define the address ranges $A(rom)$, $A(hd)$, and $A(ram)$ of ROM, hard disk, and RAM in the memory system by

$$A(rom) = [0_{32} : (4K-1)_{32}],$$
$$A(hd) = [hdbase : hdbase +_{32} (8K-1)_{32}],$$
$$A(ram) = \mathbb{B}^{32} \setminus (A(rom) \cup A(hd)).$$

For addresses $a \in \mathbb{B}^{32}$, the content of memory system $ms(d)$ at address a is then defined as

$$ms(d)(a) = \begin{cases} d.rom(a[11:0]), & a \in A(rom), \\ d.ports(a[12:0]), & a \in A(hd), \\ d.m(a), & a \in A(ram). \end{cases}$$

Depending on the mode bit we define the generalized pc as

$$gpc(d) = \begin{cases} d.pc, & \neg mode(d), \\ pma(d.pc, d), & mode(d). \end{cases}$$

We define the current instruction as

$$I(d) = ms(d)_4(gpc(d)).$$

Depending on the mode bit the generalized effective address is defined as

$$gea(d) = \begin{cases} ea(d), & \neg mode(d), \\ pma(ea(d), d), & mode(d). \end{cases}$$

We define that the disk is accessed in configuration d by

$$hdacc(d) \equiv ls(I(d)) \wedge gea(d) \in A(hd).$$

Some new software conditions arise:

- page tables should lie before the ports:

$$\langle d.pto \rangle + 4 \cdot \langle d.ptl \rangle - 1 < \langle hdbase \rangle.$$

- instructions should not be fetched from the ports:

$$gpc(d) \in A(rom) \cup A(ram).$$

- an access to a non-idle disk can only be a read of the command and status register:

$$\neg hdidle(d) \wedge hdacc(d) \rightarrow l(d) \wedge gea(d) = hdbase +_{32} (4K + 4)_{32}.$$

Note that this simple condition excludes reads of the buffer during disk reads.

With these restrictions, the semantics of the new ISA model can be defined. We use *two* transition functions δ and η to do this. Basically, as before δ models a processor step and the new function η models the completion of a hard disk access.

- Processor steps: let

$$d' = \delta(d, eev)$$

be the result of a processor step in configuration d and assume the absence of interrupts, i.e.,

$$\neg jisr(d, eev).$$

Then the effect of load instructions is defined by

$$l(d) \rightarrow d'.gpr(rt(d)) = ms(d)_4(gea(d)).$$

The semantics of store instructions is defined by

$s(d) \rightarrow$

$$d'.m(x) = \begin{cases} byte(i, d.gpr(rt(d))), & x = gea(d) +_{32} i_{32} \wedge i < 4 \wedge \\ & x \in A(ram), \\ d.m(x), & \text{otherwise,} \end{cases}$$

$$d'.hd.ports(x) = \begin{cases} byte(i, d.gpr(rt(d))), & x = gea(d)[12:0] +_{13} i_{13} \wedge i < 4 \wedge \\ & x \in A(hd), \\ d.hd.ports(x), & \text{otherwise,} \end{cases}$$

which of course implies

$s(d) \rightarrow$

$$ms(d')(x) = \begin{cases} byte(i, d.gpr(rt(d))), & x = gea(d) +_{32} i_{32} \wedge i < 4 \wedge \\ & x \in A(ram) \cup A(hd), \\ ms(d)(x), & \text{otherwise.} \end{cases}$$

- Hard disk steps. These steps can only be executed when the disk is busy. Thus let
$$d' = \eta(d) \wedge \neg hdidle(d).$$

We only specify components of the configuration which change. In d' the status register is cleared:
$$cms(d') = 00.$$

In read operations, the buffer is overwritten by the page in memory addressed by the swap page address:
$$hdr(d) \rightarrow buf(d') = d.sm(spa(d)).$$

In write operations, the buffer is copied into the page in swap memory addressed by the swap page address:
$$hdw(d) \rightarrow d'.sm(spa(d)) = buf(d).$$

Processor + disk computations are sequences (d^i). In each step either a processor step or a disk step is performed. We use
$$O = \{p, d\}$$

as the set of oracle inputs for the nondeterministic choices, where p indicates a processor step and d indicates a disk step. The sequence of oracle inputs is collected into a stepping function
$$s : \mathbb{N} \rightarrow \{p, d\}.$$

We count the number of processor instructions completed up to step i by a function ic:

$$ic(0) = 0,$$

$$ic(i+1) = \begin{cases} ic(i)+1, & s(i) = p, \\ ic(i), & \text{otherwise.} \end{cases}$$

If (eev_{isa}^j) is the input sequence of external interrupt vectors, then for all j processor step j should depend on eev_{isa}^j. If in step i of the ISA computation $s(i) = p$, we perform processor step $ic(i)$, otherwise we perform a disk step. Thus, if we perform a processor step, we should consume external interrupt input $eev_{isa}^{ic(i)}$. We define

$$d^{i+1} = \begin{cases} \delta(d^i, eev_{isa}^{ic(i)}), & s(i) = p, \\ \eta(d^i), & \text{otherwise.} \end{cases}$$

This defines the computation (d^i) for input sequence (eev_{isa}^i) and stepping function s. The distinction whether step i is a disk step or a processor step is trivially made by the stepping function.

The following *liveness condition* states that disk accesses eventually terminate. It belongs to the definition of the ISA of processor + disk:

$$hdbusy(d^i) \; \rightarrow \; \exists i' > i : hdidle(d^{i'}).$$

Alternatively we can specify the computation by a transition function Δ specifying the possible next states. For this style of specification we use extended configurations

$$N = (N.d, N.ic),$$

where $N.d$ is an ISA configuration and $N.ic \in \mathbb{N}$ counts processor instructions completed so far. For extended configurations (d, ic) we now define the *set* $\Delta(d, ic)$ of possible next configurations (d', ic') in the following way: if the disk is idle, only a processor step is possible. If the disk is busy, two cases are possible: i) a processor step with interrupt vector eev_{isa}^{ic} is performed, or ii) a disk step is performed. Processor steps increase the instruction count; disk steps do not:

$$\Delta(d, ic) = \begin{cases} \{(\delta(d, eev_{isa}^{ic}), ic+1)\}, & hdidle(d), \\ \{(\delta(d, eev_{isa}^{ic}), ic+1), (\eta(d), ic)\}, & hdbusy(d). \end{cases}$$

With this formalization an extended computation for input sequence (eev^i) is a sequence $N = (N^i)$ of extended configurations satisfying for all i

$$(N^{i+1}.d, N^{i+1}.ic) \in \Delta(N^i.d, N^i.ic).$$

Note that we can reconstruct a stepping function from instruction counts: the processor is stepped if and only if the instruction count increases. We capture this in the predicates

$$pstep(N, i) \equiv N^{i+1}.ic = N^i.ic + 1,$$
$$hdstep(N, i) \equiv N^{i+1}.ic = N^i.ic.$$

We have

$$N^{i+1}.d = \begin{cases} \delta(N^i.d, eev_{isa}^{N^i.ic}), & pstep(N,i), \\ \eta(N^i.d), & hdstep(N,i). \end{cases}$$

Thus N is an extended computation for input sequence (eev_{isa}^i) if and only if the sequence of ISA configurations $(N^i.d)$ is a computation for (eev_{isa}^i) with the stepping function reconstructed from the instruction count.

In this model the liveness condition of disks is

$$hdbusy(N^i.d) \rightarrow \exists i' > i : hdidle(N^{i'}.d).$$

In contexts where only extended computations are considered we often simply call them computations.

14.3.5 Hardware Correctness

In order to show that the hardware constructed in Sect. 14.3.2 interprets the nondeterministic ISA from Sect. 14.3.4, we first modify the simulation relation in an obvious way. In the simulation relation $sim(d,h)$ used in the correctness proof, we require the following conditions to hold:

- processor registers and swap memory have identical values:

$$X \in \{pc, gpr, spr, sm\} \rightarrow d.X = h.X.$$

- RAM and ROM store corresponding data:

$$a \in A(ram) \cup A(rom) \rightarrow d.m_4(a[31:2]00) = h.m(a[31:2]).$$

- $spar$ and $cmsr$ have identical values:

$$y \in \{spar, cmsr\} \rightarrow y(d) = y(h).$$

- except during disk reads the buffers store corresponding data:

$$\neg hdr(d) \rightarrow buf(d) = buf(h).$$

  During these operations, the hardware buffer is left unspecified (it is gradually filled with data from swap memory) and the ISA buffer has the old value. This does not cause problems, because by the software conditions the buffer of busy disks is never read by the processor.

As usual, one constructs from the initial hardware configuration h^0 an initial ISA configuration d^0 satisfying $sim(d^0, h^0)$. From the sequence (eev_h^i) of external hardware event signals and the hardware computation (h^t) we define the sequence $(le(i))$ of lengths of processor instructions, the corresponding numbers $(t(i))$ of hardware cycles and the external interrupt event signals (eev_{isa}^i) as in Sect. 14.2. The definition of the ISA computation (d^i) now becomes very slightly more technical.

For hardware cycles t, we define a predicate $pend(t)$ (end of processor step) stating that a processor step is completed in cycle t by

$$pend(t) \leftrightarrow \exists i : t = t(i)$$

and we count the number $ne(t) \in [0 : 2]$ of disk and processor steps completed in cycle t by

$$ne(t) = hdend(t) + pend(t).$$

The number of ISA steps completed before cycle t is defined by

$$NE(0) = 0,$$
$$NE(t+1) = NE(t) + ne(t).$$

For each hardware cycle t where any ISA step ends, i.e., with $ne(t) \neq 0$, we extend the definition of the stepping function $s(i)$ for arguments

$$i \in [NE(t), NE(t+1)).$$

If $ne(t) = 1$, then only a processor step or a disk step is completed and the choice is obvious

$$s(NE(t)) = \begin{cases} p, & pend(t), \\ d, & hdend(t). \end{cases}$$

If $ne(t) = 2$, then a processor step and a disk step both end in cycle t, but in the hardware the processor does not see the effect of the disk step. Thus in the ISA the processor step should be performed before the disk step, and we define

$$s(NE(t)) = p,$$
$$s(NE(t)+1) = d.$$

Because for all i processor step i is completed in cycle $t(i)$ we obtain the following by induction on i.

Lemma 131.
$$ic(NE(t(i))) = i.$$

By induction on t, one now establishes the simulation relation with the hardware computation.

Lemma 132.
$$sim(d^{NE(t)}, h^t).$$

14.3.6 Order Reduction

We collect the processor components and the hard disk components of ISA configurations d into parts

$$proc(d) = (d.pc, d.gpr, d.spr, d.m),$$
$$hd(d) = (d.ports, d.sm)$$

and collect for later reference a few simple observations.

Lemma 133.

1. *Disk steps do not change the processor part:*

$$proc(\eta(d)) = proc(d).$$

2. *The effect of processor steps which do not access the disk depends only on the processor part:*

$$\neg hdacc(d) \wedge proc(d) = proc(e) \;\rightarrow\; proc(\delta(d, eev)) = proc(\delta(e, eev)).$$

3. *Processor steps which do not access the disk do not change the disk part:*

$$\neg hdacc(d) \;\rightarrow\; hd(\delta(d, eev)) = hd(d).$$

4. *The effect of disk steps on the disk part does not depend on the processor part:*

$$hd(d) = hd(e) \;\rightarrow\; hd(\eta(d)) = hd(\eta(e)).$$

The nondeterministic programming model forces the programmer to produce programs which work for all permitted orders of processor steps and disk steps. If we could reduce the set of orders we have to consider, programming might become easier. Indeed we will define a much simpler and smaller set of *normal* computations and then show an order reduction theorem stating that, for any computation, there is a normal computation producing exactly the same results as far as the processor parts of configurations are concerned. First we observe that disk steps are always separated by processor steps which access the disk, because i) they can only occur when the disk is busy, ii) they clear *cmsr*, and iii) the disk can only become busy if the processor writes *cmsr*.

Let $N = (N^i)$ be an extended computation. We call a computation step i in computation N *normal* if it is a processor step or if it is a disk step immediately followed by a processor step that accesses the disk:

$$n\text{-}step(N, i) \;\leftrightarrow\; pstep(N, i) \vee hdstep(N, i) \wedge hdacc(N^{i+1}.d).$$

We call computation N normal if all steps are normal, and we call it k-normal if all steps up to and including step k are normal:

$$normal(N) \leftrightarrow \forall i : n\text{-}step(N,i),$$
$$k\text{-}normal(N) \leftrightarrow \forall i \leq k : n\text{-}step(N,i).$$

Note that all computations are 0-normal, because after reset the disk is idle and thus no disk step is possible in step 0.

We call nondeterministic computations N and M *processor equivalent* and write $N \equiv_p M$ if configurations with equal instruction count have equal processor parts.

$$N \equiv_p M \leftrightarrow \forall i,j : N^i.ic = M^j.ic \rightarrow proc(N^i.d) = proc(M^j.d).$$

An easy exercise shows the following.

Lemma 134. $\equiv_p$ *is an equivalence relation.*

Proof. Symmetry is trivial. For the proof of reflexivity consider i and j such that

$$N^i.ic = N^j.ic.$$

If $i = j$ there is nothing to show, otherwise assume without loss of generality $i < j$. Then step i is a disk step (which does not change the processor state by Lemma 133.1) and we have

$$proc(N^i.d) = proc(N^{i+1},d) = proc(N^j.d).$$

For the proof of transitivity assume $N \equiv_p M$ and $M \equiv_p Q$ and let

$$N^i.ic = Q^k.ic = n.$$

Because instruction counts in computations are consecutive and start with zero, we find an index j with

$$M^j.ic = n.$$

Using processor equivalence of computations we conclude immediately

$$proc(N^i.d) = proc(M^j.d) = proc(Q^k.d).$$

The main result of this section is the following order reduction theorem.

Lemma 135. *For every nondeterministic computation N, there exists a processor equivalent normal computation M:*

$$\forall N \exists M : normal(M) \wedge M \equiv_p N.$$

We proceed to prove the result in two steps. First we successively switch disk steps which are not directly followed by a hard disk access with the processor step to the right. This allows us to transform computation N for any k into k-normal computations $N(k)$ which are processor equivalent to computation N. In a second step we construct from this sequence of computations a single computation M by taking the i'th extended configuration from the i-normal computation $N(i)$ for all i

$$M^i = N(i)^i$$

and show that it is processor equivalent to N and normal.

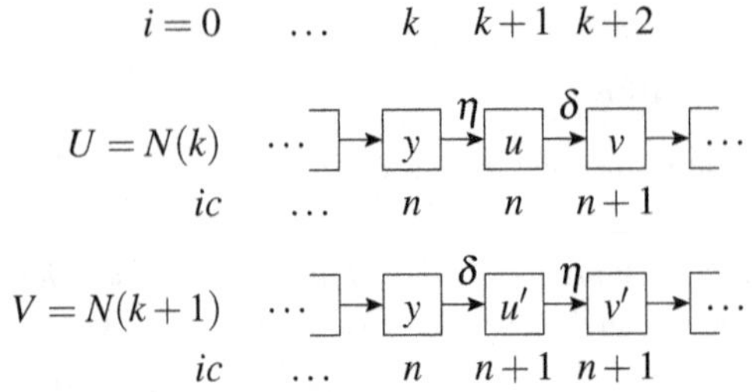

$$i = 0 \quad \ldots \quad k \quad k+1 \ \ k+2$$

Fig. 224. Exchanging a disk step and a processor step which does not access the disk in computations $U = N(k)$ and $V = N(k+1)$

Lemma 136. *For all k there is a computation $N(k)$ which is k-normal and processor equivalent to N:*

$$\exists N(k) : k\text{-}normal(N(k)) \wedge N(k) \equiv_p N.$$

Proof. by induction on k. For $k = 0$ we simply take $N(0) = N$. In the induction step from k to $k+1$ there are two cases. If step $k+1$ in computation $N(k)$ is already normal we simply take $N(k+1) = N(k)$, i.e.,

$$k+1\text{-}normal(N(k)) \ \rightarrow \ N(k+1) = N(k).$$

Otherwise, step $k+1$ of computation $U = N(k)$ is a disk step, and step $k+2$ is a processor step which does not access the disk, as illustrated in Fig. 224.

We abbreviate

$$n = U^k.ic = U^{k+1}.ic,$$
$$y = U^k.d,$$
$$u = U^{k+1}.d = \eta(y),$$
$$v = U^{k+2}.d = \delta(u, eev^n).$$

We construct computation V by interchanging these steps. Outside of steps $k+1$ and $k+2$ we leave computation U unchanged:

$$i \notin [k+1 : k+2] \rightarrow V^i = U^i.$$

In steps k and $k+1$ we continue the computation with

$$V^{k+1}.d = \delta(y, eev^n) = u',$$
$$V^{k+1}.ic = V^k.ic + 1 = n+1,$$
$$V^{k+2}.d = \eta(u') = v',$$
$$V^{k+2}.ic = V^{k+1}.ic = n+1.$$

That the new sequence (V^k) is a computation follows if we can show

$$v = v'.$$

Using Lemma 133, this can be easily shown. For the processor part we conclude successively

$$proc(y) = proc(u) \quad \text{(part 1)},$$
$$proc(u') = proc(v) \quad \text{(part 2)},$$
$$proc(u') = proc(v') \quad \text{(part 1)},$$

which shows

$$proc(v') = proc(v).$$

For the disk part we conclude successively

$$hd(y) = hd(u') \quad \text{(part 3)},$$
$$hd(u) = hd(v') \quad \text{(part 4)},$$
$$hd(u) = hd(v) \quad \text{(part 3)},$$

which shows

$$hd(v) = hd(v').$$

Computation V is obviously $(k+1)$-normal. By induction hypothesis we have

$$U \equiv_p N.$$

By the transitivity of processor equivalence it suffices to show

$$V \equiv_p U.$$

Thus we have to show

$$V^i.ic = U^j.ic \;\rightarrow\; proc(V^i.d) = proc(U^j.d).$$

This is done by a simple case split on i:

- if $i \neq k+1$ we have
$$V^i = U^i$$
and the claim follows from reflexivity of processor equivalence.
- if $i = k+1$ we have
$$V^i.ic = V^{k+1}.ic = n+1 = U^{k+2}.ic = U^j.ic.$$

Using the reflexivity of processor equivalence and the identities shown above we conclude

$$proc(U^j.d) = proc(U^{k+2}.d)$$
$$= proc(v)$$
$$= proc(u')$$
$$= proc(V^{k+1}.d)$$
$$= proc(V^i.d).$$

By Lemma 136 we can reorder the original computation N for any k into a processor-equivalent computation $N(k)$ whose first k steps are normal. If we choose k larger than the number of computation cycles possible (there is a maximal number of such cycles [Pau78] until the end of the solar system), this solves the problem for practical purposes. However, with a few extra lines we can complete the proof of the desired reorder theorem. Consider Fig. 225. There we have drawn an infinite matrix of extended configurations, where for each k row k contains the extended configurations of computation $N(k)$. Thus in every row we know for all i:

$$N(k)^{i+1} \in \Delta(N(k)^i).$$

We construct a new sequence M of extended configurations with the configurations along the diagonal of the matrix, i.e., we set for all k

$$M^k = N(k)^k.$$

This sequence is a computation, because for all k we have by construction of the sequences $N(k+1)$

$$M^k = N(k)^k = N(k+1)^k.$$

Because $N(k+1)$ is an extended computation we get

$$M^{k+1} = N(k+1)^{k+1} \in \Delta(N(k+1)^k).$$

Hence

$$M^{k+1} \in \Delta(M^k).$$

It remains to show that computation M is processor equivalent to the original computation N.

Thus let

$$M^i.ic = N(i)^i.ic = N^j.ic.$$

From

$$N(i) \equiv_p N$$

and the construction of M we immediately get

$$proc(M^i.d) = proc(N(i)^i.d) = proc(N^j.d)$$

and the reduction theorem is proven.

14.3.7 Disk Liveness in Reordered Computations

Formulating disk liveness for reordered computations is not entirely trivial. Indeed in the reordered computation M a disk access can disappear. Fortunately this will turn out to be harmless. We first observe a few details about the reordering process. We call a step i in a computation N a processor-disk access if in configuration N^i the disk is accessed and the processor is stepped:

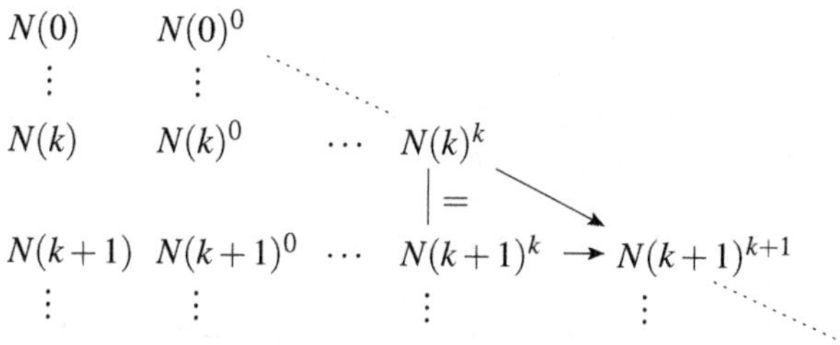

$$N(0) \qquad N(0)^0$$

$$\vdots \qquad \vdots$$

$$N(k) \qquad N(k)^0 \qquad \cdots \qquad N(k)^k$$

$$\big| = $$

$$N(k+1) \quad N(k+1)^0 \quad \cdots \quad N(k+1)^k \;\rightarrow\; N(k+1)^{k+1}$$

$$\vdots \qquad \vdots \qquad \qquad \vdots \qquad \qquad \vdots$$

Fig. 225. Constructing a new computation M along the diagonal of the infinite matrix whose rows are the computations $N(k)$

$$pdacc(N,i) \equiv hdacc(N^i.d) \wedge pstep(N,i).$$

Such steps are never reordered:

$$pdacc(N(k),i) \equiv pdacc(N(k+1),i).$$

By induction we get

$$pdacc(N,i) \;\rightarrow\; \forall k : pdacc(N(k),i),$$

which implies that the processor-disk accesses are still in their place in the reordered sequence M:

$$pdacc(N,i) \;\rightarrow\; pdacc(M,i).$$

We define a finite or infinite sequence of step numbers $x(j)$ where the j'th processor-disk access occurs in all the computations under consideration:

$$x(1) = \min\{i \mid pdacc(N,i)\},$$

$$x(j+1) = \begin{cases} \min\{i > x(j) \mid pdacc(N,i)\}, & \exists i > x(j) : pdacc(N,i), \\ \infty, & \text{otherwise.} \end{cases}$$

We consider the intervals $X(j)$ of steps where disk steps can occur. If $x(j)$ is defined, we define

$$X(j) = (x(j) : x(j+1)).$$

By the operating conditions for disks at most one disk step (indicating the end of a disk access) can occur in each interval $X(j)$, because a second access can only end after it was started, and this involves a processor-disk access:

$$\#\{i \in X(j) \mid hdstep(N,i)\} \in \{0,1\}.$$

Disk steps are not reordered across processor-disk accesses:

$$\forall k : \#\{i \in X(j) \mid hdstep(N,i)\} = \#\{i \in X(j) \mid hdstep(N(k),i)\}.$$

Because disk steps are not reordered across processor-disk accesses and because the processor is stepped in these steps, configurations at steps $x(j)$ and $x(j)+1$ are identical (for $x(j) \in \mathbb{N}$):

$$i \in \{x(j), x(j)+1\} \;\rightarrow\; \forall k : N^i = N(k)^i = M^i. \tag{11}$$

Busy signals can only become active by processor-disk accesses at steps $x(i)$. Thus within intervals they are non-increasing:

$$i \in X(j) \;\rightarrow\; hdbusy(N^i.d) \geq hdbusy(N^{i+1}.d) \wedge hdbusy(M^i.d) \geq hdbusy(M^{i+1}.d). \tag{12}$$

Because disk steps (rendering this disk idle) are reordered to the right, the disk is also busy more often as reordering progresses:

$$hdbusy(N(k)^i.d) \leq hdbusy(N(k+1)^i.d).$$

Thus

$$\forall k : hdbusy(N^i.d) = hdbusy(N(0)^i.d) \leq hdbusy(N(k)^i.d)$$

and

$$hdbusy(N^i.d) \leq hdbusy(N(i)^i.d) = hdbusy(M^i.d). \tag{13}$$

In the final reordered sequence M the disk steps are reordered to the right border of the intervals $X(j)$ if this border exists. If there are only $p \in \mathbb{N}$ processor-disk accesses, then we have

$$X(p) = (x(p), \infty)$$

and a disk access ending after step $x(p)$ is visible in every computation $N(k)$ but not in M. In this situation disk liveness is lost in the reordered computation M. We prevent this situation with a software condition on the reordered computation M stating that:

If a disk is busy in step $x(j)$ or a disk access was started in this step, then the disk must be accessed by the processor in a later step:

$$y \in \{x(j), x(j)+1\} \wedge hdbusy(M^y.d) \;\rightarrow\; \exists t > y : pdacc(M,t).$$

With this software condition we can show liveness for the reordered computation.

Lemma 137.

$$hdbusy(M^i.d) \;\rightarrow\; \exists t > i : hdidle(M^t.d).$$

Moreover t is the index of a processor-disk access

$$\exists q : t = x(q).$$

Proof. We proceed in several steps that are illustrated in Fig. 226.

1. By hypothesis we have

$$hdbusy(M^i.d).$$

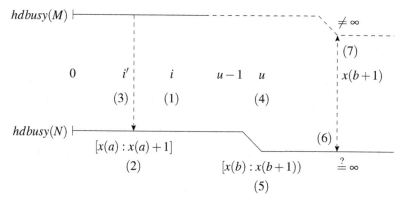

Fig. 226. Illustration of the order of arguments in the proof of disk liveness after reordering. Step 6 requires the software condition

2. Index i lies in some interval

$$i \in [x(a) : x(a+1)).$$

We introduce

$$i' = \begin{cases} x(a), & i = x(a), \\ x(a)+1, & i > x(a). \end{cases}$$

By hypothesis and in the second case by monotonicity of busy signals (Eq. 12) we conclude

$$hdbusy(M^{i'}.d).$$

3. By Eq. 11 configurations of computations M and N are equal at interval boundaries. Hence

$$hdbusy(N^{i'}.d).$$

4. By disk liveness for the original computation N there is an index $u > i'$ such that the disk is idle in step u of the original computation. For the smallest such index the disk is busy in the previous step:

$$\exists u > i' : hdidle(N^u.d) \wedge hdbusy(N^{u-1}.d).$$

Because reordering makes the disk busy more often (Eq. 13) we also have

$$hdbusy(M^{u-1}.d).$$

5. This index u lies in some set $[x(b) : x(b+1))$. Because $u > i' \geq x(a)$ we have $b \geq a$. In the figure we illustrate the case $b > a$, but $b = a$ is possible too:

$$\exists b : b \geq a \wedge u \in [x(b) : x(b+1)).$$

If $u = x(b)$ we have by Eq. 11

$$hdbusy(M^u.d) = hdbusy(N^u.d) = 0$$

and we are done with

$$t = u = x(b).$$

6. Otherwise, note that $u > x(b) + 1$, or equivalently $x(b) + 1 \leq u - 1$, because the processor step at $x(b)$ cannot turn the disk idle. We have by Eqs. 11 and 12

$$hdbusy(M^{x(b)+1}.d) = hdbusy(N^{x(b)+1}.d) \geq hdbusy(N^{u-1}.d) = 1$$

and get by the software condition that the disk will be accessed again in the future, thus $x(b+1)$ is finite:

$$x(b+1) \in \mathbb{N}.$$

7. Again using Eqs. 11 and 12 we conclude

$$0 = hdbusy(N^u.d) \geq hdbusy(N^{x(b+1)}.d) = hdbusy(M^{x(b+1)}.d)$$

and we are done with

$$t = x(b+1).$$

14.3.8 $C0$ + Assembly + Disk + Interrupts

We proceed to define and justify the semantics of C + assembly for the case in which the system contains disks and external interrupts are allowed. Recall that configurations have the form $y = d$ or $y = (d, c)$, where d is an ISA configuration and c is a $C0$ configuration. We have to deal with several technical issues:

- We have now two sources of nondeterminism: i) the choice of a C abstraction after running assembly code, and ii) the interleaving of disk steps and processor steps.
- C+A computations proceed with different granularities: i) with $C0$ steps as long as there is a C abstraction, and ii) with ISA steps otherwise. External interrupt signals come originally with the granularity of hardware cycles defined as (eev_h^t). In the hardware correctness proofs of Sects. 14.1.6 and 14.2.2 we already abstracted a sequence of interrupt event signals

$$eev_{isa} = (eev_{isa}^i)$$

which change in the absence of devices at the granularity of ISA instructions. After the introduction of disks, only the processor steps of the ISA consumed inputs eev_{isa}^i. In Sect. 14.3.4 we accounted for this by using extended configurations of the form (d, ic), where the new component ic counted the instructions completed. The set of possible next configurations was defined as

$$\Delta_{ISA}(d, ic, eev_{isa}) = \begin{cases} \{(\delta_M(d, eev_{isa}^{ic}), ic+1)\}, & hdidle(d), \\ \{(\delta_M(d, eev_{isa}^{ic}), ic+1), (\eta(d), ic)\}, & \text{otherwise.} \end{cases}$$

Thus, the external interrupt to be consumed in configuration (d, ic) was determined by the instruction count component. In the sequel we abstract from sequence (eev_{isa}^i) a sequence (eev_{c+a}^i) where i counts C+A instructions. We will do this by extending C+A configurations y with an instruction count component $ic \in \mathbb{N}$ counting the instructions completed at the ISA level so far. Thus we will have C+A configurations of the form

$$y = (d, ic) \quad \text{or} \quad y = (d, c, ic).$$

To address the components of a C+A configuration we also use the notation

$$y = (y.d, y.ic) \quad \text{resp.} \quad y = (y.d, y.c, y.ic).$$

- While translated C code is running we would like to ignore possible disk steps occurring in that period. Indeed this can be justified. By Lemma 135 we can restrict our attention to C+A computations, whose underlying ISA computation is normal: as long as the disk is not accessed normal computations are deterministic, because disk steps simply don't occur.[2] We introduce a new software condition requiring that the address range of $C0$ variables should not overlap with the disk ports. As $C0$ variables occupy addresses below $hmax$ and ports occupy addresses starting at $hdbase$ this amounts to software condition

$$\langle hmax \rangle \leq \langle hdbase \rangle.$$

This of course implies that translated $C0$ statements cannot access the ports:

$$transc(d) \rightarrow \neg hdacc(d),$$

and hence no disk steps occur (in the semantics) while translated C code is executed.

In the sequel only normal ISA steps are considered. Assuming the above software conditions we extend the definition of C+A semantics of Sect. 13.3.3. We always refer to the next ISA configuration reached by a processor step by

$$d' = \delta_M(d, eev_{isa}^{ic})$$

and we always refer to the set of possible next ISA configurations by

$$D' = \Delta_{ISA}(d, ic, eev_{isa}).$$

For a given C+A configuration $y = (d, ic)$ or $y = (d, c, ic)$ and a sequence (eev_{isa}^i) we define the set Y' of possible next C+A configurations y'

$$Y' = \Delta_{C+A}(y)$$

by an obvious case split using notation from Sect. 13.3:

---

[2] This does not mean that disk steps of the hardware do not occur while the disk is not accessed. It simply means that the programmer can ignore disk steps for the time being and treat them as if they happened immediately before the disk is accessed.

- running outside of translated C and staying outside of translated C:

$$y = (d, ic) \wedge \neg transc(d) \wedge \neg transc(d').$$

We simply make a possible step of the ISA computation and increase the instruction count:

$$Y' = \{(d', ic+1) \mid d' \in D'\}.$$

Note that this definition covers the occurrence of interrupts $(jisr(d, eev_{isa}^{ic}))$.

- running translated $C0$ code (extended or not):

$$y = (d, c, ic) \wedge transc(d).$$

The ISA computation is normal and due to the software condition above does not contain disk steps; thus it is deterministic. We consider the sequence (q^i) of next ISA configurations after d:

$$q^0 = d,$$
$$q^{i+1} = \delta_M(q^i, eev_{isa}^{ic+i}).$$

We stop the sequence in two cases:

(a) an interrupt occurs:

$$jisr(q^i, eev_{isa}^{ic+i}).$$

We execute the interrupt, drop the C abstraction, and increase the instruction count by $i+1$:

$$Y' = \{(q^{i+1}, ic+i+1)\}.$$

(b) a consistency point is reached without interrupt:

$$cpoint(q^{i+1}) \wedge \neg jisr(q^i, eev_{isa}^{ic+i}).$$

We execute a C step and increase the instruction count by $i+1$:

$$Y' = \{(q^{i+1}, \delta_C(c), ic+i+1)\}.$$

- switching to assembly:

$$y = (d, c, ic) \wedge \neg transc(d').$$

We simply drop the high-level language abstraction and proceed only with the low-level computation:

$$Y' = \{(d', ic)\}.$$

Note again, that this covers the case of interrupts.

- returning to C:

$$y = (d, ic) \wedge transc(d').$$

The ISA computation is normal. We construct the sequence (q^i) of next ISA configurations as above and stop it in two cases:

(a) an interrupt occurs:

$$jisr(q^i, eev_{isa}^{ic+i}).$$

Then

$$Y' = \{(q^{i+1}, ic+i+1)\}.$$

(b) a C abstraction can be reconstructed without interrupt:

$$\exists c : wfc(c) \wedge consis(q^{i+1}, c) \wedge \neg jisr(q^i, eev_{isa}^{ic+i}).$$

We reconstruct a well-formed consistent C configuration and increase the instruction count by $i+1$:

$$Y' = \{(q^{i+1}, c, ic+i+1) \mid wfc(c) \wedge consis(q^{i+1}, c)\}.$$

14.3.9 Hard Disk Driver

In what follows we will often define functions which do not return a value. Such functions are called *procedures*. Their declaration has the form

$$void \quad f\,(\,t_1 \quad x_1\,, \,\ldots\,,\, t_p \quad x_p\,)\,\{\,t_{p+1} \quad x_{p+1}\,;\,\ldots\,;\, t_r \quad x_r\,;\, body\,\}$$

In contrast to functions, procedures are not called as right-hand sides of assignments. A call of a procedure is a statement and has the form

$$f(e_1, \ldots, e_p)$$

Extending the grammar of $C0$ and the compiler to handle this is easy but not very instructive in lectures. Here we treat procedures simply as *syntactic sugar*, i.e., as a shorthand for something known. We introduce a global integer variable WOV (write-only variable), which is never read, and we use declarations and calls of functions f as introduced above simply as shorthands for

$$int \quad f\,(\,t_1 \quad x_1\,, \,\ldots\,,\, t_p \quad x_p\,)\,\{\,t_{p+1} \quad x_{p+1}\,;\,\ldots\,;\, t_r \quad x_r\,;\, body\,\}$$

and

$$WOV = f(e_1, \ldots, e_p).$$

In order to write a simple hard disk driver, we define a few auxiliary C+A procedures. We define their semantics in terms of the C+A configurations

$$y = (d, c) \quad \text{and} \quad y' = (d', c')$$

before and after execution of the procedures.

1. $e_1 = readms(e_2)$
   Specification:

$$c'.m(lv(e_1, c)) = d.ms_4(va(e_2, c)_{32}).$$

Reads into the $C0$ variable given by e_1 a word from the main memory at the address given by e_2.
Implementation:

```
int readms (uint a)
{
   int tmp;
   gpr (1)  = a;
   asm (
      lw    2  1  0
      sw    2 29 -4
   );
   return tmp
};
```

For an explanation, observe that while this function is running in configuration $(\tilde{d}, \tilde{c}, ic)$ we have:

$$ba(tmp, \tilde{c}) = \tilde{d}.spr(29_5) -_{32} 4_{32}.$$

An alternative implementation is

```
int readms (uint a)
{
   int tmp;
   gpr (1)  = a;
   gpr (2)  = tmp& {1};
   asm (
      lw    3 1 0
      sw    3 2 0
   );
   return tmp
};
```

This violates the very conservative context condition 16 that we used to show the general absence of dangling pointers. However, no dangling pointer is produced here.

2. $writems(e_1, e_2)$
   Specification:

   $$d'.ms_4(va(e_2, c)_{32}) = va(e_1, c)_{32}.$$

   Writes the value of the $C0$ variable given by e_1 to the main memory at the address given by e_2.
   Implementation:

```
void writems (uint x,  uint a)
{
   gpr (1)  = x;
   gpr (2)  = a {1};
   asm( sw    1 2 0 )
};
```

3. $copyms(e_1, e_2, e_3)$

Specification:

$$d'.ms_{4 \cdot va(e_3, c)}(va(e_2, c)) = d.ms_{4 \cdot va(e_3, c)}(va(e_1, c)).$$

Copies $va(e_3, c)$ words from the main memory starting at the address given by e_1 to the main memory starting at the address given by e_2.

Implementation:

```
void copyms (uint a,  uint b,  uint L)
{
   gpr (1)  = a;
   gpr (2)  = b {1};
   gpr (3)  = L {1,2};
   asm (
      blez 3 7          // until L <= 0 {
      lw   4 1 0        //
      sw   4 2 0        //    m(b)  = m(a)
      addi 1 1 4        //    a += 4
      addi 2 2 4        //    b += 4
      addi 3 3 -1       //    L -= 1
      blez 0 -6         // }
   )
};
```

That the implementation of the last three procedures satisfies the desired specification follows directly from the C+A semantics, and straightforward induction on $\langle d.gpr(3_5)\rangle$ in the case of $copyms$.

4. $readdisk(e_1, e_2)$

Specification:

$$hdidle(d') \wedge d'.ms_{4K}(va(e_1, c)_{20} \circ 0^{12}) = d.sm_{4K}(va(e_2, c)_{28} \circ 0^{12}).$$

A single page of memory is copied to the main memory address given by e_1 from the swap memory page address given by e_2. We require the disk to be idle after the execution of the procedure.

We assume that the disk is idle before execution of the procedure and that the values of e_1 and e_2 are not too large:

$$hdidle(d) \wedge (va(e_1, c) < 2^{20}) \wedge (va(e_2, c) < 2^{28}).$$

Implementation:

```
void readdisk (uint ppx,  uint spx)
{
   int y;
   writems (spx, hdbase+4*K);    // spa(d)  = spx
```

```
writems(1,hdbase+4*K+4);     // issue read access
                             // via cmsr(d)
y = 1;
while y!=0
{
   y = readms(hdbase+4*K+4) // polling
};
copyms(hdbase,ppx*4*K,K)
};
```

Here, we issue a hard disk read request by writing to the command and status register $cmsr(d)$ of the hard disk after writing the swap memory page address to the address port of the hard disk. We poll the command and status register until its value turns zero. After the hard disk turns idle, the requested data is found in the data buffer of the hard disk; we copy it from there to the designated main memory page.

The correctness proof for this program has one new aspect: we have to argue when disk steps (signaling the completion of a disk access) can occur. Because idle disks do not make steps, there is no disk step before the access is started with the write to the command and status register. Using C+A semantics we are looking for execution of two *writems* procedures in a configuration $y^{j_0} = (d^{j_0}, c^{j_0})$ satisfying

$$spa(d^{j_0}) = va(spx, c)_{28},$$
$$cmsr(d^{j_0}) = 1_{32},$$

hence

$$hdbusy(d^{j_0}).$$

The ISA portion of the extended C+A computation is normal. Thus the disk step signalling the completion of the disk access is reordered to occur before an assembly instruction accessing the disk. During the polling the only such instruction is the load instruction

```
lw    3 2 0
```

belonging to the loop body, which samples the command and status register. For $j \geq 1$ let $z(j)$ be the extended C+A configuration, in which this lw instruction is executed in iteration j of the loop. We denote the corresponding instruction count as

$$x(j) = z(j).ic.$$

If the disk is found to be busy another iteration of the loop follows. Thus the software condition for the liveness of disks in reordered computations from Sect. 14.3.6 is fulfilled. We can apply Lemma 137 and conclude that for some iteration j_k of the loop the disk is idle in ISA configuration $z(j_k).d$ and the loop is left after a successful disk operation. That the desired data is read from the disk follows from the ISA semantics for disks.

5. $writedisk(e_1, e_2)$
   Specification:

$$d'.sm_{4K}(va(e_2, c)_{28} \circ 0^{12}) = d.ms_{4K}(va(e_1, c)_{20} \circ 0^{12}).$$

We leave the implementation of this function and the correctness proof as an exercise. Do not forget to poll after the write access to the disk is started; otherwise you cannot conclude that the buffer is copied to the swap memory. Of course one can increase efficiency (and make correctness proofs slightly more involved) by polling only before one starts the next disk access.

14.4 Final Remarks

In this chapter we have made the step from user programming to system programming, and we recommend to remember quite a few key technical points:

- about interrupts:
  - the new data structures visible to the system programmer: interrupt event signals and their names, special purpose registers and their names.
  - the new instructions (*sysc*, *eret* and *move*).
  - semantics of *jisr* and the new instructions.
  - in the hardware the special purpose registers are implemented by an SPR-RAM. This allows us to translate the specification of *jisr* and *eret* into hardware by appropriate wiring of the individual data inputs in a straightforward way.
  - in our hardware design the duration of instruction execution is variable: a non-masked interrupt in the fetch cycle triggers *jisr* immediately and the hardware proceeds to the execution of the next instruction. Technically we introduce hardware cycle $t(i)$ when execution of ISA instruction i ends.
  - external interrupts eev_{isa}^i visible in ISA are abstracted in a nontrivial way from the hardware signals eev_h^t with the help of the cycles $t(i)$.

  In light of the rumor that interrupts are supposed to be very complicated, this is not too bad (at least for sequential processors).
- about address translation:
  - the partition of virtual addresses into page index $va.px$ and byte index $va.bx$.
  - the specification of signals $ptea$, pte, ppx, v, pma, and the various page fault signals.
  - the extension of the instruction set, where translated or untranslated pc and ea are used depending on the *mode*.
  - in the processor design there are up to four memory accesses per instruction in translated mode. Thus we split cycles *fetch* and *execute* into two phases when the hardware is running in translated mode. This obviously affects the definitions of cycles $t(i)$ and signals eev_{isa}^i.

- when execution of a new instruction starts (in a cycle $t(i) + 1$), the auxiliary registers like I (instruction register) contain 'stale' data from the previous instruction. Care has to be taken that these stale data do not trigger internal event signals.

- about disks:
  - the new components of configurations: i) swap memory and ii) ports (buffer, command and status, swap page address).
  - the semantics and the operating conditions (only poll status when the disk is busy).
  - minor nuisance: I/O-ports in hardware are word addressed, I/O-ports in ISA are byte addressed. Hence the disk model differs (in a quite trivial way) in hardware and in ISA.
  - construction of a hardware memory system with ROM, RAM, and disk ports is straightforward.
  - ISA of hardware + disk is nondeterministic. Disk steps — when disk accesses end — are interleaved with processor steps which is deterministic (due to physics) but not revealed to the programmer.
  - in the hardware, disk accesses and processor steps can both end in the same hardware cycle t. In ISA one has to produce two ISA steps for this cycle, and the order is important (disk step second; the processor does not notice it due to register semantics). Technically the number of ISA steps $ne(t)$ ending in hardware cycle t and the number of ISA steps $NE(t)$ ended before cycle t are introduced to handle this.
  - only processor steps i of the ISA consume external interrupt signals. Their number is tracked with function $ic(i)$. If we view the sequence (eev_{isa}^i) as an input tape, then function $ic(i)$ gives the position of a read head on that tape before execution of (disk or processor) instruction i. For processor steps one defines
  $$d^{i+1} = \delta_M(d^i, eev_{isa}^{ic(i)}).$$
  - if a tree falls in a forest and nobody hears it, does its fall produce a sound? In the same philosophical spirit you can ask: if a disk access finishes and the processor does not poll the disk (and no interrupt is produced), has the access already finished? Clearly it might have finished, but you can safely assume that it has not. That is the order reduction theorem. The single exchange steps are basically trivial, but reordering an infinite sequence requires a diagonalization argument.
  - liveness in reordered sequences is problematic, because if you start a disk access and then never access the disk, the disk step ending the access (which exists by hardware disk liveness) is reordered into nirvana. One can show liveness if the processor always polls again at some point in the future, if the access has not yet finished. Instead of remembering our proof we recommend to try to find a simpler one.

- the semantics of 'C+A + disk + interrupts' is defined with reordered sequences. Because translated C code does not access I/O-ports, disk steps can be ignored during C steps.
- except for disk liveness nothing of the above is complicated. With these models in place writing a disk driver *and* proving its correctness is indeed completely straightforward.

Hardware designs of pipelined DLX processors with out-of-order execution and the interrupt and address translation mechanisms presented here have been formally verified [BJK$^+$03, DHP05]. The verified processor design with address translation from [DHP05] is unfortunately somewhat inefficient: *pc* and *ea* can be translated in parallel by two memory management units, but *these units are in the same pipeline stage as the subsequent translated memory access*, and this is bad for the cycle time. Adding translation lookaside buffers and splitting address translation and memory access into different pipeline stages makes constructions, the specification of ISA, and proofs considerably harder. We have outlined a paper and pencil solution in a recent lecture [Pau13]; working out the details is current research.

A nondeterministic ISA for a pipelined processor with interrupts and devices has been formally verified [HT09]; the pipelining introduces some very serious complications into the proof. A formal correctness proof of essentially the reorder theorem and the disk driver of this text is reported in [AH08].

There are presently no formally verified hardware designs for multi-core processors. A gate-level design of such a processor together with a paper and pencil correctness proof has been produced only very recently [KMP14]; the design has a cache coherent sequentially consistent shared memory and pipelined processors, but no store buffers, MMUs, interrupt mechanism, and devices yet. Producing such a design and correctness proof is current research. Note that without such a proof it is very difficult to identify the set of *all* software conditions which must be satisfied by an ISA program for it to run correctly on the hardware. Without a list of software conditions which is *known* to be complete one obviously has no guarantee that any ISA program runs correctly on the processor hardware: your favorite program just might violate a condition which is as yet unknown to you. As all programs are written in ISA or are translated into it, one cannot guarantee that *any* program runs correctly.

The manuals of current industrial multi-core processors are thousands of pages long. Spread out over the text they *do* list numerous software conditions that must be obeyed. But the technology to show that this list is complete is presently not there. Fortunately the people who build real-time systems controlling cars, airplanes, elevators, weapons, etc. are ordinary engineers and *much* more conservative than average programmers and software engineers. When they use multi-core processors in real-time systems, they first turn off almost all parallel features in the processor that can be turned off. Of course at first sight this looks completely unreasonable. It is not: for an ordinary engineer — whose mistakes can kill people — a modern processor manual is just not precise enough. And a feature that is turned off cannot produce an undocumented software condition.

14.5 Exercises

1. Computation of the repeat signal:
   a) an n-bit find-first-one circuit n-$FF1$ has inputs $X[n-1:0] \in \mathbb{B}^n$ and outputs $Y[n-1:0] \in \mathbb{B}^n$ satisfying

   $$Y[i] \equiv X[i] \wedge \bigwedge_{j<i} \neg X[i].$$

   Construct such a circuit with cost $O(n)$ and delay $O(\log(n))$.
   Hint: first compute the parallel prefix OR $Z = PP_\vee(X)$ of X.
   b) for all i compute the interrupt-level signals $IL[i]$ signalling that the interrupt level equals i:

   $$IL[i](h, eev) \equiv il(h, eev) = i.$$

   c) compute $repeat(h, eev)$.
2. Prove the correctness of the hardware MMU in user mode, i.e., prove that for MIPS + Interrupts configurations d and hardware configurations h, if

   $$d \sim h \quad \text{and} \quad h.mode[0],$$

   then

   a) $$h.E \rightarrow ma(h) = \begin{cases} ptea(d, d.pc), & \neg h.phase, \\ pma(d, d.pc), & h.phase. \end{cases}$$

   b) $$h.E \wedge ls(d) \rightarrow ma(h) = \begin{cases} ptea(d, ea(d)), & \neg h.phase, \\ pma(d, ea(d)), & h.phase. \end{cases}$$

   You can assume the correctness of the old hardware design (without the MMU).
3. Write a C+A function

   $$uint \quad va2pa(uint\ a),$$

   which runs in *untranslated mode* and when started in configuration

   $$y = (d, c)$$

   returns

   $$\langle pma(d, va(a, c)_{32}) \rangle.$$

   Hint: fill in the following code:

```
uint va2pa (uint a)
{
    uint px;
    uint bx;
    uint ptea;
    uint pte;
    uint ppx;
    uint pa;
```

```
bool v;

px   =                         //
bx   =                         //
ptea =                         //
pte  =                         //
v    =                         //
if v
{
   ppx =                       //
   pa  =
}
else
{
   pa  =                       // return 2^32 - 1
}
return pa
};
```

4. Consider the processor design with address translation.
   a) Optimize the design such that in translated mode only three cycles are needed for the execution of instructions which do not perform loads or stores.
   b) What happens in the correctness proof for (a) with the definitions of $t(i)$ and eev^i_{isa}?
5. Optimize the same processor design further.
   a) All instructions (user or system mode, interrupted or not) should always be executed in one cycle.
      Hint: first determine how many ports you need in the hardware memory $h.m$.
   b) What happens in the correctness proof for (a) to the definitions of $t(i)$ and eev^i_{isa}?
6. Several disks.
   a) Define hardware and ISA models of a processor with two disks d_1 and d_2. For $x \in \{0, 1\}$ use components like $h.d_x.ports$ and $h.d_x.sm$ in the configurations.
   b) State the software conditions.
      Hint: make sure the ports in the memory system don't overlap.
   c) Generalize the order reduction theorem.
   d) Show that steps of disk 2 can be reordered behind the steps of a driver run for disk 1.
      Hint: you need the above software condition.
7. One disk with an interrupt output.
   a) Modify the disk definition such that bit $cmsr[2]$ is turned on when an access ends and make this signal external interrupt signal $eev[2]$ of the processor. It can only be turned off by a write to $cmsr$.
   b) Adjust the order reduction theorem and prove it.

Hint: you had better mask $eev[2]$.

c) Formulate and prove an order reduction theorem for a system with two such disks, of course with different interrupt event signals $eev[2]$ and $eev[3]$.

d) Show that disk steps of disk 2 can be reordered behind the steps of a driver run of disk 1.

8. Consider a system with a processor and one disk which produces interrupts.

a) Write a 'C+A + disk + interrupts' program which starts a disk access and subsequently does not poll the disk. Instead it should be interrupted once interrupt signal $eev[2]$ is activated.

b) What happens in reaction to the activation of the interrupt signal?

c) Can we handle interrupts in any meaningful way without knowledge of the ROM? If yes explain how. If no give a content of the ROM where things fall apart.

9. Timer specification. Define a hardware device with the following properties:

a) I/O-ports: i) $counter \in \mathbb{B}^{64}$, ii) command and status register $cmsr \in \mathbb{B}^{32}$.

b) two bits of $cmsr$ are meaningful: i) $cmsr[0]$: ticking and ii) $cmrs[1]$: interrupt. While the timer is ticking, the counter is decreased in every hardware cycle. Interrupt is activated when the counter reaches zero, at which time the timer stops ticking, and can only be cleared by a write to $cmsr$.

10. Give a gate-level design of

a) a timer.

b) a processor of the main text with a timer.

11. Processor + timer.

a) Specify an ISA model for processor + timer.
Attention: the number of cycles for the execution of instructions varies. Just formulate liveness: at some instructions the timer stops ticking.

b) Suppose you start the timer (by writing to $cmsr$) with value $T = \langle counter \rangle$. Give upper and lower bounds for the number of instructions that can be executed before the timer stops ticking.

c) Assume the hardware is clocked at 1 GHz. How much time passes until the interrupt?

d) Specify and program a driver for the timer.

12. Order reduction for processor + timer. Consider a system of processor + disk + timer.

a) Formulate and prove an order reduction theorem.

b) Conclude that — with proper disabling of interrupts — timer interrupts can be reordered behind driver runs of the disk. Thus in particular if you write a block to the disk, it is there after the run.

c) Now connect the timer to a bomb which destroys the disk while the driver is running. Obviously the block is not written to the disk; the disk is not even there any more. In what line of the definitions or the proof did we make a mistake?
Hint: you can model things by *not* writing something down. Example: every breath is followed by the next. This models — because breaths have a minimal length — immortality.

d) Look up what is an information side channel. Did we model the world outside of processors and devices?

e) Can the infotainment system of a car be ignored by the motor control?
   Hint: infotainment controls climate. What does heating or cooling the car do to the range of electric cars (or others)?

Einstein said: make everything as simple as possible but not simpler. Order reduction is great in the absence of information side channels. And information side channels are real.

A Generic Operating System Kernel

Joint Work with Jonas Oberhauser

Finally, we specify and implement what we call a generic operating system kernel. In a nutshell a kernel has two tasks:

- to simulate on a single physical processor multiple virtual processors of almost the same ISA;
- to provide services to these virtual processors via system calls.

As long as no system call is executed, a program on a virtual machine is executed as if the program were running alone on its machine. The physical machine is also called the *host* whereas the virtual machines are called *users* or *guests*. Usually, the physical or *host* machine is slightly more powerful than the virtual machines. In particular, in a classical kernel the host can run in translated/system mode whereas the guests can only run in untranslated mode. However, some kernels — called *hypervisors* — permit guests to run in translated mode. Thus the guests of hypervisors can themselves be kernels which in turn have user processes. Under a hypervisor therefore one can run several kernels or several operating systems on the same hardware. Here we restrict our attention to classical kernels.

The two main tasks of a kernel mentioned above suggests splitting the kernel implementation into two layers, one for each of the tasks: i) a lower virtualization layer and ii) an upper layer which includes the scheduler and the handlers for system calls. In this text we treat the lower layer and leave the programmer the freedom to program the upper layer in any way she or he wishes. With all the computational models developed so far in place and justified, we follow our usual pattern of specification, implementation, and correctness proof. In Sect. 15.1 we explain the differences between physical and virtual processors: roughly speaking virtual processors cannot use address translation and not all interrupts are visible. In particular the virtual machine sees page faults only if it tries to access pages that are not allocated by it. Page

© Springer International Publishing Switzerland 2016
W.J. Paul et al., *System Architecture*, DOI 10.1007/978-3-319-43065-2_15

faults due to invalid page table entries are invisible and must be handled transparently by the implementation of the kernel.

In Sect. 15.2 we formally specify a computational model called CVM (Communicating Virtual Machines), which is realized by the virtualization layer. The user of the virtualization layer simply sees a number of virtual user machines communicating with a so-called *abstract kernel*. The latter is an arbitrary (!) C0 program which is allowed to call a very small number of special functions that we call *CVM primitives*. The specification of the CVM model including the semantics of all special functions is extremely concise. No reference to inline assembly is necessary in the specification. Only the semantics of ISA, of C0, and the parallelism of CVM are used.

CVM is implemented by what we call a concrete kernel written in 'C+A + disk + interrupts'. The concrete kernel and its main data structures are introduced in Sect. 15.3. We identify the simulation relation we have to maintain and then hope that this will guide the implementation in a straightforward way. It turns out that we have to maintain three simulation relations:

- between the physical machine configuration d and the virtual user machines vm. This involves address translation and process control blocks.
- between the C abstraction k of the concrete kernel configuration (if present) and the physical machine d. This is handled basically by $consis(k,d)$, but if users are running we must maintain a substitute for this C abstraction.
- between the C abstraction k of the concrete kernel configuration and the configuration c of the abstract kernel. Because the code of the abstract kernel is a part of the concrete kernel this requires a very rudimentary theory of linking and gives rise to a simulation relation k-$consis$.

Some complexity is added, because each of the relations involved changes slightly when the simulated machine (user, concrete kernel, or abstract kernel) starts or stops running. Identifying these relations is the heart of the matter. From then on things become easier.

In Sect. 15.4 we introduce a top-level flow chart for the CVM primitive *runvm*, which starts user processes, handles page faults transparently, and saves user processes when they return to the kernel. We annotate it with a nontrivial number of invariants, but identifying these invariants is not very hard: they are just the invariants needed to maintain the three simulation relations. Programming the building blocks of the flow chart such that the invariants are maintained is done in the subsequent sections in a fairly straightforward way.

Section 15.5 deals with booting and the simulation of CVM steps by the abstract kernel. Section 15.6 deals with page fault handling. Section 15.7 provides implementations of CVM primitives that can be used by the abstract kernel, and explains how to write a dispatcher of the abstract kernel. The programs involved in these last three sections are fairly straightforward, and with all computational models in place correctness proofs mostly boil down to bookkeeping.

In the classroom this chapter takes much less time than one would expect: the difference between virtual and physical machines is easily explained. The CVM model is simple. The concrete kernel has only very few data structures. *Then* one has to invest time to explain the three simulation relations and their variants and the overall correctness theorem. The only serious piece of programming is the flow chart for the *runvm* primitive. Gaining the (numerous) annotations from the simulation relations is already a fairly routine process. We explain how a few of them were obtained; the others can be looked up in the text. The remaining programming is mostly routine work: no complex manipulations of data structures, simple loops, and routine correctness proofs. Most of it is better given as a reading assignment.

There are however two details which do deserve attention in the classroom: i) dropping C abstractions and ii) gaining them back.

15.1 Physical and Virtual Machines

15.1.1 Physical Machines

Our physical machines are the full MIPS machines with operating system support defined in Chap. 14 with a 1 terabyte disk and interrupts. In order to illustrate the power of paging we will however assume that the hardware memory of the physical machine has less than the full 2^{20} pages that could possibly be addressed by the 32-bit architecture. Instead we assume that the memory of the physical machine, resp. the physical memory, has only

$$P < 2^{20}$$

pages. For any number $X < 2^{20}$ of pages we denote the set $A(X)$ of addresses a whose page index (interpreted as a binary number) is less than X by

$$A(X) = \{a \in \mathbb{B}^{32} \mid \langle a.px \rangle < X\}.$$

The memory component $d.m$ of physical machine configurations is thus restricted to

$$d.m : A(P) \to \mathbb{B}^8$$

and accessing in untranslated mode a location which is not physically present results in a page fault.

$$\neg mode(d) \wedge d.pc \notin A(P) \to pff(d),$$
$$\neg mode(d) \wedge ls(d) \wedge ea(d) \notin A(P) \to pfls(d).$$

Modifying the hardware to generate these interrupts is an easy exercise.

15.1.2 Virtual Machines

We restrict the configurations d of virtual machines and define their next state function

$$d' = \delta_{VM}(d, eev).$$

- At any time address space is restricted to a certain number of pages X. Thus

$$d.m : A(X) \to \mathbb{B}^8.$$

Pages can be freed and more pages can be allocated by system calls. Note that on the physical machine a very large disk is available. One can use this large swap memory to simulate virtual memories which are larger than the physical memory, i.e., we explicitly allow $X > P$.

- The machines can only run in untranslated mode. Therefore special purpose registers $mode$, pto, and ptl are absent. Accessing a location which is not in the available address space results in a page table length exception, and this is the only source of page faults:

$$pff(d) \equiv ptlef(d)$$
$$\equiv d.pc \notin A(X),$$
$$pfls(d) \equiv pflels(d)$$
$$\equiv ls(d) \wedge ea(d) \notin A(X).$$

- If an interrupt occurs in a virtual machine, i.e., if

$$jisr(d, eev),$$

then control will be transferred to the kernel. The status register sr is frozen. Note that in a physical machine the frozen value is saved into esr:

$$d'.sr = \delta_M(d, eev).esr = d.sr.$$

Therefore register esr is absent in virtual machines. Depending on the type of interrupt the pc is immediately updated to the address where computation should resume after the run of the handler in the kernel; this is exactly the address saved by the physical machine in register epc:

$$d'.pc = \delta_M(d, eev).epc.$$

The remaining special purpose registers eca and $edata$ are updated in the usual way:

$$R \in \{eca, edata\} \to d'.R = \delta_M(d, eev).R.$$

- Direct accesses to the special purpose registers and $eret$ instructions are illegal:[1]

$$move(d) \vee eret(d) \to ill(d).$$

The set of configurations of virtual machines with X memory pages is denoted by $K_{VM}(X)$. The set of all virtual machine configurations is

$$K_{VM} = \bigcup_X K_{VM}(X).$$

---

[1] Note that indirect access to these registers is possible for virtual machines via system calls.

We define two processor configurations of virtual machines to be equivalent if they have the same pc, general-purpose registers, and status register:

$$\begin{aligned} proc(d) \sim_{VM} proc(d') \equiv \ & d.pc = d'.pc \ \wedge \\ & d.gpr = d'.gpr \ \wedge \\ & d.sr = d'.sr. \end{aligned}$$

We define two virtual machine configurations to be equivalent if their processor configurations are equivalent and their memories are equal:

$$\begin{aligned} d \sim d' \equiv \ & proc(d) \sim_{VM} proc(d') \ \wedge \\ & d.m \ = \ d'.m. \end{aligned}$$

Obviously this is an equivalence relation. We compare computation steps of virtual machines which start from equivalent configurations

$$d \sim d'$$

with identical input eev. Whether an interrupt occurs depends only on a single special purpose register: the status register sr containing the masks. Hence

$$jisr(d, eev) \equiv jisr(d', eev).$$

If no interrupt occurs, then the special purpose register file is not updated, because $movg2s$ instructions produce illegal instruction interrupts. Updates to the remaining components do not depend on the special purpose register file, because $movs2g$ and $eret$ instructions produce an illegal interrupt. Hence

$$\neg jisr(d, eev) \ \rightarrow \ \delta_{VM}(d, eev) \sim \delta_{VM}(d', eev).$$

If an interrupt occurs, then changes to the configuration depend on sr and on no other special purpose registers. In a physical machine all special purpose registers except ptl and pto are updated when an interrupt occurs. In virtual machines, registers pto and ptl are not present, hence at an interrupt all special purpose registers of a virtual machine are updated. Thus after the interrupt configurations will be equivalent:

$$jisr(d, eev) \ \rightarrow \ \delta_{VM}(d, eev) \sim \delta_{VM}(d', eev).$$

For later reference we summarize as follows.

Lemma 138. *Suppose*

$$d \sim d'.$$

Then

1. $$jisr(d, eev) \equiv jisr(d', eev),$$
2. $$\delta_{VM}(d, eev) \sim \delta_{VM}(d', eev).$$

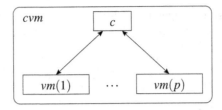

Fig. 227. Illustration of a CVM configuration. Virtual MIPS machines with configurations $vm(u)$ communicate with an abstract kernel with $C0$ configuration c

15.2 Communicating Virtual Machines

15.2.1 Configurations

In what follows, we define a model of a generic operating system kernel and its user processes. We refer to this model by the name *communicating virtual machines* (CVM). A CVM configuration *cvm* with $p \in \mathbb{N}$ users is illustrated in Fig. 227. For simplicity we will keep the number p of users fixed. Enriching the model and its implementation such that user processes can be created and terminated is not very difficult.

As the name suggests, CVM is a model of *virtual machines* (also called *users* or *user processes*) as defined above which may communicate by means of services provided through the *abstract kernel*. The abstract kernel is a $C0$ program that uses a small number of *CVM primitives* (also called *special functions*), which perform low-level functions such as switching control to a specific user process, reading registers of a virtual machine, increasing the address range of a virtual machine, etc. We collect their names in a set

$$SPF = \{\ runvm,$$
$$alloc, free,$$
$$readgpr, writegpr,$$
$$readspr, writespr,$$
$$readpage, writepage\ \}.$$

A statement s is a call of a CVM primitive if is has the form

$$f()\quad \text{or}\quad f(e_1, \ldots, e_{np})$$

for a special function name $f \in SPF$ and parameters $e_i \in L(Pa)$ for $i \in [1 : np]$.

$$spfcall(s) \equiv \exists f \in SFP : (s = f() \lor \exists ps \in L(PaS) : s = f(ps)).$$

These external functions are used like predefined operators:

- they can be called with the syntax of ordinary $C0$ function calls,
- they have a predefined semantics that we will specify below in the CVM model.

The point of using external functions is that the entire kernel and its interaction with the user processes can be specified entirely without '$C0$ + assembly'. Only

- *C0* for the abstract kernel,
- MIPS ISA for the virtual machines, and
- the combination of these machines in the parallel CVM model for the specification of the CVM primitive functions

are used. In the implementation, the use of 'C0 + assembly' can of course not be avoided, but its use is restricted to the implementation of the few CVM primitives.

The abstract kernel has to fulfill a single software condition:

There is a global variable

```
uint CP
```
---

which is used by the abstract kernel to track which user runs next or has been running before control returned to the kernel.

Formally a CVM configuration *cvm* has the following components:

- the current process identifier

$$cvm.cp \in [0 : p].$$

Here, we use $cvm.cp = 0$ to denote that the kernel is currently being executed; $cvm.cp = u \in [1 : p]$ denotes that virtual machine u is being executed.
- the page table length of every user process

$$cvm.ptl : [1 : p] \rightarrow \mathbb{N}.$$

The page table length is the number of pages that are allocated to virtual machine u by the kernel. Thus, virtual machine u can access addresses in the range $A(cvm.ptl(u))$.
- the virtual machine configurations

$$cvm.vm : [1 : p] \rightarrow K_{VM}.$$

For all users u we require

$$cvm.vm(u) \in K_{VM}(cvm.ptl(u)).$$

- the *C0* configuration of the abstract kernel

$$cvm.c.$$

15.2.2 Semantics

In a nutshell CVM executes:

1. user machines using the MIPS ISA semantics δ_{VM} for virtual machines,

2. the $C0$ kernel program by using $C0$ semantics, as long as no CVM primitives are executed,
3. CVM primitives using their semantics which have to be specified for each primitive separately,
4. a switch from a user process u to the kernel as a reaction to an interrupt on virtual machine u,
5. a switch from the kernel to user u by execution of the CVM primitive *runvm*.

Formally the semantics of CVM

$$cvm' = \delta_{CVM}(cvm, eev)$$

is defined by a straightforward case split. We only list components that change: all components which are not mentioned in a specific step remain unchanged.

1. User u is running and there is no interrupt:

$$cvm.cp = u \in [1:p] \quad \text{and} \quad \neg jisr(cmv.vm(u), eev).$$

The model performs a regular MIPS step of virtual machine u:

$$cvm'.vm(u) = \delta_M(cvm.vm(u), eev).$$

2. The kernel is running and $hd(cvm.c.pr)$ is **not** a CVM primitive:

$$cvm.cp = 0 \quad \text{and} \quad \neg spfcall(hd(cvm.c.pr)).$$

The model performs a regular $C0$ step of the kernel:

$$cvm'.c = \delta_C(cvm.c).$$

3. The kernel is running and $hd(cvm.c.pr)$ is a CVM primitive:

$$cvm.cp = 0 \quad \text{and} \quad spfcall(hd(cvm.c.pr)).$$

For expressions $e \in L(E)$ we use the shorthand

$$\tilde{e} = va(e, cvm.c).$$

For any CVM primitive, we have

$$cvm'.c.pr = tail(cvm.c.pr).$$

Since the specific effect depends on the CVM primitive in question, we perform a case split on $hd(cvm.c.pr)$:

a) *runvm*(): Starts the user process $\widetilde{CP}$:

$$cvm'.cp = \widetilde{CP}.$$

b) $alloc(u,x)$: Allocates $\tilde{x}$ new pages for user process $\tilde{u}$:

$$cvm'.ptl(\tilde{u}) = cvm.ptl(\tilde{u}) + \tilde{x}.$$

New addresses $a \in A(cvm'.ptl(\tilde{u})) \setminus A(cvm.ptl(\tilde{u}))$ are initialized with zeros:

$$cvm'.vm(\tilde{u}).m(a) = 0^8.$$

c) $free(u,x)$: Frees $\tilde{x}$ pages of user process $\tilde{u}$:

$$cvm'.ptl(\tilde{u}) = cvm.ptl(\tilde{u}) - \tilde{x}.$$

d) $readgpr(e,u,i)$: Reads general-purpose register $\tilde{i}$ of user process $\tilde{u}$ to the C0 variable $e*$:

$$va(e*, cvm'.c) = \langle cvm.vm(\tilde{u}).gpr(\tilde{i}_5) \rangle.$$

e) $writegpr(u,i,e)$: Writes the value of C0 expression e to general-purpose register $\tilde{i}$ of user process $\tilde{u}$:

$$cvm'.vm(\tilde{u}).gpr(\tilde{i}_5) = \tilde{e}_{32}.$$

f) $readspr(e,u,i)$ and $writespr(u,i,e)$ for special purpose registers are defined in a completely analogous way.

g) $readpage(e,u,i)$: Reads page $\tilde{i}$ of user $\tilde{u}$ into the C0 variable $e*$ of array type $uint[K]$:

$$\forall k \in [0:K-1] : va(e*[k], cvm'.c) = \langle cvm.vm(\tilde{u}).m_4(i_{20} \circ k_{10} \circ 00) \rangle.$$

h) $writepage(u,i,e)$: Writes the C0 variable $e*$ of array type $uint[K]$ into page $\tilde{i}$ of user $\tilde{u}$:

$$\forall k \in [0:K-1] : cvm'.vm(\tilde{u}).m_4(i_{20} \circ k_{10} \circ 00) = (\widetilde{e*[k]})_{32}.$$

The last two primitives are illustrated in Fig. 228. With the composition of the primitives

```
readpage(e,u,i);
writepage(v,j,e);
```

one can obviously copy page $\tilde{i}$ of user $\tilde{u}$ into page $\tilde{j}$ of user $\tilde{v}$.

Note that this is not necessarily an exhaustive list of CVM primitives. We could extend this list later on.

4. User u is running and an interrupt occurs:

$$cvm.cp = u \in [1:p] \quad \text{and} \quad jisr(cvm.vm(u), eev).$$

The virtual machine makes a step as specified in Sect. 15.1.2 and control is transferred to the kernel:

$$cvm'.vm(u) = \delta_{VM}(cvm.vm(u), eev),$$
$$cvm'.cp = 0.$$

$$cvm.vm(\tilde{u}).m$$

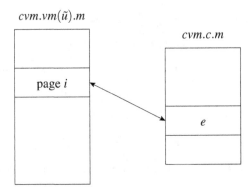

Fig. 228. Copying a page between $C0$ array e of the kernel and page i of virtual machine u in CVM

In order to better understand the given specification of CVM, consider an abstract kernel that is structured roughly as follows:

```
[...]                 // initialize data structures
while true
{
   scheduler();
   runvm();
   dispatcher()
}
```

After initialization the kernel is in a simple endless loop. In the scheduler a new value of a variable CP is computed. This is the number of the user process to be started next. The CVM primitive *runvm* suspends the kernel and starts running the user process. Control can return to the kernel only via an interrupt. After an interrupt the dispatcher can inspect all registers of the interrupted process using the *readgpr* and *readspr* primitives, determine the cause of the interrupt, and call the appropriate handler. In case of a syscall interrupt, the user could for instance specify the nature of the service requested in general-purpose register number 1 (which will be saved into the process control block of the interrupted process), and it can pass parameters to the kernel via further general-purpose registers.

15.3 Concrete Kernel

We proceed to implement the computational model of CVM by what we call the *concrete kernel*, a program written in $C0$ + assembly and running on a physical machine. It is obtained from the abstract kernel by adding:

- data structures and functions supporting the simulation of multiple virtual machines,

PCB

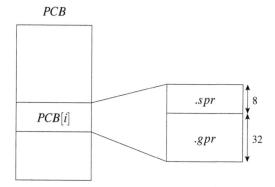

Fig. 229. *PCB* is an array of process control blocks. For each user process *u* process control block *PCB[u]* contains an array *PCB[u].gpr* of general-purpose registers and an array *PCB[u].spr* of special purpose registers

- implementations of the special functions.

So far we have not put any restrictions on the abstract kernel except the presence of a global variable *CP* which keeps track of the current process. In what follows we will make a trivial assumption: names of types and variables introduced by us in the new data structures of the concrete kernel should not be used in the abstract kernel. But even this assumption can easily be removed: in case the new names clash with names in an existing abstract kernel, we could as well rename the new types and variables.

We aim at a simulation of CVM computations with configurations *cvm* by C+A computations with configurations $y = (d, c, ic)$ if translated *C* code is running or $y = (d, ic)$ otherwise. In case no translated *C* code is running we will substitute the missing *C* abstraction *c* of *y* by another *C0* configuration *k*. For the proof that the concrete kernel provides this simulation, we will couple CVM configurations *cvm* and configurations (d, c, ic) or (d, k, ic) by numerous invariants *inv-X*(d, c, cvm). The first one of these invariants states that variable *CP* has to track the current process *cvm.cp* of CVM, while the kernel is not running.

Invariant 11

$$inv\text{-}cp(c, cvm) \equiv (cvm.cp > 0 \rightarrow va(CP, c) = cvm.cp).$$

15.3.1 Data Structures

In the kernel implementation we add to the abstract kernel several data structures (see Figs. 229 and 230):

1. *PCB_t PCB* – a global array of process control blocks, one for every virtual machine and one for the kernel itself.

   The type of process control blocks *PCB_t* is declared as follows:

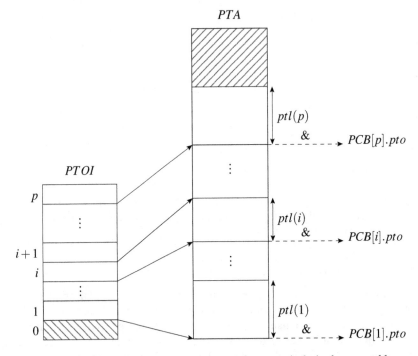

Fig. 230. Page tables for all user processes are stored consecutively in the page table array *PTA*. The index at which the page table for user *u* starts is stored in element *PTOI*[*u*] of the page table origin index array. The base address of *PTA*[*PTOI*[*u*]] is stored at component $pto = spr(5_5)$

```
typedef uint[32] u;
typedef uint[8] v;
typedef struct {u gpr; v spr} pcb;
typedef pcb[p+1] PCB_t;
```

where *p* is the number of CVM user processes. *PCB*[*u*] holds the general- and special purpose registers of virtual machine *u* (or of the kernel in case $u = 0$).

2. *PTA⊥ PTA* – a global array of page table entries that support the address translation for the users.

Here, *PTA⊥* is the array type:

```
typedef uint[PTASIZE] PTA_t;
```

where *PTASIZE* is the maximal number of words occupied by the page tables of all virtual machines.

3. *PTOI⊥ PTOI* – a global array of indices of page table array (*PTA*) elements.

Here, $PTOI\_t$ is the array type:

```
typedef uint[p+1] PTOI_t;
```

The page table for user u starts in array PTA at index $PTOI[u]$.

More data structures will be added later for the page fault handler. The above data structures suffice to formulate the intended simulation relations between physical machine configurations d and CVM configurations cvm.

15.3.2 Virtual Address Translation via C Data Structures

For configurations c of the concrete kernel and users $u \in [1 : p]$ we define the current page table origin index $ptoi(c, u)$ by

$$ptoi(c, u) = va(PTOI[u], c).$$

For each user we will store the page table for this user at indices

$$[ptoi(c, u) : ptoi(c, u) + cvm.ptl(u)).$$

Thus we require the following.

Invariant 12

$$inv\text{-}ptoi(c, cvm) \equiv \forall u \in [1 : p-1] : ptoi(c, u) + cvm.ptl(u) < ptoi(c, u+1) \land$$
$$ptoi(c, p) + cvm.ptl(p) < PTASIZE \land$$
$$ptoi(c, 1) = 0.$$

The base address of the first page table array element $PTA[PTOI[u]]$ for user u is stored in the control block $PCB[u]$ for user u at component $pto = spr(5_5)$.

Invariant 13

$$inv\text{-}pto(c) \equiv \forall u \in [1 : p] : va(PCB[u].spr[5], c)_{32} = ba(PTA[PTOI[u]], c).$$

It is tempting to achieve this with the assignment

```
PCB[u].spr[5] = PTA[PTOI[u]]&;
```

but the left-hand side has type *uint*, and the right-hand side has type *uint*∗. Thus type correctness of assignments would be violated. However, using extended $C0$ we can achieve the type cast with the two statements

```
gpr(1) = PTA[PTOI[u]]&;
PCB[u].spr[5] = gpr(1);
```

The page table length for user u is stored in the process control block $PCB[u]$ for user u at component $ptl = spr(6_5)$.

Invariant 14

$$inv\text{-}ptl(c,cvm) \equiv \forall u \in [1:p] : va(PCB[u].spr[6],c) = cvm.ptl(u).$$

We define translation of virtual addresses $a \in \mathbb{B}^{32}$ in terms of the $C0$ data structures. Below *ptei* is used as an abbreviation for 'page table entry index':

$$
\begin{aligned}
ptei(c,u,a) &= ptoi(c,u) + \langle a.px \rangle, \\
pte(c,u,a) &= va(PTA[ptei(c,u,a)],c)_{32}, \\
v(c,u,a) &= pte(c,u,a)[11], \\
ppx(c,u,a) &= pte(c,u,a)[31:12], \\
pma(c,u,a) &= ppx(c,u,a) \circ a.bx,
\end{aligned}
$$

$$
\begin{aligned}
ptle(c,u,a) &\equiv \langle a.px \rangle \geq va(PCB[u].spr[6],c), \\
ipf(c,u,a) &\equiv \neg ptle(c,u,a) \wedge \neg v(c,u,a), \\
pf(c,u,a) &\equiv ptle(c,u,a) \vee ipf(c,u,a).
\end{aligned}
$$

In order to prevent user u from tampering with the memory of the kernel and other users u' we maintain an invariant stating that different valid page table entries map addresses to different pages and not into the program or data region of the kernel.

Invariant 15

$$
\begin{aligned}
inv\text{-}ppx(c) \equiv \forall u,u' \in [1:p] \; \forall a,a' \in \mathbb{B}^{32} : \\
(v(c,u,a) \rightarrow \langle ppx(c,u,a) \rangle \geq \langle hmax \rangle) \wedge \\
(u \neq u' \wedge v(c,u,a) \wedge v(c,u',a') \rightarrow ppx(c,u,a) \neq ppx(c,u',a')).
\end{aligned}
$$

We collect the above invariants into a single invariant.

Invariant 16

$$
\begin{aligned}
inv\text{-}pt(c,cvm) \equiv \; &inv\text{-}ptoi(c,cvm) \wedge \\
&inv\text{-}ptl(c,cvm) \wedge \\
&inv\text{-}pto(c) \wedge \\
&inv\text{-}ppx(c).
\end{aligned}
$$

It states that the page tables in $C0$ configuration c are well-formed and can be used for encoding CVM configuration cvm.

15.3.3 Simulation Relation for Virtual Machines

We proceed to specify how a CVM configuration cvm is encoded by an ISA configuration d together with a consistent $C0$ configuration c, or — seen from the side of CVM — how to abstract user machine configuration $vm(d,c,u)$ from an ISA configuration d and a $C0$ configuration c. The abstraction is done in different ways depending on whether u is running or suspended:

$$vm(d,c,u) = \begin{cases} vmrun(d,c,u), & u = va(CP,c), \\ vmsus(d,c,u), & \text{otherwise.} \end{cases}$$

We can assume that the page table invariant $inv\text{-}pt(c,cvm)$ and the current process invariant $inv\text{-}cp(c,cvm)$ hold.

Registers and PC

Let

$$RF \in \{gpr, spr\}.$$

If user machine $vm(d,c,u)$ is running, then each register $RF(i_5)$ is found in the corresponding register of the host. The program counter is found in the pc of the host:

$$vmrun(d,c,u).RF(i_5) = d.RF(i_5),$$
$$vmrun(d,c,u).pc = d.pc.$$

If user machine $vm(d,c,u)$ is suspended, then each general-purpose register $gpr(i_5)$ is found in the corresponding entry of its process control block:

$$vmsus(d,c,u).gpr(i_5) = va(PCB[u].gpr[i],c)_{32}.$$

Special purpose registers $eca = spr(2_5)$ and $edata = spr(4_5)$ are abstracted in the same way:

$$vmsus(d,c,u).eca = va(PCB[u].spr[2],c)_{32},$$
$$vmsus(d,c,u).edata = va(PCB[u].spr[4],c)_{32}.$$

In contrast, the status register and program counter of machines which are not running are stored in components $esr = spr(1_5)$ and $epc = spr(3_5)$ of the process control block:

$$vmsus(d,c,u).sr = va(PCB[u].spr[1],c)_{32},$$
$$vmsus(d,c,u).pc = va(PCB[u].spr[3],c)_{32}.$$

Memory

For each user u and page index

$$px < cvm.ptl(u)$$

in its address range we also need a swap memory address $sma(u,px)$, where the memory page px of user u is stored in case it is not in the physical memory. We define this mapping here in a brute force way. For each user process we reserve 4 Gbyte, which is 2^{20} pages, of disk space. For all p processes we use the last $p \cdot 2^{20}$ pages. Swap memory space for user u then starts at pages

$$smbase(u) = \begin{cases} (2^{28} - p \cdot 2^{20})_{28}, & u = 1, \\ smbase(u - 1) +_{28} (2^{20})_{28}, & u > 1. \end{cases}$$

Starting at these page addresses we reserve space for the user pages in the obvious order:

$$sma(u, px) = smbase(u) +_{28} (0^8 \circ px).$$

If the valid bit for address a of user u is on, then the memory content $cvm.vm(u).m(a)$ is stored in the physical memory $d.m$ at the physical memory address obtained from the translation:

$$v(c, u, a) \rightarrow cvm.vm(u).m(a) = d.m(pma(c, u, a)). \tag{14}$$

If the valid bit for address a of user u is off, then the memory content $cvm.vm(u).m(a)$ is stored at byte $\langle a.bx \rangle$ of the disk page with address $sma(u, a.px)$:

$$\neg v(c, u, a) \rightarrow cvm.vm(u).m(a) = byte(\langle a.bx \rangle, d.sm(sma(u, a.px))). \tag{15}$$

With Eqs. 14 and 15 we define the memory of the abstracted virtual machine

$$vmrun(d, c, u).m = vmsus(d, c, u).m = vm(d, c, u).m$$

both for running and suspended processes u as

$$vm(d, c, u).m(a) = \begin{cases} d.m(pma(c, u, a)), & v(c, u, a), \\ byte(\langle a.bx \rangle, d.sm(sma(u, a.px))), & \text{otherwise.} \end{cases}$$

Invariant $inv\text{-}vm(d, c, cvm)$ then states that all virtual machines $cvm.vm(u)$ can be abstracted from (d, c) with one exception: for the current process we can only abstract a configuration which is equivalent to $cvm.vm(u)$.

Invariant 17 *Invariant*

$$inv\text{-}vm(d, c, cvm)$$

holds iff for all $u > 0$

$$u \neq va(CP, c) \rightarrow vm(d, c, u) = cvm.vm(u),$$
$$u = va(CP, c) \rightarrow vm(d, c, u) \sim cvm.vm(u).$$

The special case for the running process comes from the use of the exception registers during the transparent handling of page fault interrupts.

15.3.4 Encoding an Abstract Kernel by a Concrete Kernel

We also have to define a relation $k\text{-}consis(k, c)$ specifying when a C0 configuration k of the concrete kernel encodes a C0 configuration c of the abstract kernel. The only new data structures introduced in the concrete kernel were in the global memory, and this will also be the case for the remaining data structures of the concrete kernel.

We require that all components of the configurations (recursion depth, stack, heap, program rest, ...) except the global memory content are the same:

$$X \neq m \rightarrow c.X = k.X,$$

and that subvariables x of the abstract kernel configuration have the same value in the concrete kernel configuration:

$$x \in SV(c) \rightarrow c.m(x) = k.m(x).$$

If the concrete kernel used the heap, the situation would be slightly more complicated. Using methods from Sect. 13.1 one would require that the heap of c is isomorphic to the portion of the heap of k which is reachable from the subvariables belonging to the abstract kernel.

Using the $C0$ semantics one infers the following by lengthy bookkeeping.

Lemma 139. *Executing a C0 step in k-consistent configurations k and c preserves k-consistency:*

$$k\text{-}consis(k,c) \wedge \neg spfcall(hd(c.pr)) \rightarrow k\text{-}consis(\delta_C(k), \delta_C(c)).$$

We would like to extend the definition of k-consistency to situations in which k has a larger recursion depth than c, i.e.,

$$r = k.rd - c.rd > 0.$$

Intuitively, the situation $r > 0$ arises if the abstract kernel calls a special function, i.e.,

$$spfcall(hd(c.pr))$$

and the concrete kernel is in the process of running an implementation of that function. In this situation we should be able to return to the trivial case, if we drop

- the head from $c.prn$,
- the first r elements from the stack of k,
- the nodes up to and including the r'th return from the start of $k.prn$.

Formally we define for sequences prn of program nodes the index of the first node generating a return statement:

$$fret(prn) = \min\{x \mid \ell(prn(x)) = rSt\}.$$

We define $TAIL$ (with capital letters) of prn by

$$TAIL(prn) = prn(fret(prn) + 1 : |prn|).$$

Taking $TAIL$ iteratively r times is defined by

$$TAIL'(prn, 0) = prn,$$
$$TAIL'(prn, r+1) = TAIL(TAIL'(prn, r)).$$

For configurations k, c, and recursion depth difference r we define the relation

$$chop(k,c,r)$$

stating that the stack of c is obtained by chopping off r elements from the stack of k and that the heap as well as the value of common subvariables is the same:

$$c.rd = k.rd - r,$$
$$X \in \{st, clr, rds\} \;\rightarrow\; c.X(y) = k.X(y) \quad \text{for} \quad y \le c.rd,$$
$$X \in \{nh, ht\} \;\rightarrow\; c.X = k.X,$$
$$x \in SV(c) \;\rightarrow\; c.m(x) = k.m(x).$$

Obviously we have

$$k\text{-}consis(k,c) \;\equiv\; chop(k,c,0) \wedge c.prn = k.prn.$$

For $r > 0$ we define

$$k\text{-}consis(k,c,r) \;\equiv\; chop(k,c,r) \wedge tail(c.prn) = TAIL'(k.prn,r).$$

For computations of the concrete and the abstract kernel which are nicely 'in sync' the above definition is fine. For very technical reasons however, this will not always be the case: there will be situations where the abstract kernel c drops a call of a special function from the program rest before the concrete kernel returns from the implementation. For these situations we define

$$k\text{-}consis'(k,c,r) \equiv chop(k,c,r) \wedge c.prn = TAIL'(k.prn,r).$$

An important special case of $k\text{-}consis'(k,c,r)$ is the situation when the program rests differ exactly by a 'dangling' return statement of k:

$$danglingr(k,c) \;\equiv\; (k.pr = return\ 1; c.pr).$$

When k and c are almost in sync except for a 'dangling' return statement, then only this statement has to be executed in k in order to reach ordinary consistency:

$$k\text{-}consis'(k,c,1) \wedge danglingr(k,c) \;\rightarrow\; k\text{-}consis(\delta_C(k),c).$$

15.3.5 Simulation Relation for the Abstract Kernel of CVM

The abstract kernel component $cvm.c$ is encoded differently by a $C0$ configuration k of the concrete kernel and an ISA configuration d of the physical machine. If the abstract kernel is running $(cvm.cp = 0)$, life is very simple: d has to encode k via $consis(k,d)$, and k has to encode $cvm.c$ via $k\text{-}consis(k,cvm.c)$ as illustrated in Fig. 231.

Invariant 18

$$inv\text{-}crun(d,k,cvm) \;\equiv\; consis(k,d) \wedge k\text{-}consis(k,cvm.c).$$

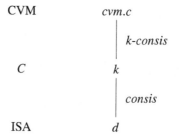

Fig. 231. The kernel is running. The C+A computation has a C abstraction k. It codes $cvm.c$ via $k\text{-}consis(k, cvm.c)$, and $consis(k, d)$ holds by the C+A semantics

When the abstract kernel is suspended and a user process is running ($cvm.cp > 0$), things are more complex. This situation will arise only during the execution of CVM primitive *runvm* in the concrete kernel while user code is running and before the return statement of the *runvm* primitive is executed.

Clearly, during the context switch to the user program, the C abstraction is lost at some point in the C+A computation. But in the correctness proof we will have to provide a substitute for it. The obvious substitute k is the C abstraction of the last configuration K before C abstractions are lost, and except for the program rest nodes and the program rest this will work just fine. For the portion of the program rest belonging to the body of function *runvm* we will not really attempt to maintain consistency. Instead we will keep in k only the return statement from this portion. Let m be the node in the derivation tree generating this return statement. We will later define

$$k.prn = m; TAIL(K.prn).$$

In the ISA configurations d arising during execution of user code, we cannot possibly hope to have consistency with k for two reasons:

- the stack base and heap pointers of k have been overwritten;
- the hardware *pc* $d.pc$ is completely unrelated to the program rest of k.

Fortunately, the pointers are saved in the process control block $PCB[0]$ of the kernel, and for *pc* consistency with k we obviously need a *pc* pointing to $start(m)$. Now *if* we were to context switch from d to a configuration

$$q = cswk(d, k),$$

where these values are restored, we should regain compiler consistency $consis(q, k)$. Guided by these observations we define

$$cswk(d, k).m = d.m,$$

$$cswk(d, k).RF(i_5) = \begin{cases} va(PCB[0].RF[i], k)_{32}, & RF = gpr \wedge i \in [28:30], \\ d.RF(i_5), & \text{otherwise,} \end{cases}$$

$$cswk(d, k).pc = start(m).$$

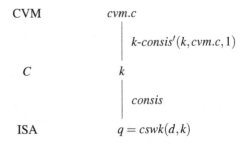

Fig. 232. A user is running. The C+A computation does not have a C abstraction. Instead a substitute k is used for this C abstraction. If in configuration d one restores the pointers from $PCB[0]$ and sets the pc to $start(m)$, one obtains configuration $q = cswk(d,k)$, which is consistent with k. In configuration $cvm.c$ the call of special function $runvm$ is completely gone, whereas in configuration k the return statement is left dangling. Configurations $cvm.c$ and k are out of sync, and we have $k\text{-}consis'(k, cvm.c, 1)$

It turns out that the $C0$ configuration $cvm.c$ of the suspended abstract kernel will be one $C0$ step ahead of the (substituted) $C0$ configuration k of the concrete kernel, which will still have the dangling return statement and a function frame for $runvm$. Thus we will have

$$k\text{-}consis'(k, cvm.c, 1) \wedge danglingr(k, cvm.c).$$

As illustrated in Fig. 232 we now define the following invariant.

Invariant 19

$$
\begin{aligned}
inv\text{-}csus(d, k, cvm) \equiv\ & consis(k, cswk(d,k))\ \wedge \\
& k\text{-}consis'(k, cvm.c, 1)\ \wedge \\
& danglingr(k, cvm.c).
\end{aligned}
$$

A simple lemma shows that address translation as defined via the $C0$ data structures in $C0$ configuration k for user u is performed by memory management units in an ISA configuration d satisfying:

- $consis(k, cswk(d,k))$, i.e., d is consistent with k except for the pc and the three pointers.
- The pto and ptl registers of the hardware contain their counterparts from the process control block of user u.

Lemma 140. *Let*

$$consis(k, cswk(d,k))$$

and

$$
\begin{aligned}
d.pto &= va(PCB[u].spr[5], k), \\
d.ptl &= va(PCB[u].spr[6], k).
\end{aligned}
$$

Then

$$pte(k,u,a) = pte(d,a),$$
$$v(k,u,a) = v(d,a),$$
$$ppx(k,u,a) = ppx(d,a),$$
$$pma(k,u,a) = pma(d,a),$$

$$ptle(k,u,a) = ptle(d,a),$$
$$ipf(k,u,a) = ipf(d,a),$$
$$pf(k,u,a) = pf(d,a).$$

Proof. This would be a bookkeeping exercise if $consis(k,d)$ held. Moreover, one observes that computations of the memory management unit only depend on pto, ptl, and m, and that these components are identical for d and $cswk(d,k)$. Thus the MMU computes the same values in configurations d and $cswk(d,k)$.

15.3.6 Technical Invariants

We are aiming at a correctness theorem stating that the C+A computation of the concrete kernel simulates the CVM computation. This simulation theorem can be expressed with the definitions made so far. In order to prove it by induction we have to augment the simulation invariants by additional *technical invariants*. In the end these technical invariants are collected into a single invariant $inv\text{-}T(d,c,cvm)$. In a proof by induction one usually writes down the entire induction hypothesis first; then one proceeds to show base case and induction step. On the other hand it is much more comfortable to introduce technical invariants in the place where they are needed. Thus if we prove the entire theorem in I steps, then in each such step i additional technical invariants $inv\text{-}t(i,d,c,cvm)$ might be introduced and we have

$$inv\text{-}T(d,c,cvm) \equiv \bigwedge_{i=0}^{I-1} inv\text{-}t(i,d,c,cvm).$$

It is possible to develop a proof by induction in this way. However, when new invariants $inv\text{-}t(i,d,c,cvm)$ are introduced at step i, one has to make sure that previous portions of the proof also would have worked if this new invariant had been in place right from the start, i.e., one has to verify in step i that invariant $inv\text{-}t(i,d,c,cvm)$ is not broken in steps $j < i$.

We start out with $inv\text{-}cp(c,cvm)$, $inv\text{-}pt(c,cvm)$, and two more simple invariants:

- if a CVM user is running $cvm.cp > 0$, the underlying hardware runs in translated mode ($mode(d)$); if the abstract kernel of CVM is running, the hardware is running in untranslated mode.

Invariant 20

$$inv\text{-}mode(d,cvm) \equiv (cvm.cp > 0 \leftrightarrow mode(d)).$$

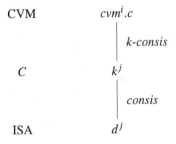

Fig. 233. The kernel is running. Step $j = s(i)$ of the C+A computation corresponds to step i of the CVM computation. The C+A computation has a C abstraction k^j. It codes $cvm^i.c$ via $k\text{-}consis(k^j, cvm.c^i)$, and $consis(k^j, d^j)$ holds by the C+A semantics

- interrupts cannot be enabled while translated C is running. The reason for this restriction is that we have not given a semantics for interrupts in this situation.

Invariant 21
$$inv\text{-}sr(d) \;\equiv\; (transc(d) \to d.sr = 0_{32}).$$

Then we summarize as follows.

Invariant 22
$$inv\text{-}t(0,d,c,cvm) \;\equiv\; \begin{aligned}&inv\text{-}cp(c,cvm) \;\wedge\\ &inv\text{-}pt(c,cvm) \;\wedge\\ &inv\text{-}mode(d,cvm) \;\wedge\\ &inv\text{-}sr(d).\end{aligned}$$

15.3.7 Formulating the Correctness Theorem

For C+A configurations y we define a predicate stating that y has a C abstraction by

$$hasc(y) \equiv \exists d,k,ic : y = (d,k,ic).$$

For ISA configurations d, C0 configurations k, and numbers ic we say that (d,k,ic) completes y if d and ic just extract the d and ic components of y and k extracts the C abstraction of y in case y has such an abstraction:

$$y \sim (d,k,ic) \equiv \begin{aligned}&y.d = d \;\wedge\\ &y.ic = ic \;\wedge\\ &hasc(y) \to y.c = k.\end{aligned}$$

Thus, if y has a C abstraction, then $y = (d,k,ic)$. Otherwise, $y = (d,ic)$ and k provides the missing C abstraction.

To formulate the correctness theorem we start with a C+A computation (y^j) and a sequence of external interrupt event vectors (eev^j_{isa}). We will construct for all i and j

- a CVM computation (cvm^i),

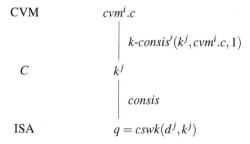

Fig. 234. A user is running. Step $j = s(i)$ of the C+A computation corresponds to step i of the CVM computation. $C0$ configuration $cvm^i.c$ is one $C0$ step ahead of k^j; thus we have $k\text{-}consis'(k^j, cvm^i.c, 1)$ and configuration $q = cswk(d^j, k^j)$ is consistent with k^j

- a sequence of external interrupt event vectors (eev^i_{cvm}),
- a sequence of step numbers $(s(i))$,
- a sequence (d^j, k^j, ic^j),

such that for all steps j of the C+A computation and all steps i of the CVM computation the following holds:

$$y^j \sim (d^j, k^j, ic^j),$$

i.e., the sequence (d^j, k^j, ic^j) provides C abstractions for the original C computations. Moreover, if $s(i) = j$, i.e., the simulation of step i of the CVM computation (cvm^i) starts with step j of the C+A computation, then

- the technical invariants $inv\text{-}T(d^j, k^j, cvm^i)$ hold,
- the ISA component d^j of the C+A computation together with k^j codes the user processes,

$$inv\text{-}vm(d^j, k^j, cvm^i)$$

- if the kernel is running:

$$cvm^i.cp = 0,$$

then we require

$$inv\text{-}crun(d^j, k^j, cvm^i)$$

to hold. Fig. 231 is thus instantiated by Fig. 233.
- if a user is running:

$$cvm^i.cp > 0,$$

then we require

$$inv\text{-}csus(d^j, k^j, cvm^i)$$

to hold. Fig. 232 is thus instantiated by Fig. 234.

In a nutshell the substituted $C0$ configurations k^j are needed to reconstruct a $C0$ configuration k^r if we run translated $C0$ code again in a later C+A step $r > j$.

We formulate this correctness statement first in nonconstructive form. A constructive formulation of the step numbers $s(i)$ can be extracted from the proof. We collect the invariants of the correctness statement into a single invariant.

Invariant 23

$$inv\text{-}kernel(d,k,cvm) \equiv \quad \begin{aligned} & inv\text{-}T(d,k,cvm) \wedge && \textit{(technical invariants)} \\ & inv\text{-}vm(d,k,cvm) \wedge && \textit{(user processes simulated)} \\ & (cvm.cp = 0 \to inv\text{-}crun(d,k,cvm)) \wedge && \textit{(running kernel simulated)} \\ & (cvm.cp > 0 \to inv\text{-}csus(d,k,cvm)) && \textit{(suspended kernel simulated)}. \end{aligned}$$

In the remainder of this chapter we prove by induction on i the following.

Lemma 141.

$$\begin{aligned} & \forall(y^j) && \textit{(C+A computation)} \\ & \forall(eev^j_{isa}) && \textit{(ISA inputs)} \\ & \exists(cvm^i) && \textit{(CVM computation)} \\ & \exists(eev^i_{cvm}) && \textit{(CVM inputs)} \\ & \exists(s(i)) && \textit{(step numbers)} \\ & \exists(d^j,k^j,ic^j) && \textit{(sequence of concrete kernel configurations)} \\[4pt] & \forall i,j: && \\ & y^j \sim (d^j,k^j,ic^j) \wedge && \textit{(providing C abstractions)} \\ & (j = s(i) \to && \textit{(indices of corresponding configurations)} \\ & inv\text{-}kernel(d^j,k^j,cvm^i)) && \textit{(kernel invariants)}. \end{aligned}$$

We treat the start of the induction when we treat interrupts, because initialization is performed as a reaction to the *reset* interrupt. In the induction step we conclude from i to $i+1$. We start with

$$j = s(i)$$

and for

$$j' = j+1, j+2, \ldots$$

we inspect C+A configurations

$$y^{j'} \sim (d^{j'}, k^{j'}, ic^{j'})$$

until we can establish

$$inv\text{-}kernel(d^{j'}, k^{j'}, cvm^{i+1})$$

at which point we set

$$s(i+1) = j'.$$

15.4 The *runvm* Primitive

In the induction step the obvious case split mirrors the different cases of the CVM semantics. Before we treat these cases in detail we give an overview over the implementation of the *runvm* primitive.

15.4.1 Overview of the *runvm* Primitive

We classified page faults as

- *ptle*: page table length exceptions.

  They are visible in CVM and are handled by the abstract kernel.
- *ipf*: invalid page exceptions ($\neg ptle$ and the requested page is not in the physical memory).

  If no page table length exception occurs, this exception is not visible in CVM, and the concrete kernel has to handle it transparently.

The tasks of the *runvm* primitive are:

1. process switch from concrete kernel to user after the start of the *runvm* primitive by macro *restore*,
2. saving a user task and exiting to the concrete kernel after an interrupt by macro *save*,
3. testing after an interrupt whether an *ipf*-page fault has to be handled by the concrete kernel by macro *compute-ipf*,
4. transparent handling of *ipf*-page faults by function call *ipf-handler*,
5. process switch from kernel to user after the handling of an *ipf*-page fault by macro *restore*,
6. exiting to the kernel after the suspension of a user task.

In order to navigate between these tasks and to identify the current user the concrete kernel uses two global variables:

- *int CP*: if a user is running it keeps track of the current user process. An invariant for it has already been defined as

$$inv\text{-}cp(c, cvm) \equiv (cvm.cp > 0 \rightarrow va(CP, c) = cvm.cp).$$

  While the kernel is running it is used i) after a scheduler run to specify the user to be started next and ii) after an interrupt to determine for which user the interrupt has to be handled.
- *bool ipf*: it is used locally in the implementation of the *runvm* primitive to signal the occurrence of an *ipf*-page fault.

The body of special function *runvm* is

```
restore;
save;
compute-ipf;
if ipf then {ipf-handler(); restore};
return 1
```

A flow diagram of the implementation of the *runvm* primitive is shown in Fig. 235. Numbers in the figure correspond to the enumeration of the tasks of the primitive given above. There are two entry points: α) from the abstract kernel and γ) from the user processes. Exit to the abstract kernel is via the return statement in (6). Exit to a user process is via an *eret* instruction in the implementations of 'restore' in (1) and (5).

15.4.2 Program Annotations

For the labels

$$x \in \{\alpha, \ldots, \xi\}$$

denoting certain points in the flow chart we would like to formulate invariants of the form

$$inv_x(d^j, k^j, cvm^i),$$

where

- invariants inv_x should hold when the computation reaches label x,
- i is an index in the CVM computation, and
- $j \in [s(i) : s(i+1) - 1]$ denotes steps of the C+A computation simulating CVM step i.

Indeed for each label x we could write invariant inv_x like a comment in the program text. A program obtained in this way is called an *annotated program* and invariants inv_x are therefore also called *program annotations*. Program annotations will be proven by induction on the CVM steps i. For each i all invariants $inv_x(d^j, k^j, cvm^i)$ are proven jointly by induction on $j \in [s(i) : s(i+1)]$. With step $j = s(i+1)$ one completes the induction step from i to $i+1$. Thus, for the proof of invariant $inv_x(d^j, k^j, cvm^i)$ we can assume that *all* invariants $inv_y(d^{j'}, k^{j'}, cvm^i)$ hold including $y = x$ provided $j' < j$. The situation where $y = x$ occurs naturally because the control flow of programs contains cycles.

We denote steps of the C+A computation when label x is reached by j_x and formulate the invariants inv_x in the form

$$inv(d^{j_x}, k^{j_x}, cvm^i).$$

In addition we refer to C+A configurations where virtual machine steps are executed by index j_λ.

In the sequel we present the annotations for the maintenance of the three simulation relations *k-consis*, *consis*, and *inv-vm*. In general we will say little about the technical invariants. Most technical invariants are maintained most of the time simply because the data structures on which they depend are not modified. Hence we mention technical invariants only in situations where the data structures on which they depend are modified.

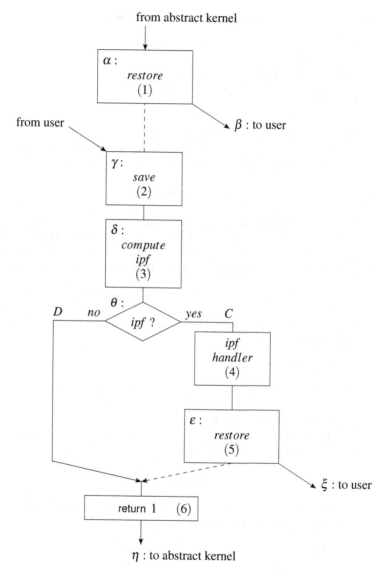

Fig. 235. Flow chart for the implementation of the *runvm* primitive. Labels with names x correspond to steps j_x in the C+A computation. The dotted arrows of the flow chart are never taken, because the computation exits to a user process. From there it returns to label γ

15.4.3 Maintaining *k-consis*

Below we introduce invariants which are needed to prove that *k-consis* is maintained by the execution of *runvm*:

- j_α: when the flow chart is entered from the abstract kernel after a call of the *runvm* primitive. We have

$$j_\alpha = j + 1$$

and from *C0* semantics and *inv-kernel*(d^j, k^j, cvm^i) we directly infer

Invariant 24

$$k\text{-}consis(k^{j_\alpha}, cvm^i.c, 1).$$

The abstract and concrete kernels are still in sync.

- j_β: when 'restore' is left after it was entered from the abstract kernel. This step completes the simulation of the call of *runvm* in CVM. Thus we have

$$j_\beta = s(i + 1).$$

The *C* abstraction is lost in the C+A computation. Thus we need later to define a substitute k^{j_β} for it. In the abstract kernel configuration $cvm^{i+1}.c$ the call of *runvm* is dropped, whereas in concrete kernel configuration k^{j_α} the return statement at the bottom of the flow chart is left dangling. In Section 15.5.4 we show

$$k\text{-}consis'(k^{j_\beta}, cvm^{i+1}.c, 1) \wedge danglingr(k^{j_\beta}, cvm^{i+1}.c).$$

This is one of the cases when an induction step for i is completed. The invariant maintained is

Invariant 25

$$k\text{-}consis'(k^{j_\beta}, cvm^i.c, 1) \wedge danglingr(k^{j_\beta}, cvm^i.c).$$

This is shown in Sect. 15.5.4.

- j_δ: after a user process was saved. This happens after an interrupt during the execution of a user step in a CVM configuration cvm^i. We have

$$s(i) < j_\delta < s(i + 1).$$

A *C* abstraction k^{j_δ} is regained. A user process is running, thus the abstract and concrete kernels are out of sync, and we get

Invariant 26

$$k\text{-}consis'(k^{j_\delta}, cvm^i.c, 1).$$

This is shown in Sect. 15.5.7.

- j_ε: when 'restore' is entered after branch *C* was taken, i.e., after a run of the page fault handler. From Invariant 26 as well as from *C0* semantics we conclude

Invariant 27

$$k\text{-}consis'(k^{j_\varepsilon}, cvm^i.c, 1).$$

- j_ξ: when 'restore' is left after it was entered by branch C. This does **not** complete the implementation of the call of *runvm*. It continues the simulation of a virtual machine step i after the handling of an *ipf*-page fault. This is transparent for the user, and the simulation of the same step continues. We have

$$s(i) < j_\xi < s(i+1).$$

As in step j_β the return statement remains dangling in the program rest of the concrete kernel while we are continuing the simulation of CVM step i.

Invariant 28

$$k\text{-}consis'(k^{j_\xi}, cvm^i.c, 1) \wedge danglingr(k^{j_\xi}, cvm^i.c).$$

This is shown in Sect. 15.5.4.

- j_η: after leaving the flow chart to the abstract kernel. This happens via branch D and completes the simulation of a CVM step i. Thus

$$j_\eta = s(i+1).$$

This happens only in case of a non-*ipf* interrupt in user mode. From Invariant 26 and $C0$ semantics we infer for the C+A step $s(i+1) - 1$ before the execution of the return statement that k-consistency is reached up to the dangling return statement, i.e.,

$$k\text{-}consis'(k^{s(i+1)-1}, cvm^{i+1}.c, 1) \wedge danglingr(k^{s(i+1)-1}, cvm^{i+1}.c).$$

Execution of the return statement gets the concrete kernel back in sync with the abstract kernel:

$$k\text{-}consis(k^{s(i+1)}, cvm^{i+1}.c).$$

This is a case where the induction step for i is completed. The invariant maintained is

Invariant 29

$$k\text{-}consis(k^{j_\eta}, cvm^{i+1}.c).$$

- j_λ: for CVM step i, which is a user step. We have

$$s(i) \leq j_\lambda$$

and require

Invariant 30

$$k\text{-}consis'(k^{j_\lambda}, cvm^i.c, 1) \wedge danglingr(k^{j_\lambda}, cvm^i.c).$$

If $j_\lambda = s(i)$, i.e., after a user was started or after a previous user step, this directly follows by induction hypothesis from $inv\text{-}csus(d^{s(i)}, k^{s(i)}, cvm^i)$. For $j_\lambda > s(i)$ the user step is repeated after page fault handling, and the invariant follows from Invariant 28.

15.4.4 Maintaining *consis*

The structure of the corresponding arguments for compiler consistency is much simpler:

- j_α: when the flow chart is entered from the abstract kernel the C+A computation keeps the C abstraction and we get the trivial

  Invariant 31
  $$consis(k^{j_\alpha}, d^{j_\alpha}).$$

- j_δ: after a user process is saved a C abstraction must be gained. Thus we need

  Invariant 32
  $$consis(k^{j_\delta}, d^{j_\delta}).$$

  This invariant is shown in Sect. 15.5.7.

- j_ε, j_η: For any step j' when 'restore' is entered after page fault handling or after the return statement, consistency holds.

  Invariant 33
  $$j' \in \{j_\varepsilon, j_\eta\} \rightarrow consis(k^{j'}, d^{j'}).$$

  Proof. By Invariant 32 we have a C abstraction in step j_δ. A C abstraction is obviously maintained by $C0$ statements in the flow chart. That a C abstraction holds after the '*ipf*-test' and after the '*ipf*-handler' is implicit in the arguments of Sect. 15.6, where we always reestablish this C abstraction immediately after the inline assembly portions.

- j_β, j_ξ: for any step j' after 'restore' is left, the kernel is suspended, the C abstraction is substituted by a suitable concrete kernel configuration $k^{j'}$, and the consistency relation for suspended kernels has to hold.

  Invariant 34
  $$j' \in \{j_\beta, j_\xi\} \rightarrow consis(k^{j'}, cswk(d^{j'}, k^{j'})).$$

  The definition of $k^{j'}$ as well as the proof of the invariant is given in Sect. 15.5.4.

- j_λ: for CVM step i, which is a user step. We have

  $$s(i) \le j_\lambda$$

  and require

  Invariant 35
  $$consis(k^{j_\lambda}, cswk(d^{j_\lambda}, k^{j_\lambda})).$$

  If $j_\lambda = s(i)$, i.e., after a user was started or after a previous user step, this directly follows by induction hypothesis from $inv\text{-}csus(d^{s(i)}, k^{s(i)}, cvm^i)$. For $j_\lambda > s(i)$ the user step is repeated after page fault handling, and the invariant follows from Invariant 34.

15.4.5 Maintaining *inv-vm*

This is slightly involved, because *ipf*-page faults are invisible in CVM. In contrast, when such an interrupt occurs in the ISA configuration, then the special purpose registers other than *pto* and *ptl* are overwritten. For ISA configurations d and inputs *eev* we define invalid page faults on fetch, invalid page faults on load/store, and general invalid page faults by

$$ipff(d,eev) \equiv ipf(d,d.pc) \wedge il(d,eev) = 17,$$
$$ipfls(d,eev) \equiv ipf(d,ea(d)) \wedge il(d,eev) = 20,$$
$$ipf(d,eev) \equiv ipff(d,eev) \vee ipfls(d,eev).$$

- j_α: when the flow chart is entered from the abstract kernel, no process is running, and we get from the induction hypothesis

Invariant 36

$$\forall u > 0 : vmsus(d^{j\alpha}, k^{j\alpha}, u) = cvm^i.vm(u).$$

- j_δ: after a process save. If a user process u was saved before step $j_\delta > s(i)$, then there was an interrupt at a step j' which also belongs to the simulation of CVM step i, i.e.,

$$s(i) \le j' < j_\delta \quad \text{and} \quad jisr(d^{j'}, eev^{j'}_{isa}).$$

Consider the machine $vmsus(d^{j\delta}, k^{j\delta}, u)$ which is abstracted after a 'process save' and which is now suspended. We must split cases depending on the occurrence of an invalid page fault. Invalid page faults are handled transparently and the saved machine is equivalent to the current configuration $cvm^i.vm(u)$ of the simulated machine. If a different interrupt occurs, the simulated machine makes a step, and the saved configuration should be the next configuration of this machine.

We define

$$u = va(CP, k^{j\delta}) = va(CP, k^{j\alpha}),$$
$$vm = vmsus(d^{j\delta}, k^{j\delta}, u),$$
$$j' = \max\{x < j_\delta \mid jisr(d^x, eev^x_{isa})\},$$

and state

Invariant 37

$$ipf(d^{j'}, eev^{j'}_{isa}) \rightarrow vm \sim cvm^i.vm(u),$$
$$\neg ipf(d^{j'}, eev^{j'}_{isa}) \rightarrow vm = cvm^{i+1}.vm(u).$$

This invariant is shown in Sect. 15.5.7.

- j_θ: after 'compute ipf'. The decision whether branch D or branch C is taken is based on the value of the Boolean variable ipf computed in 'compute ipf'. Let j_θ be the C+A step in which this flag is tested. Then

Invariant 38

$$va(ipf, k^{j_\theta}) \equiv ipf(d^{j'}, eev_{isa}^{j'}).$$

This invariant is shown in Sect. 15.6.1.

- j_η: the flow chart is left. This happens after branch D is taken. On this branch configuration vm is not changed.

From Invariants 37 and 38 we conclude

$$vmsus(d^{j_\eta}, k^{j_\eta}, u) = vmsus(d^{j_\delta}, k^{j_\delta}, u)$$
$$= cvm^{i+1}.vm(u).$$

This completes the simulation of CVM step i. We set

$$s(i+1) = j_\eta$$

and conclude

$$inv\text{-}vm(d^{s(i+1)}, k^{s(i+1)}, cvm^{i+1}).$$

- j_ε: after the 'ipf-handler'. Let j_ε be the C+A step after completion of the 'ipf-handler'. Then the abstracted user machine configuration vm for which the page fault was handled did not change. Let j_δ be the last step before step j_ε, when label δ was reached. Then

Invariant 39

$$vmsus(d^{j_\delta}, k^{j_\delta}, u) = vmsus(d^{j_\varepsilon}, k^{j_\varepsilon}, u).$$

This invariant is shown in Sects. 15.6.3 and 15.6.4.

When 'restore' is entered after branch C was taken, we conclude from Invariants 37, 38, and 39 for j_ε:

$$cvm^i.vm(u) \sim vmsus(d^{j_\delta}, k^{j_\delta}, u)$$
$$= vmsus(d^{j_\varepsilon}, k^{j_\varepsilon}, u).$$

Thus we get

Invariant 40

$$cvm^i.vm(u) \sim vmsus(d^{j_\varepsilon}, k^{j_\varepsilon}, u).$$

- j_β, j_ξ: when 'restore' is left in step j_β after branch A or in step j_ξ after branch C, then the suspended machine of user u was successfully restored to the registers from its process control block.

Invariant 41

$$vmrun(d^{j_\beta}, k^{j_\beta}, u) = vmsus(d^{j_\alpha}, k^{j_\alpha}, u),$$
$$vmrun(k^{j_\xi}, k^{j_\xi}, u) = vmsus(d^{j_\varepsilon}, k^{j_\varepsilon}, u).$$

This invariant is shown in Sect. 15.5.4.

Together with Invariants 36 and 40 we get

Invariant 42

$$x \in \{j_\beta, j_\xi\} \rightarrow cvm^i.vm(u) \sim vmrun(d^x, k^x, u).$$

- j_λ: for CVM step i, which is a user step. We have

$$s(i) \leq j_\lambda$$

and require for the running user u

Invariant 43

$$cvm^i.vm(u) \sim vmrun(d^{j_\lambda}, k^{j_\lambda}, u).$$

If $j_\lambda = s(i)$, i.e., after a user is started or after a previous user step, this follows directly by induction hypothesis from $inv\text{-}vm(d^{s(i)}, k^{s(i)}, cvm^i)$. For $j_\lambda > s(i)$ the user step is repeated after page fault handling, and the invariant follows from Invariant 42.

15.5 Simulation of CVM Steps

15.5.1 Scratch Memory

Before we proceed we have to introduce some new technical invariants.

We need to have a look at a preliminary memory map of the kernel implementation as shown in Fig. 236. We are presently only interested in the first two pages:

- ROM on page 0. After an interrupt, program execution always continues at the first address of the ROM. We elaborate on the role of the ROM when we treat the handling of interrupts by the concrete kernel.
- *scratch* on page 1. This page covers byte addresses from $4K$ to $8K - 1$.

The code region $[a : b]$ and the data region $[sbase : hmax)$ for the compiled kernel both lie above the scratch memory:

$$8K \leq \langle a \rangle < \langle b \rangle < \langle sbase \rangle < \langle hmax \rangle.$$

We will use the first two words of scratch memory to store the base addresses of two process control blocks, and introduce

Invariant 44 *Invariant*

$$inv\text{-}scratch(d, k)$$

holds iff

- *for the kernel:*

$$d.m_4((4K)_{32}) = ba(PCB[0], k),$$

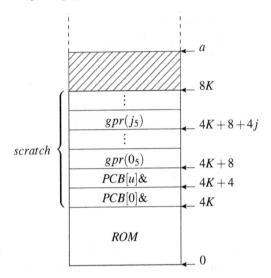

Fig. 236. Memory map for the lower portion of physical memory. The first two pages are ROM and scratch memory. Above lies the code region $[a : b]$ of the compiled kernel. The first two words of scratch memory store the base addresses of $PCB[0]$ and $PCB[u]$. The next words are reserved for the temporary storage of the general-purpose registers

- *for the current user process if a user is running:*

$$cvm.cp = u > 0 \rightarrow d.m_4((4K+4)_{32}) = ba(PCB[u],k).$$

We include these predicates among the technical invariants:

$$inv\text{-}t(2,d,k,cvm) \equiv inv\text{-}scratch(d,k).$$

The next 32 words are used as temporary storage for the general-purpose registers of the processor. We always store the register j at the address $(4K+8+4j)_{32}$. Thus, in configurations d after saving $gpr(j_5)$ we want to have

$$d.m_4((4K+8+4j)_{32}) = d.gpr(j_5).$$

This use of the scratch page is illustrated in Fig. 236. As long as d and k are consistent the base addresses stored in the scratch memory can easily be computed and written with C+A programs. Initialization of the first word in scratch memory is achieved by

```
gpr(1) = PCB[0]&;
asm( sw 1 0 4K );
```

and if

$$u = va(CP,k)$$

then initialization of the second word in scratch memory is achieved by

```
gpr(1)  =  PCB[CP]&;
asm( sw 1 0 4K+4 );
```

These data are stored in scratch memory because we will also need these data when $consis(k,d)$ does not hold.

In order to store and restore the value of register r to/from the scratch page, we define assembler macros $ssave(r)$ and $srestore(r)$:

```
ssave(r):
   sw  r  0  4K+8+4r

srestore(r):
   lw  r  0  4K+8+4r
```

Observe that both macros are implemented using direct addressing mode, i.e., the effective address is determined using general-purpose register $rs = 0_5$ which is tied to zero by construction.

We leave space for storing general-purpose registers because we often need to perform certain auxiliary computations *before* saving registers into process control blocks. Because the auxiliary computations overwrite certain registers, we have to save the original values of these registers into the scratch, so that we can save them to the process control blocks later.

15.5.2 $C0$ Step of the Kernel

From the induction hypothesis we have

$$k\text{-}consis(k^j, cvm^i.c) \quad \text{and} \quad consis(k^j, d^j).$$

We perform a single step in the C+A computation and the CVM computation:

$$s(i+1) = s(i) + 1 = j + 1.$$

Then k-consistency is preserved by Lemma 139 and consistency is preserved by the C+A semantics. The page table invariant is preserved because steps of the abstract kernel do not access the data structures of the concrete kernel. User simulation

$$inv\text{-}vm(d^{j+1}, k^{j+1}, cvm^{i+1})$$

and the technical invariants are preserved because steps of the abstract kernel

- do not access i) the process control blocks, ii) the disk, iii) the pages of users (by Invariant 15), and iv) the scratch memory;
- do not change the mode of the machine;
- do not change $cvm.cp$ (which remains zero). Invariant 11 therefore holds trivially, even if variable CP is changed.

15.5.3 ISA Step of a User Without Interrupt

If this happens in a C+A step j during the simulation of CVM step i, then we have

$$j \geq s(i)$$

where $j = s(i)$ occurs in situations where the user machine was running before or was just started. If execution of CVM step i is repeated in step j after the handling of a page fault, then $j > s(i)$ holds. In any case we perform a single step in the C+A computation and the CVM computation; the abstract kernel configuration stays the same:

$$s(i+1) = j+1,$$
$$k^{j+1} = k^j.$$

Let

$$cvm^i.cp = u$$

be the current user process.

If $s(i) = j$, from the induction hypothesis we have

$$mode(d^j) \quad \text{and} \quad consis(k^j, cswk(d^j, k^j)) \quad \text{and} \quad inv\text{-}vm(d^j, k^j, cvm^i).$$

If $s(i) < j$, then $j \in \{j_\beta, j_\xi\}$ and we infer consistency and user simulation from Invariants 34 and 42. That user mode is active in d^j follows from the semantics of *eret*, which is the last instruction in the implementation of 'restore'.

From Invariant $inv\text{-}vm(d^j, k^j, cvm^i)$ we infer that the processor portion of the ISA configuration is similar to the processor portion of the running user:

$$proc(d^j) \sim_{VM} proc(cvm^i.vm(u)).$$

In particular we get

$$d^j.pc = cvm^i.vm(u).pc.$$

The instruction is fetched in translated mode. The page table entry accessed is valid, because otherwise a page fault interrupt would occur. Abbreviating the current user configuration by

$$d_u = cvm^i.vm(u)$$

we get from the ISA semantics, Lemma 140, and $inv\text{-}vm(d^j, k^j, cvm^i)$:

$$\begin{aligned}
I(d^j) &= d^j.m_4(pma(d^j, d^j.pc)) \\
&= d^j.m_4(pma(k^j, u, d^j.pc)) \\
&= d_u.m_4(d_u.pc) \\
&= I(d_u).
\end{aligned}$$

Thus the hardware executes the currently simulated instruction. Now three cases are possible:

1. No load or store instruction is executed. The next processor configuration depends only on the instruction executed:

$$proc(d^{j+1}) = proc(\delta_M(d^j))$$
$$\sim_{VM} proc(\delta_M(d_u)) \qquad \text{(Lemma 138).}$$

The memory is not changed, thus for $a \in A(cvm^i.ptl(u))$ we get:

$$\delta_M(d^j).m(pma(d^j,a)) = d^j.m(pma(d^j,a))$$
$$= d^j.m(pma(k^j,u,a))$$
$$= d_u.m(a)$$
$$= \delta_M(d_u).m(a).$$

Thus $inv\text{-}vm$ is preserved.

2. A load is executed. From

$$proc(d^j) \sim_{VM} proc(d_u) \quad \text{and} \quad I(d^j) = I(d_u)$$

we get

$$ea(d^j) = ea(d_u).$$

As for the instruction fetch we conclude for the load results

$$lres(d^j) = lres(d_u),$$

which implies

$$proc(\delta_M(d^j)) \sim_{VM} proc(\delta_M(d_u)).$$

As in case 1 the memory is not changed.

3. A store is executed. Let

$$a = ea(d^j) = ea(d_u)$$

be the common effective address and

$$s = d^j.gpr(rt(d^j)) = d_u.gpr(rt(d_u))$$

be the common word which is stored. For address a we have to show separately that $inv\text{-}vm$ is preserved. From the ISA semantics and Lemma 140 we get:

$$\delta_M(d^j).m_4(pma(d^j,a)) = s$$
$$= \delta_M(d_u).m_4(a).$$

Invariants for the page tables, the simulation for the abstract kernel, and the technical invariants are preserved because by invariant $inv\text{-}ppx(k^j)$ the physical memory addresses used are not below address $hmax$ and thus are outside the kernel data structures and the scratch memory.

15.5.4 Restoring a User Process

$C0$ macro 'restore' is entered in C+A step j during simulation of CVM step i for some user u. We have

$$j > s(i) \quad \text{and} \quad u = va(CP, k^j).$$

If the macro was entered from the abstract kernel in a step $j = j_\alpha$, then the concrete and abstract kernel are still in sync, and by Invariant 24 we have

$$k\text{-}consis(k^j, cvm^i.c, 1).$$

If it was entered after handling of a page fault via branch C in a step $j = j_\varepsilon$, then the concrete and abstract kernel are already out of sync and by Invariant 27 we have

$$k\text{-}consis'(k^j, cvm^i.c, 1).$$

Thus in particular the program rests of abstract and concrete kernel are related by

$$TAIL(k^j.prn) = \begin{cases} tail(cvm^i.c.prn), & j = j_\alpha, \\ cvm^i.c.prn, & j = j_\varepsilon \end{cases}$$

and the stack of $cvm^i.c$ is obtained from the stack of k^j by chopping off the top elements of the stack of k^j:

$$chop(k^j, cvm^i.c, 1).$$

In both cases by Invariants 33, 36, and 40 we have

$$consis(k^j, d^j) \quad \text{and} \quad vmsus(d^j, k^j, u) \sim cvm^i.vm(u).$$

We specify the implementation of macro 'restore-user' in several steps:

- First we save the base address of $PCB[u]$ into the second word of scratch memory via register 1.

```
gpr(1) = PCB[CP]&;
asm( sw 1 0 4K+4 );
```

For the resulting configuration d^{j_1} this updates the second word of scratch memory as required by $inv\text{-}scratch(d^{j_1}, k^{j_1})$:

$$k^{j_1}.m_4((4K+4)_{32}) = ba(PCB[u], k^j)$$
$$= ba(PCB[u], k^{j_1}).$$

From $inv\text{-}scratch(d^j, k^j)$ for the first word of scratch memory we know:

$$k^{j_1}.m_4((4K)_{32}) = k^j.m_4((4K)_{32})$$
$$= ba(PCB[0], k^j)$$
$$= ba(PCB[0], k^{j_1}).$$

- We load this value in register 1 and save counters bpt, spt and hpt into $PCB[0]$.

```
asm(
   lw    1  0 4K          // gpr(1) = PCB[0].gpr&
   sw    28 1 4*28        // store kernel bpt, spt
   sw    29 1 4*29        // and hpt into PCB[0]
   sw    30 1 4*30
);
```

In the resulting configuration k^{j_2} we still have consistency

$$consis(k^{j_2}, d^{j_2})$$

and register 1 has value

$$d^{j_2}.gpr(1_5) = ba(PCB[0], k^{j_1})$$
$$= ba(PCB[0], k^{j_2}).$$

The intended registers have been saved, i.e., for indices $l \in [28 : 30]$ we also have

$$va(PCB[0].gpr[l], k^{j_2}) = d^j.gpr(l_5). \tag{16}$$

- In order to restore the registers of user u we load the base address of $PCB[u]$ into register 1. We restore registers

$$esr = spr(1_5), \qquad pto = spr(5_5),$$
$$epc = spr(3_5), \qquad ptl = spr(6_5)$$

via register 2.

```
asm(
   lw    1 0 4K + 4       // gpr(1) = PCB[u]&
   lw    2 1 4*(32+1)     // load user's spr(1)
   movg2s 1 2
   lw    2 1 4*(32+3)     // load user's spr(3)
   movg2s 3 2
   lw    2 1 4*(32+5)     // load user's spr(5)
   movg2s 5 2
   lw    2 1 4*(32+6)     // load user's spr(6)
   movg2s 6 2
);
```

In the resulting configuration k^{j_3} we have restored the special purpose registers:

$$z \in \{1, 3, 5, 6\} \rightarrow d^{j_3}.spr(z_5) = va(PCB[u].spr[z], k^j).$$

- We restore user registers $gpr(1_5), \ldots, gpr(31_5)$; $gpr(1_5)$ is restored last.

```
asm (
    lw      31 1 4*31        // load user's gpr(31)
    [...]
    lw      1 1 4*1          // load user's gpr(1)
);
```

In the resulting configuration d^{j4} registers $gpr(1_5), \ldots, gpr(31_5)$ are restored:

$$z \in [1:31] \rightarrow d^{j4}.gpr(z_5) = va(PCB[u].gpr[z], k^j).$$

Consistency is lost because the pointers $gpr(28_5)$, $gpr(29_5)$, and $gpr(30_5)$ have been overwritten. We define a partial context switch function $cswp$, which restores only the pointers but not the pc:

$$cswp(d,k).X = \begin{cases} cswk(d,k).X, & X \neq pc, \\ d.pc, & X = pc. \end{cases}$$

We define a substitute k^{j4} for a C abstraction by dropping the head of the program rest and treating the last asm-statement as a noop (i.e., a statement which changes nothing) on the $C0$ data structures. Thus, for $C0$ configurations c we define:

$$\delta_{C\text{-}noop}(c).X = \begin{cases} tail(c.X), & X \in \{pr, prn\}, \\ c.X, & \text{otherwise.} \end{cases}$$

We define the substitute as

$$k^{j4} = \delta_{C\text{-}noop}(k^{j3}).$$

Because the last asm-statement did not update the $C0$ data structures we conclude as follows.

Lemma 142.
$$consis(k^{j4}, cswp(d^{j4}, k^{j4})).$$

- We complete the 'restore' portion of the flow chart with a 'return from exceptions' statement.

```
asm ( eret );
```

Let d^{j5} be the ISA configuration after execution of the $eret$ instruction. We define a substitute k^{j5} for the C abstraction. For components X other than the program rest nodes and the program rest we define this again as

$$X \notin \{prn, pr\} \rightarrow k^{j5}.X = \delta_{C\text{-}noop}(k^{j4}).X.$$

Let m be the node in the derivation tree generating the return statement of special function $runvm$. Then we define the prn component of k^{j5} as

$$k^{j_5}.prn = \begin{cases} m; tail(cvm^i.c.prn), & j = j_\alpha, \\ m; cvm^i.c.prn, & j = j_\varepsilon. \end{cases} \qquad (17)$$

Because the *eret* instruction does not modify $C0$ data structures we infer consistency except for the pc if the full context switch function *cswk* is applied:

$$consis'(k^{j_5}, cswk(d^{j_5}, k^{j_5})).$$

We can also conclude that virtual machine u is properly simulated, i.e.,

$$vmrun(d^{j_5}, k^{j_5}, u) = vmsus(d^j, k^j, u).$$

This implies Invariant 41. By Equation 17 we have

$$hd(k^{j_5}.prn) = m.$$

By the definition of *cswk* the program counter in d^{j_5} points to the start address of the code generated for m:

$$d^{j_5}.pc = start(m).$$

Hence we have

$$consis(k^{j_5}, cswk(d^{j_5}, k^{j_5})).$$

With C+A step j_5 the computation has left the 'restore' macro. If the macro was entered from the abstract kernel in $j = j_\alpha$, then the call of the *runvm* primitive in step i of the abstract kernel is completed in step j_5 and we define

$$s(i+1) = j_\beta = j_5.$$

Otherwise we have reached label ξ, i.e.,

$$j_\xi = j_5.$$

In both cases we can conclude Invariant 34. In both cases the lower stack frames were not accessed, thus relations

$$chop(k^{j_\beta}, cvm^{i+1}, 1) \quad \text{and} \quad chop(k^{j_\xi}, cvm^i, 1)$$

are maintained.

In the first case the call of the *runvm* primitive is dropped from the program rest of the abstract kernel. From Equation 17 we obtain

$$k^{j_\beta}.pr = return\ 1; tail(cvm^i.c.pr)$$
$$= return\ 1; cvm^{i+1}.c.pr,$$

i.e., we have

$$danglingr(k^{j_\beta}, cvm^{i+1}.c)$$

and we conclude Invariant 25. In the second case Equation 17 directly gives

$$danglingr(k^{j_\xi}, cvm^i.c)$$

and we conclude Invariant 28.

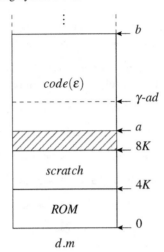

d.m

Fig. 237. Memory map for the code region [a : b] of the translated kernel in physical memory. It lies above the ROM and the scratch page. Code for initialization of the concrete kernel starts at address *a*. Address *γ-ad* corresponds to entry point *γ* in the flow chart of the *runvm* primitive (Fig. 235)

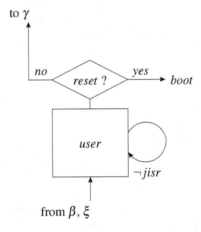

Fig. 238. Flow chart for the simulation of user steps

15.5.5 Testing for Reset

Fig. 238 illustrates the simulation of user steps. The flow chart shown there is entered after execution of the 'restore' macro in steps j_β or j_ξ. Execution of user steps that do not produce interrupts has already been treated in Sect. 15.5.3. An interrupt occurs in step j of a C+A computation in two situations:

- after *reset*. Then the translated kernel has to be loaded from the disk into addresses $[a : b]$ of physical memory. This is done by a program called the *boot loader*. Then execution of the translated kernel is started at address a.
- after an interrupt, other than *reset*. Then the user process has to be saved; control is transferred to the kernel at an address γ-*ad* corresponding to label γ in the flow chart of the *runvm* primitive (Fig. 235), where the macro 'save' begins.

Suppose that

$$jisr(d, eev)$$

holds, i.e., an interrupt which causes a jump to the interrupt service routine occurred in configuration d of the physical machine, and let

$$d' = \delta_M(d, eev).$$

Note that the machine now runs in untranslated mode:

$$mode(d') = 0.$$

For all interrupts the routine starts at address 0^{32}. This address lies in the ROM for a very good reason. This is illustrated in Fig. 237.

- In case of a *reset* interrupt, i.e., if $d'.eca[0]$, the ROM is the only memory with known content in the physical machine. In this case the C+A computation starts with

$$d' = d^0.$$

  The translated kernel has to be loaded by the *boot loader* from the disk to addresses $[a : b]$ of the physical memory. Then execution of the translated kernel is started at address a. There the kernel must be initialized such that all invariants hold.
- If a different interrupt occurred, i.e., if $\neg d'.eca[0]$, then the interrupt occurred during simulation of a CVM step i and a C+A step $j \geq s(i)$. The invariants can be assumed:

$$d = d^j \wedge inv\text{-}csus(d^j, k^j, cvm^i) \wedge inv\text{-}vm(d^j, k^j, cvm^i) \wedge inv\text{-}T(d^j, k^j, cvm^i).$$

  In this case the software in the ROM is used to jump to entry point γ of the implementation of the *runvm* primitive as described in Sect. 15.4.1 in order to save the current user process $u = cvm^i.cp$.

Obviously, the interrupt service routine in the ROM first has to test for the occurrence of a *reset* interrupt by inspecting the exception cause register $d'.eca$ in the special purpose register file. For this purpose eca has to be moved to the general-purpose register file by a *movs2g* instruction, where it overwrites the destination register of this instruction. In case we start the interrupt service routine in this way and discover that no *reset* occurred, we have a problem: the destination register was overwritten and cannot be saved to the process control block any more.

This problem is easily solved. Prior to the *movs2g* instruction we save a register (here $gpr(1_5)$) into the scratch area.[2] Then we move *eca* to $gpr(1_5)$ and perform the necessary testing. In case of a *reset* interrupt we execute the boot loader, which has a length of *lb* words, and then jump to an address *a* where the code for initialization of the concrete kernel starts. In case no *reset* interrupt occurred we jump to label γ of the implementation of *runvm*, where the code for saving the user and restoring the kernel starts.

```
ssave(1)              // store gpr(1)
movs2g 2 1            // gpr(1) = spr(2)
andi 1 1 1
blez 1 (lb+2)        // if not reset, jump to continue

bootloader:
  [...]               // load operating system code
                      // from hard disk
  j     0ba[27:2]    // jump to address a

continue:
  j     gamma-ad     // jump to address of label gamma
```

15.5.6 Boot Loader and Initialization

We still have to treat the case that an interrupt was discovered in ISA configuration *d*, and in the first four instructions of the code from Sect. 15.5.5 it was found that it was a *reset*. In this case the interrupt started the C+A computation, thus

$$d' = d^0.$$

Then the boot loader has to copy the translated concrete kernel code $code(\varepsilon)$ from the swap memory of the disk to addresses $[a : b]$ of the physical memory.

Let

$$lk = \langle b \rangle - \langle a \rangle + 4$$

be the length of the translated concrete kernel measured in bytes, and let

$$LK = \lceil lk/4K \rceil$$

be the number of pages occupied by it. We assume that this code is originally stored on the first *LK* pages of the swap disk:

---

[2] As announced in Sect. 8.2.6 this is the first place where we *need* the direct addressing mode used in the implementation of *ssave*. If $gpr(0_5)$ were not tied to zero and we could only use relative addressing, we would first need to set $gpr(0_5)$ to zero in order to address the scratch page. However, then we would not be able to save the value of $gpr(0_5)$ after a non-reset interrupt. Because $gpr(0_5)$ is tied to zero there is no need to save it, and it already has the desired value.

$$d^0.sm_{LK}(0^{28})[0:lk-1] = code(\varepsilon).$$

Finally we assume that address a is aligned to a page boundary:

$$a.bx = 0^{12}.$$

The boot loader in the ROM is the translated code of a C+A program which simply calls the appropriate disk driver from Sect. 14.3.9 LK times.

```
int SPX;
int PPX;
int L;

SPX = 0;        // start swap page address on the disk
PPX = ⟨a⟩;      // start physical page index in RAM
L = LK;         // number of pages to be loaded
while L>0
{
  readdisk(PPX, SPX);
  SPX = SPX + 1;
  PPX = PPX + 1;
  L = L - 1
}
```

Using the specification of the disk driver and C+A semantics we find for the resulting ISA configuration d^{j_1}:

$$d^{j_1}.m_{lk}(a) = code(\varepsilon).$$

After execution of the subsequent jump instruction

j	0ba[27:2]

the computation is in ISA configuration d^{j_1+1} and the initialization of the concrete kernel starts. Presently we cannot write this code, because we have not yet introduced all data structures of the concrete kernel. We will leave writing this code as an easy exercise. Note that the translated code will start with some instructions which set up pointers bpt, spt, and hpt. After execution of these instructions the C+A computation is in a configuration

$$y^{j_2} = (d^{j_2}, k^{j_2}),$$

which has a C abstraction. The initialization code has to be written such that in the C+A configuration (d^{j_3}, k^{j_3}) after its execution we have

$$inv\text{-}vm(d^{j_3}, k^{j_3}, cvm^0) \wedge inv\text{-}crun(d^{j_3}, k^{j_3}, cvm^0) \wedge inv\text{-}T(d^{j_3}, k^{j_3}, cvm^0).$$

With

$$s(0) = j_3$$

this then establishes the base case of the correctness proof for the concrete kernel, i.e., the proof of Lemma 141.

15.5.7 Process Save

Suppose $C0$ macro 'save' is entered in C+A step $j > s(i)$ during simulation of CVM step i with current user

$$u = va(CP, k^j) > 0.$$

Then an interrupt occurred at a preceding step j'

$$j' = \max\{x < j \mid jisr(d^x, eev_{isa}^x)\}$$

and

$$s(i) \leq j' < j.$$

Because the test for *reset* failed, the system remains initialized and we can assume that the invariants hold. From Invariant 30 we get

$$k\text{-}consis'(k^{j'}, cvm^i.c, 1) \land danglingr(k^{j'}, cvm^i.c).$$

From Invariant 35 we get

$$consis(k^{j'}, cswk(d^{j'}, k^{j'}))$$

and for the running machine

$$vm = vmrun(d^{j'}, k^{j'}, u)$$

we get from Invariant 43

$$vm \sim cvm^i.vm(u).$$

Now in the correctness proof — not in the programming — we must split cases depending on the visibility of the interrupt in the CVM computation.

- The interrupt is visible if

$$\neg ipf(d^{j'}, eev_{isa}^{j'}).$$

  In this case the simulation of CVM step i has succeeded at the ISA level.

  We only have to save configuration $cvm^{i+1}.vm(u)$ in order to complete the CVM step. We identify the input sampled in CVM step i as

$$eev_{cvm}^i = eev_{isa}^{j'}$$

and conclude with Lemma 138

$$jisr(cvm^i, eev_{cvm}^i).$$

We almost have

$$cvm^{i+1}.vm(u) \sim vmrun(d^{j'+1}, k^{j'+1}, u)$$

but not quite. The necessary conditions are fulfilled for memory and general-purpose registers:

$$cvm^{i+1}.vm(u).m = vm(d^{j'+1}, k^{j'+1}, u).m,$$
$$cvm^{i+1}.vm(u).gpr = d^{j'+1}.gpr.$$

Moreover, we even have

$$R \in \{eca, edata\} \;\to\; cvm^{i+1}.vm(u).R = d^{j'+1}.R.$$

However, after an interrupt we have in the physical machine

$$R \in \{pc, sr\} \to d^{j'+1}.R = 0_{32}$$

but we find the new pc of the virtual machine in the exception pc register

$$cvm^{i+1}.vm(u).pc = d^{j'+1}.epc$$

and the frozen status register in the exception status register

$$cvm^{i+1}.vm(u).sr = d^{j'+1}.esr.$$

- The interrupt is invisible if
$$ipf(d^{j'}, eev_{isa}^{j'}).$$

In this case we have to save a configuration equivalent to $cvm^i.vm(u)$ before we handle the page fault.

We almost have

$$cvm^i.vm(u) \sim vmrun(d^{j'+1}, k^{j'+1}, u)$$

but not quite. The necessary conditions are fulfilled for memory and general-purpose registers:

$$cvm^i.vm(u).m = vm(d^{j'+1}, k^{j'+1}, u).m,$$
$$cvm^i.vm(u).gpr = d^{j'+1}.gpr.$$

We almost certainly have

$$R \in \{eca, edata\} \to cvm^i.vm(u).R \neq d^{j'+1}.R$$

because these registers have been overwritten. Fortunately, this does not matter, because we only aim at saving an equivalent virtual machine configuration. Because a page fault has the resume type *repeat*, we find the old pc of the virtual machine in the exception pc register

$$cvm^i.vm(u).pc = d^{j'+1}.epc$$

and the old status register in the exception status register

$$cvm^i.vm(u).sr = d^{j'+1}.esr.$$

Now we can argue about the code of the inline assembly portion 'save' of *runvm*. First, observe that register $gpr(1_5)$ was saved in scratch memory by the code of the test for *reset*:

$$d^{j'+1}.gpr(1_5) = d^{j'+2}.m_4((4K+12)_{32}) \quad \text{(effect of } ssave(1)\text{)}$$
$$= d^{j}.m_4((4K+12)_{32}) \quad \text{(scratch memory unchanged)}.$$

We proceed to save the registers of the user process into process control block $PCB[u]$. By invariant $inv\text{-}scratch(d^j, k^j)$ the base address of this process control block is conveniently stored in the second word of scratch memory:

$$d^j.m_4((4K+4)_{32}) = ba(PCB[u], k^j).$$

From there we load it into $gpr(1_5)$ and save registers $gpr(2_5), \dots, gpr(31_5)$ using addressing relative to $gpr(1_5)$. In the end we save the value for $gpr(1_5)$ as stored in the scratch memory.

```
lw    1 0 4K+4          // gpr(1) = PCB[u] &
sw    2 1 4*2           // save register gpr(2)
[...]
sw    31 1 4*31         // save register gpr(31)
add   2 0 1             // gpr(2) = PCB[u] &
srestore(1)            // restore gpr(1) from scratch
sw    1 2 4*1           // save register gpr(1)
```

Next, the special purpose registers are saved via register 1. Note that we do not need to save the values of status and mode registers because they are known from the semantics of *jisr*, i.e., $d^j.sr = 0^{32}$ and $mode(d^j) = 0$. For the correctness proof the corresponding *PCB* components are irrelevant as well. Registers *pto* and *ptl* cannot be accessed by the user, hence we do not update the *PCB* for them either.

```
movs2g 1 1             // save register spr(1)
sw     1 2 4*(32+1)
[...]
movs2g 4 1             // save register spr(4)
sw     1 2 4*(32+4)
```

In the configuration d^{j_1} after this, all user registers are saved in a way simulating the next or old CVM configuration $cvm^i.vm(u)$ as required by Invariant 37:

$$\neg ipf(d^{j'}, eev_{isa}^{j'}) \rightarrow vmsus(d^{j_1}, k^{j_1}, u) = cvm^{i+1}.vm(u),$$
$$ipf(d^{j'}, eev_{isa}^{j'}) \rightarrow vmsus(d^{j_1}, k^{j_1}, u) \sim cvm^i.vm(u).$$

Here k^{j_1} is obtained from $k^{j'}$ by updating the parts of the process control block that were changed by the inline assembly routine:

$$X \neq m \;\rightarrow\; k^{j_1}.X = k^{j'}.X,$$

$$k^{j_1}.m(x) = \begin{cases} \langle d.gpr(i_5) \rangle, & x = PCB[u].gpr[i] \wedge i \in [1:31], \\ \langle d.spr(i_5) \rangle, & x = PCB[u].spr[i] \wedge i \in [1:4], \\ k^{j'}.m(x), & \text{otherwise.} \end{cases}$$

With the intention of regaining a C abstraction in the C+A computation (which means that we can think in terms of $C0$ programs) we restore pointers

$$(bpt, spt, hpt) = (gpr(28_5), gpr(29_5), gpr(30_5))$$

from process control block $PCB[0]$. By invariant $inv\text{-}scratch(d^j, k^j)$ its base address is found in the first word of the scratch memory.

```
lw    1 0 4K           // gpr(1) = PCB[0]&
lw    28 1 4*28        // restore register gpr(28)
lw    29 1 4*29        // restore register gpr(29)
lw    30 1 4*30        // restore register gpr(30)
```

Note that in the resulting configuration d^{j_2} we have regained consistency with some concrete kernel configuration k^{j_2} except for the program rest:

$$\exists k^{j_2} : consis'(k^{j_2}, d^{j_2}).$$

Let n be the node in the derivation tree of the concrete kernel generating the first statement of macro *compute-ipf*. By compiler construction (translation of statement sequences), we know that in configuration d^{j_2} the pc points to the first byte in the code of n

$$d^{j_2}.pc = start(n)$$

and hence

$$j_2 = j_\delta.$$

We construct a well-formed and consistent C abstraction k^{j_2}. Except for the program rest nodes and the program rest we of course copy the substitute configuration k^{j_1}:

$$X \neq \{prn, pr\} \rightarrow k^{j_2}.X = k^{j_1}.X.$$

We reconstruct the program rest nodes $k^{j_2}.prn$ starting with node n. We continue this sequence by taking successor nodes as defined by function *succ* until we hit a return statement (it will be the statement at the bottom of Fig. 235):

$$k^{j_2}.prn(1) = n,$$
$$\ell(k^{j_2}.prn(x)) \neq rSt \rightarrow k^{j_2}.prn(x+1) = succ(k^{j_2}.prn(x)).$$

We complete $k^{j_2}.prn$ with the 'TAIL' of $k^{j_1}.prn$:

$$TAIL(k^{j_2}.prn) = TAIL(k^{j_1}.prn).$$

The resulting configuration k^{j_2} is well-formed and consistent with d^{j_2}:

$$wfc(k^{j_2}) \wedge consis(k^{j_2}, d^{j_2}).$$

For the time j' of the interrupt we concluded

$$k\text{-}consis'(k^{j'}, cvm^i.c, 1) \wedge danglingr(k^{j'}, cvm^i.c).$$

The 'TAIL' of the program rest nodes component of the substituted C abstractions and the lower portions of the stack did not change since step j', i.e.,

$$TAIL(k^{j_2}.prn) = TAIL(k^{j'}.prn)$$

and

$$chop(k^{j_2}, cvm^i.c, 1) \wedge chop(k^{j'}, cvm^i.c, 1).$$

We conclude

$$TAIL(k^{j_2}.prn) = cvm^i.c.prn$$

and

$$k\text{-}consis'(k^{j_\delta}, cvm^i.c, 1)$$

as required for Invariant 26 with $j_\delta = j_2$.

15.6 Page Fault Handling

15.6.1 Testing for ipf Page Fault

The statements of the following code start at label δ. All declarations of variables are parts of the global variable declaration sequence of the concrete kernel.

- Detecting a page fault:

$$il = il(d^j, eev).$$

The interrupt level $(il > 0)$ is determined by the following obvious procedure.

```
uint i;
uint IL;
uint found;
[...]
IL = 0;
i = 1;
found = 0;
while found==0
{
   IL = IL + 1;
   i = 2 * i;
```

```
    gpr(1) = i;
    gpr(2) = PCB[CP].spr[2] {1};
    asm( and  3 1 2 );
    found = gpr(3)
};
```

For potential interrupt levels

$$il = 1, 2, \ldots$$

it successively generates masks

$$(2^{il})_{32} = 0^{31-il} 10^{il}$$

and ANDs them with the exception cause register of the interrupted process until the desired interrupt level is found.

Let y^{j_3} be the C+A configuration after execution of this code. Then

$$va(IL, k^{j_3}) = il(d^{j_3}, eev).$$

- Identifying the 'exception' virtual address eva causing the interrupt. For C+A configurations (d, k) and CVM configurations cvm this is

$$eva(d, k, cvm) \equiv \begin{cases} cvm.vm(u).pc, & pff(d), \\ ea(cvm.vm(u)), & pfls(d). \end{cases}$$

Overloading notation we set

$$eva = eva(d^j, k^j, cvm^i).$$

Page fault on fetch and load/store are recognized by interrupt levels 17 and 20 resp. In the first case the exception virtual address is found in the process control block of the current user in the exception pc register $(spr(3_5))$, while in the second case it is contained in the exception data register $(spr(4_5))$:

$$il = 17 \rightarrow pff(d^j) \wedge eva = va(PCB[CP].spr[3], k^{j_3})_{32},$$
$$il = 20 \rightarrow pfls(d^j) \wedge eva = va(PCB[CP].spr[4], k^{j_3})_{32}.$$

In all other cases it does not matter, and we set the corresponding C0 variable EVA to 0.

```
uint EVA;
[...]
EVA = 0;
if il==17 {EVA = PCB[CP].spr[3]};
if il==20 {EVA = PCB[CP].spr[4]};
```

Let y^{j_4} be the C+A configuration after execution of this code. Then

$$va(EVA, k^{j_4})_{32} = eva.$$

- We compute the page index

$$evpx = eva.px$$

of the exception virtual address by a right shift.

```
uint EVPX;
[...]
gpr(1) = EVA;
asm( srl  1 1 12 );
EVPX = gpr(1);
```

Let y^{j_5} be the C+A configuration after execution of this code. Then

$$va(EVPX, k^{j_5})_{20} = evpx.$$

- We check for a page table length exception, which occurs if the exception virtual address *eva* has a page index exceeding the length of the page table allocated for the user.

```
bool ptle;
[...]
ptle = ( EVPX>=PCB[u].spr[6] );
```

Let y^{j_6} be the C+A configuration after execution of this code. Then *ptle* is true iff a page table length exception occurred in the simulated virtual machine:

$$va(ptle, k^{j_6}) = 1 \equiv ptlef(cvm^i.vm(u)) \lor ptlels(cvm^i.vm(u)).$$

- The decision to branch to the page fault handler is based on flag *ipf*.

```
bool ipf;
[...]
ipf = ( il==17 || il==20 ) && !ptle;
```

After execution of this statement we are in step j_7 of the C+A computation with

$$va(ipf, k^{j_7}) = 1 \equiv ipf(d^{j'}, eev_{isa}^{j'}),$$

i.e., Invariant 38 holds with

$$j_\theta = j_7.$$

If no *ipf*-page fault is handled branch D is followed in the flow chart of *runvm*. Before execution of the return statement we are in C+A configuration y^{j_8} satisfying

$$consis(k^{j_8}, d^{j_8})$$

and

$$k^{j_8}.pr = return\ 1; cvm^i.c.pr.$$

Execution of the return statement returns control to the abstract kernel and leads to configuration y^{j_9} satisfying

$$inv\text{-}crun(d^{j_9}, k^{j_9}, cvm^i).$$

In the CVM semantics an interrupt from a user in step i does not change the abstract kernel component

$$cvm^{i+1}.c = cvm^i.c.$$

Hence with

$$s(i+1) = j_9$$

we complete the induction step for this case with

$$inv\text{-}crun(d^{s(i+1)}, k^{s(i+1)}, cvm^{i+1}).$$

15.6.2 Auxiliary Data Structures

In order to treat page fault handling we obviously have to know where user pages are stored in physical memory. We reserve

$$nup \geq 2$$

physical user pages for this purpose in physical memory as indicated in Fig. 239. Any physical user page can serve to store any page for any user depending on the page tables. Space for the pages starts at address *upbase* which is page aligned:

$$upbase.bx = 0^{12}.$$

Space for physical user page i starts at base address

$$upba(i) = upbase +_{32} (4K \cdot i)_{32}$$

and the page index of such an address is abbreviated as

$$upx(i) = upba(i).px.$$

The current content of such a page in ISA configuration d is

$$up(d, i) = d.m_{4K}(upba(i)).$$

Physical user pages lie above *hmax* and below the base address *hdbase* of the disk:

$$\langle hmax \rangle \leq \langle upbase \rangle < \langle upbase \rangle + 4K \cdot nup - 1 < \langle hdbase \rangle.$$

Valid physical page indices of page table entries always point into this region.

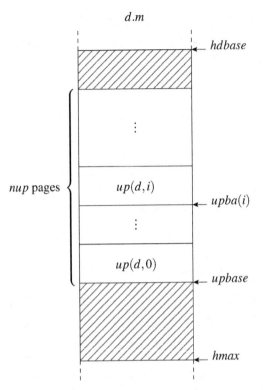

Fig. 239. Memory map for physical user pages in physical memory. They lie above the upper border *hmax* of the heap and below the base address *hdbase* of the disk ports

Invariant 45

$$inv\text{-}ppxrange(c,cvm) \equiv \forall u \, \forall a \in A(cvm.ptl(u)) : v(c,u,a) \rightarrow$$
$$\langle ppx(c,u,a) \rangle - \langle upbase.px \rangle \in [0 : nup-1].$$

A physical user page i is called full if some valid page table entry points to it (see Fig. 240):

$$upfull(c,i) \equiv \exists u,a : v(c,u,a) \wedge ppx(c,u,a) = upx(i).$$

By invariant *inv-ppx(c)* this page table entry and the pair (u,a) are unique.

We introduce variables which — among other things — keep track of full and free physical user pages:

- *bool psfull*: a flag indicating that all physical user pages are full. If this flag is off a page fault can simply be handled by swapping the missing user page from the disk to a free physical user page. Otherwise a 'victim page' has first to be swapped out to the disk, and then the missing page is swapped in.

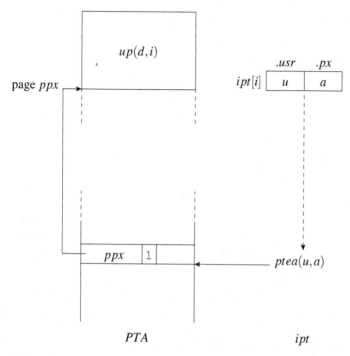

Fig. 240. A physical user page i is full if the ppx component of some valid page table entry $pte(u,a)$ points to it. In this case the inverse page table $ipt[i]$ stores u and $a.px$

- *uint nextup*: a variable storing the index of a physical user page to be swapped in next. If this page is already occupied by a user page, this victim page has to be swapped out before swapping in a new page.

```
typedef struct {uint usr; uint px} auxrec;
typedef auxrec[nup] ipt_t;
```

- *ipt_t ipt*: a variable $ipt[i]$ keeps track of the process number *usr* and the page index *px*, s.t. page *px* of user *usr* is currently held by physical user page i. The corresponding array *ipt* of structs is in a sense an inverse page table.

We require that after startup all physical swap pages are free:

$$va(psfull, k^0) = 0,$$
$$va(nextup, k^0) = 0.$$

We maintain the invariant that a physical user page i is full if all physical user pages are full or — while there are still free physical swap pages — if it lies below the *nextup* index.

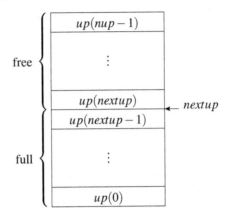

Fig. 241. Free physical user pages are filled from bottom to top. Pages below *nextup* are already filled. Once all physical user pages are filled, variable *nextup* is increased modulo *nup* after each page fault. This effectively turns the physical user pages into a FIFO queue

Invariant 46

$$inv\text{-}psfull(c) \equiv \forall i : upfull(c,i) \leftrightarrow (va(psfull,c) = 1 \lor i < va(nextup,c)).$$

Thus physical user pages are filled from bottom to top. This is illustrated in Fig. 241. After all physical user pages are full, we will increment variable *nextup* modulo *nup* after each page fault. In this way among all physical user pages we will always swap out the page which was swapped in first. Data structures which store and evict elements in this order are called *FIFO queues*, where FIFO abbreviates 'first in first out'.

Invariant 47

$$inv\text{-}ipt(c) \equiv \forall u,a,i : (v(c,u,a) \land ppx(c,u,a) = upx(i)) \rightarrow \\ (va(ipt[i].usr,c) = u \land va(ipt[i].px,c)_{20} = a.px).$$

This is illustrated in Fig. 240. The newly introduced invariants are collected into the following technical invariant.

Invariant 48
$$inv\text{-}t(3,d,c,cvm) \equiv inv\text{-}ppxrange(c,cvm) \land \\ inv\text{-}psfull(c) \land \\ inv\text{-}ipt(c).$$

The invariant holds after initialization. The steps previously described maintain it, because they do not write to the data structures on which this invariant depends.

If an *ipf*-page fault was detected in the test of Sect. 15.6.1, then branch *C* is followed. After the assignment to *ipf*, a C+A configuration y^{j8} is reached. This configuration satisfies

$$k\text{-}consis'(k^{j8}, cvm^i.c, 1)$$

because *k-consis′* already holds in C+A step j_2 and is not broken by subsequent C+A steps.

15.6.3 Swapping a Page Out

Page fault handling has to split cases on the *psfull* flag. If the flag is off, one only has to swap in the missing page. If it is on, one has to swap out page *nextup* and then swap in the missing page. After that one increases *nextup* modulo *nup*. In the first case one might have to turn on the *psfull* flag.

```
void ipf-handler()
{
   if !psfull {swap_in()}
   else {swap_out(); swap_in()};

   nextup = nextup + 1;

   if nextup==nup {nextup = 0; psfull = 1}
};
```

Procedure

```
void swap_out()
{
   [...]
};
```

is only invoked if flag *psfull* is on, i.e., if no free physical user page is available

$$va(psfull, k^{j8}) = 1,$$

which implies that all physical user pages are full by Invariant 46:

$$\forall i : upfull(k^{j8}, i).$$

We introduce the following shorthands:

$ui = va(nextup, k^{j8})$	(index of physical user page swapped),
$usrout = va(ipt[nextup].usr, k^{j8})$	(user swapped out),
$pxout = va(ipt[nextup].px, k^{j8})_{20}$	(page index of page swapped out),
$ppxout = upx(ui)$	(physical page index of swapped page),
$spaout = sma(usrout, pxout)$	(swap page address of swapped page),
$pteiout = ptei(k^{j8}, usrout, pxout \circ 0^{12})$	(index of page table entry to be invalid).

By the definition of *upfull*, invariant *inv-ipt*(k^{j8}), and the definition of *ptei* we conclude for any byte index $bx \in \mathbb{B}^{12}$:

$$v(k^{j8}, usrout, pxout \circ bx) = 1,$$
$$ppx(k^{j8}, usrout, pxout \circ bx) = ppxout,$$
$$pma(k^{j8}, usrout, pxout \circ bx) = ppxout \circ bx,$$
$$ptei(k^{j8}, usrout, pxout \circ bx) = pteiout.$$

Recall that $inv\text{-}vm(d, k, cvm)$ implies

$$\forall u : vm(d, k, u).m = cvm.vm(u).m.$$

We abbreviate this part of the invariant by $inv\text{-}m(d, k, cvm)$. Since the interrupt handler so far has not touched the user memory or page tables, by induction hypothesis we have $inv\text{-}m(d^{j8}, k^{j8}, cvm^i)$ and in particular

$$d^{j8}.m_{4K}(ppxout \circ 0^{12}) = cvm^i.vm(usrout).m_{4K}(pxout \circ 0^{12}) \qquad (18)$$

using the definition of $vm(d^{j8}, k^{j8}, usrout)$. Recalling that the swap memory address for user u and page index px was defined as

$$smbase(0) = (2^{28} - p \cdot 2^{20})_{28},$$
$$sma(u, px) = smbase(0) +_{28} u_8 \circ px,$$

we abbreviate

$$UPBASE = \langle upbase \rangle,$$
$$SMBASE = \langle smbase(0) \rangle,$$

and compute successively $ppxout$, $spaout$ and $pteiout$.

```
uint PPXOUT;
uint SPAOUT;
uint PTEIOUT;
[...]
PPXOUT = UPBASE / 2^12 + nextup;
SPAOUT = SMBASE + ipt[nextup].usr * 2^20
                + ipt[nextup].px;
PTEIOUT = PTOI[ipt[nextup].usr] + ipt[nextup].px;
```

In the C+A configuration y^{j9} after execution of this code we have

$$va(PPXOUT, k^{j9})_{20} = ppxout,$$
$$va(SPAOUT, k^{j9})_{28} = spaout,$$
$$va(PTEIOUT, k^{j9}) = pteiout.$$

Physical user page $nextup$ is swapped out with a call of the hard disk driver and entry $PTA[pteiout]$ is invalidated. The latter is achieved by subtraction of a number $2^{11} = 2048$ from the value of variable $PTA[pteiout]$.

```
writedisk(SPAOUT,PPXOUT);
PTA[PTEIOUT] = PTA[PTEIOUT] - 2048;
```

In the resulting C+A configuration y^{j10} by the specification of the disk driver and C+A semantics we have:

$$
\begin{aligned}
d^{j10}.sm(sma(usrout, pxout)) &= d^{j10}.sm(spaout) \quad \text{(Def. of } spaout) \\
&= d^{j9}.m_{4K}(ppxout \circ 0^{12}) \quad \text{(driver spec.)} \\
&= cvm^i.vm(usrout).m_{4K}(pxout \circ 0^{12}). \quad \text{(Eq. 18)}
\end{aligned}
$$

For the page table entry we get:

$$
\begin{aligned}
v(k^{j10}, usrout, pxout \circ 0^{12}) &= pte(k^{j10}, usrout, pxout \circ 0^{12})[11] \\
&= va(PTA[pteiout], k^{j10})_{32}[11] \\
&= 0.
\end{aligned}
$$

This shows the following.

Lemma 143.

$$
inv\text{-}m(d^{j10}, k^{j10}, cvm^i).
$$

15.6.4 Swapping a Page In

Procedure

```
void swap_in()
{
   [...]
};
```

is called in two situations:

- if all physical user pages were full — after a run of procedure *swap_out* — starting in configuration
$$
y^{j10}.
$$

- if free physical user pages were available — immediately after the detection of an *ipf*-page fault — in configuration y^{j8}. In this case no page is swapped out, and we set
$$
y^{j10} = y^{j8}.
$$

Swapping pages in is very similar to swapping them out. The inverse page tables are not needed to determine the parameters of the disk driver, but they have to be updated. We reuse abbreviation

$$
ui = va(nextup, k^{j10}) \quad \text{(index of physical user page swapped)}
$$

and — using u, eva, and $evpx$ as defined before — introduce the abbreviations:

$$ppxin = upx(ui) \qquad \text{(physical page index of the swapped page)},$$
$$spain = sma(u,evpx) \qquad \text{(swap page address of the swapped page)},$$
$$pteiin = ptei(k^{j10},u,evpx \circ 0^{12}) \qquad \text{(page table entry index accessed)}.$$

An ipf-page fault occurs only if

$$v(k^{j10},u,eva) = 0.$$

In this case we know from invariant $inv\text{-}vm(d^{j10},k^{j10},cvm^i)$ that the page of user u with index $evpx$ that is missing in physical memory is found on the disk at the swap page address $spain$:

$$cvm^i.vm(u).m_{4K}(evpx \circ 0^{12}) = d^{j10}.sm(spain). \tag{19}$$

We compute successively $ppxin$, $spain$, and $pteiin$.

```
uint PPXIN;
uint SPAIN;
uint PTEIIN;
[...]
PPXIN = UPBASE / 2^12 + nextup;
SPAIN = SMBASE + CP * 2^20 + EVPX;
PTEIIN = PTOI[CP] + EVPX;
```

In the C+A configuration y^{j11} after execution of this code we have:

$$va(PPXIN,k^{j11})_{20} = ppxin,$$
$$va(SPAIN,k^{j11})_{28} = spain,$$
$$va(PTEIIN,k^{j11}) = pteiin.$$

Physical user page $nextup$ is filled with the missing page by a call of the hard disk driver and $PTA[pteiin]$ is updated. The latter is achieved in a similar way to the above.

```
readdisk(PPXIN,SPAIN);
PTA[PTEIIN] = PPXIN * 2^12 + 2^11;
```

In the resulting C+A configuration y^{j12} by the specification of the disk driver and C+A semantics we have:

$$d^{j12}.m_{4K}(ppxin \circ 0^{12}) = d^{j10}.sm(spain) \quad \text{(driver spec.)}$$
$$= cvm^i.vm(u).m_{4K}(evpx \circ 0^{12}). \quad \text{(Eq. 19)}$$

For the page table entry we get:

$$pte(k^{j_{12}}, u, evpx \circ 0^{12}) = va(PTA[pteiin], k^{j_{12}})_{32}$$
$$= ppxin \circ 10^{11}.$$

This shows the following.

Lemma 144.

$$inv\text{-}m(d^{j_{12}}, k^{j_{12}}, cvm^i).$$

The inverse page tables are updated by

```
ipt[nextup].usr = CP;
ipt[nextup].px = PPXIN;
```

In the resulting C+A configuration $y^{j_{13}}$ invariants

$$inv\text{-}ipt(k^{j_{13}}) \quad \text{and} \quad inv\text{-}ppxrange(k^{j_{13}}, cvm^i)$$

hold. This completes the code of procedure *swap_in*. Procedure *ipf-handler* ends with an update of variables *nextup* and *psfull*.

```
nextup = nextup + 1;
if nextup==nup {nextup = 0; psfull = true};
```

In the resulting C+A configuration $y^{j_{14}}$ invariant

$$inv\text{-}psfull(k^{j_{14}})$$

holds.

Thus after the handling of the page fault all technical invariants hold and the memories of the suspended virtual machines are still consistent with the CVM model.

Lemma 145.

$$inv\text{-}T(d^{j_{14}}, k^{j_{14}}, cvm^i) \wedge inv\text{-}m(d^{j_{14}}, k^{j_{14}}, cvm^i).$$

With

$$j_\varepsilon = j_{14}$$

and Invariant 37 this shows Invariant 40.

Note that we also have the following.

Lemma 146.

$$va(nextup, k^{j_\varepsilon}) = (va(nextup, k^{j_\delta}) + 1 \bmod nup).$$

15.6.5 Liveness

During the execution of a user step in CVM configuration cvm^i up to two ipf interrupts can occur: one on instruction fetch and — if a load/store instruction was fetched — another one during the execution of this instruction. Observe also that the page fault on load/store can occur before the page fault on fetch: this is the case if the page containing the instruction was evicted during the handling of the page fault on load/store. This raises the question of whether we can ever succeed in handling all page faults of an instruction such that finally we can execute it. Formally this is the question of the existence of step number $s(i+1)$. Indeed without the hypothesis that the number nup of user pages in the physical memory is at least 2 there would be a problem. In the — artificial — case of a single user page the code page and the data page of a load/store instruction can evict each other forever. As we have $nup \geq 2$ we can argue about the liveness of the user step simulation as follows.

If more than one ipf-page fault is handled, then the page fault handler is run twice in steps $j > s(i)$ with exception virtual addresses eva^1 and eva^2. In particular C+A steps j_δ, j_ε, and j_ξ occur twice. Thus we index them too with superscripts 1 and 2. We have

$$eva^1 \neq eva^2$$

because the page ppx needed in physical memory to avoid the page fault on exception address eva^1 was just swapped in. By Lemma 146 and because $nup \geq 2$ we have

$$va(nextup, k^{j_\delta^2}) = va(nextup, k^{j_\varepsilon^1})$$
$$= (va(nextup, k^{j_\delta^1}) + 1 \bmod nup)$$
$$\neq va(nextup, k^{j_\delta^1}).$$

This means that the page ppx which was previously swapped in is not the victim page swapped out during the handling of the next page fault. Therefore no page fault will occur in C+A step j_ξ^2 and thus with

$$s(i+1) = j_\xi^2 + 1$$

the simulation of user step i will succeed if there is no other interrupt. In case of

$$jisr(d^{j_\xi^2}, eev^{j_\xi^2})$$

the kernel is entered and we have

$$s(i+1) = j_\eta.$$

15.7 Other CVM Primitives and Dispatching

Implementations of the remaining CVM primitives can now be described very easily. Correctness proofs are left as an exercise.

15.7.1 Accessing Registers of User Processes

The four CVM primitives which permit the abstract kernel to access the registers of user processes are only executed if the kernel is running. Thus the kernel simply has to access the corresponding registers in the process control blocks.

- $readgpr(e,u,i)$ is implemented as

```
e* = PCB[u].gpr[i];
```

- $writegpr(u,i,e)$ is implemented as

```
PCB[u].gpr[i] = e;
```

- $readspr(e,u,i)$ is implemented as

```
e* = PCB[u].spr[i];
```

- $writespr(u,i,e)$ is implemented as

```
PCB[u].spr[i] = e;
```

15.7.2 Accessing User Pages

Implementation of the two primitives $readpage(e,u,k)$ and $writepage(u,k,e)$ is slightly more involved. Let

$$\tilde{k} = va(k,cvm^i.c),$$
$$\tilde{u} = va(u,cvm^i.c),$$

then the primitives should only be executed in cvm step i with at least $\tilde{k}+1$ pages in the address range of user $\tilde{u}$:

$$\tilde{k} < cvm^i.ptl(\tilde{u}).$$

One proceeds in four obvious steps.

- Compute a flag signaling that page $\tilde{k}$ of user $\tilde{u}$ is valid. Store it in a Boolean variable $validk$. Below we use a binary constant $VMASK = 0^4 10^{11}$.

```
bool validk;
[...]
gpr(1) = PTA[PTOI[u] + k];
asm ( andi 1 1 0bVMASK );
validk = ( gpr(1)>0 );
```

- If the page is not valid, handle a page fault for user $\tilde{u}$ with exception page index $\tilde{k}$. We temporarily break invariant *inv-cp* and set the current process variable CP to $\tilde{u}$.

```
uint CPSAVE;
[...]
if !validk
{
   CPSAVE = CP;             // save CP
   CP = u;
   EVPX = k;
   ipf-handler();
   CP = CPSAVE             // restore CP
};
```

- Compute the base addresses of page $\tilde{k}$ and array type variable $\tilde{e}*$ of user $\tilde{u}$ in physical memory. Store them resp. in variables *bak* and *bae* of type *uint*.

```
uint bak;
uint bae;
[...]
gpr(1) = PTA[PTOI[u] + k];    // read updated PTA entry
asm( srl  1 1 12 );           // shift right by 12 bits
bak = gpr(1);
gpr(1) = e;
bae = gpr(1);
```

- Use the *copyms* macro from Sect. 14.3.9 to do the actual copying.

  For read page:

```
copyms(bak,bae,K);
```

  For write page:

```
copyms(bae,bak,K);
```

15.7.3 *free* and *alloc*

We have implemented page tables for all user processes in an extremely brute force and simple way: we simply packed them all consecutively into the single array *PTA*. The penalty for this is that growing or shrinking page tables for a single user u forces us to move all page tables of users $u' > u$ up, if pages are allocated for u, or down, if pages of user u are freed. Restrictions for the use of CVM primitives *free* and *alloc* in CVM step i are listed below.

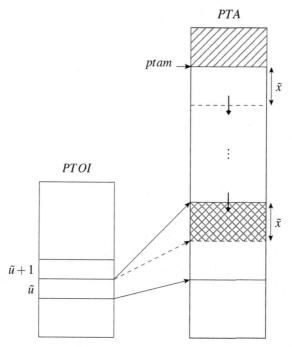

Fig. 242. Freeing $\tilde{x}$ pages of user $\tilde{u}$. The page tables of all users $\tilde{y} > \tilde{u}$ are moved down by $\tilde{x}$ in the page table array PTA. This concerns page table entry indices between the old $PTOI[\tilde{u}+1]$ and the old $ptam$. The new $PTOI[\tilde{u}+1]$ is shown with a dotted arrow

In a similar way to the above we abbreviate

$$\tilde{u} = va(e_1, cvm^i.c),$$
$$\tilde{x} = va(e_2, cvm^i.c).$$

1. *free*(e_1, e_2)

   Specification:
   $$cvm^{i+1}.ptl(\tilde{u}) = cvm^i.ptl(\tilde{u}) - \tilde{x}.$$

   Frees the number of pages given by e_2 of the user process given by e_1.
   We should only free pages that the user has, i.e.,

   $$\tilde{x} \le cvm^i.ptl(\tilde{u}).$$

   Implementation:

```
void free(uint u, uint x) {
   [...]
}
```

Freeing $\tilde{x}$ pages of user $\tilde{u}$ is illustrated in Fig. 242. We proceed in three steps.

- We decrease the page length of process $\tilde{u}$ and the page table origin indices of processes

$$\tilde{y} \in [\tilde{u}+1:p].$$

```
uint y;
[...]
PCB[u].spr[6] = PCB[u].spr[6] - x;
y = u + 1;
while y<=p
{
  PTOI[y] = PTOI[y] - x;
  y = y + 1
};
```

- We compute the maximal index *ptam* of an occupied page table array entry and copy entries $PTA[k]$ with indices $k \in [PTOI[\tilde{u}+1]:ptam]$ downward to position $(k-\tilde{x})$. We start with the lowest index.

```
uint ptam;
uint pti;
[...]
if u<p
{
  ptam = PTOI[p] + PCB[p].spr[6] - 1;
  pti = PTOI[u + 1];
  while pti<=ptam
  {
    PTA[pti] = PTA[pti + x];
    pti = pti + 1
  }
};
```

- We adjust page table origins in the process control blocks.

```
y = u + 1;
while y<=p
{
  gpr(1) = PTA[PTOI[y]]&;
  PCB[y].spr[5] = gpr(1);
  y = y + 1
};
```

2. $alloc(e_1, e_2)$

   Specification:

   $$cvm^{i+1}.ptl(\tilde{u}) = cvm^i.ptl(\tilde{u}) + \tilde{x}.$$

   Allocates the number of pages given by e_2 for the user process given by e_1.

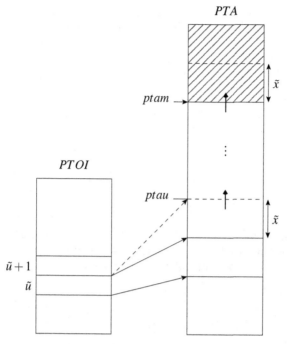

Fig. 243. Allocating $\tilde{x}$ pages to user $\tilde{u}$. The page tables of all users $\tilde{y} > \tilde{u}$ are moved up by $\tilde{x}$ in the page table array *PTA*. This concerns page table entry indices between the old $PTOI[\tilde{u}+1]$ and the old *ptam*. The new $PTOI[\tilde{u}+1] = ptau+1$ is shown with a dotted arrow. The new page table entries for user $\tilde{u}$ are initialized with zeros. This concerns indices between $ptau - \tilde{x}+1$ and *ptau*

There must be free space in the page table array for the newly allocated pages. The number of occupied page table entries in CVM configuration *cvm* is

$$npages(cvm) = \sum_u cvm.ptl(u).$$

The number of free pages in CVM configuration *cvm* is

$$nfpages(cvm) = PTASIZE - npages(cvm).$$

Thus we require

$$\tilde{x} \le nfpages(cvm^i).$$

Implementation:

```
void alloc(uint u, uint x) {
    [...]
}
```

Adding $\tilde{x}$ pages to user $\tilde{u}$ is illustrated in Fig. 243. We proceed in five steps.

- We increase the page length of process $\tilde{u}$ and the page table origin indices of processes

$$\tilde{y} \in [\tilde{u}+1:p].$$

```
uint y;
[...]
PCB[u].spr[6] = PCB[u].spr[6] + x;
y = u + 1;
while y<=p
{
  PTOI[y] = PTOI[y] + x;
  y = y + 1
};
```

- We compute the maximal index *ptam* of an occupied page table array entry and copy entries $PTA[k]$ with indices $k \in [PTOI[\tilde{u}+1]:ptam]$ upward to position $(k+\tilde{x})$. We start with the highest index.

```
uint ptam;
uint pti;
[...]
if u<p
{
  ptam = PTOI[p] + PCB[p].spr[6] - 1;
  pti = ptam;
  while pti>=PTOI[u + 1]
  {
    PTA[pti] = PTA[pti - x];
    pti = pti - 1
  }
};
```

- We adjust page table origins in the process control blocks exactly as before.

```
y = u + 1;
while y<=p
{
  gpr(1) = PTA[PTOI[y]]&;
  PCB[y].spr[5] = gpr(1);
  y = y + 1
};
```

- We compute the new maximal page table entry index *ptau* for user $\tilde{u}$ and set the new page table entries $k \in [ptau-\tilde{x}+1:ptau]$ to zero.

```
uint ptau;
uint c;
```

```
[...]
ptau = PTOI[u] + PCB[u].spr[6] - 1;
c = 0;
while c<x
{
  PTA[ptau - c] = 0;
  c = c + 1
};
```

- Making use of CVM primitive *writepage* we initialize the new pages of process $\tilde{u}$ on the swap page with all zeros.

```
typedef uint[K] upage_t;
[...]
upage_t e;
[...]
c = 0;
while c<K
{
  e[c] = 0;
  c = c + 1
};                          // creating a zero page
c = PCB[u].spr[6] - x;      // first new page
while c<PCB[u].spr[6]
{
  writepage(u,c,e&);
  c = c + 1
}; // writing x zero pages into user memory
```

15.7.4 Application Binary Interface and Dispatcher for the Abstract Kernel

We have completed construction of the concrete kernel and its correctness proofs, at least for the primitives we have mentioned. More primitives for creating and killing processes could quite easily be added. In contrast to the concrete kernel we have said very little so far about the abstract kernel. Indeed this is the point of the CVM model: it works with any abstract kernel. To program an abstract kernel one basically has to do two things:

- construct a scheduler computing new values of variable *CP* before calling the *runvm* primitive,
- define and implement services which are provided to the user processes via system calls.

Obviously the user needs to know:

- what services are provided,
- how to specify a particular service,
- how to pass parameters to the service.

The specification with this information for user processes written in ISA is called the application binary interface (ABI). If the user processes are compiled programs this interface has a counterpart on the level of the programming language used, which is called the application programming interface (API). We present here a very simple ABI.

- Formally services are simply $C0$ functions of the abstract kernel which can make use of the CVM primitives. The CVM model permits us both to specify their effect and to prove the correctness of their implementation. No assembler programming is involved any more: it is hidden in the semantics of the CVM primitives.
- Services are provided by handlers $h_1, \ldots, h_k$. The number $s \in [1:k]$ of the service h_s is passed by the user in general-purpose register $gpr(1_5)$.
- Each service has four parameters, passed in registers $gpr(2_5), \ldots, gpr(5_5)$.

The implementation of the abstract kernel will then contain a function called the *dispatcher*.

- It computes the interrupt level il from $PCB[CP].spr[2]$. We leave the programming of this test in $C0$ as an exercise.
- For each interrupt level it calls an interrupt service routine $isr\_il(\ldots)$ with appropriate parameters.
- If the interrupt level is 21, i.e., the level of the *sysc* interrupt, it inspects $PCB[CP].gpr[1]$ and calls the matching handler h_i with parameters

$$PCB[CP].gpr[2], \ldots, PCB[CP].gpr[5].$$

An obvious implementation is given below.

```
uint t_1;
[...]
uint t_5;
[...]
void dispatcher()
{
  uint il;
  [...]    // calculate interrupt level, store in 'il'
  if il==1  {isr_1(...)};
  [...]
  if il==20 {isr_20(...)};
  if il==21
  {        // call matching handler for syscall
    readgpr(t_1&,CP,1);
    [...]
    readgpr(t_5&,CP,5);
    if t_1==1 h_1(t_2,t_3,t_4,t_5);
    [...]
    if t_1==k h_k(t_2,t_3,t_4,t_5);
```

```
    };
    if il==22 {isr_22(...)};
    [...]
};
```

15.8 Final Remarks

Although this chapter is not short, you do not have to memorize terribly much:

- for virtual machines: the memory limitation and the visible interrupts.
- for the CVM model with abstract $C0$ kernel, ISA user processes, and parallelism. Very important: no assembly portions in the abstract kernel. Whatever is programmed later in the concrete kernel with assembly code can be specified by CVM primitives.
- the data structures of the concrete kernel.
- the three simulation relations involved:
  - for the user processes. Memory defined from physical memory and swap memory by address translation. Registers of running processes in physical registers. Registers of other processes in process control block *except* for *pc* and *sr* which are in components *epc* and *esr* of the process control blocks.
  - consistency of the concrete kernel: ordinary *consis* if the kernel is running. Otherwise *consis* would only hold if one restored the *pc* as well as the base, stack, and heap pointers from $PCB[0]$.
  - coupling concrete and abstract kernel: ordinary *k-consis* while the kernel is running. Dangling return while user processes are running. This is modeled by *k-consis'*.
- for the programming:
  - the flow chart of the *runvm* primitive.
  - for page fault handling: maintaining inverse page tables and free user pages. No big problem if you forget this: once you try to program it, you will see what is missing. Important however: eviction strategy first in first out (FIFO) and at least two physical user pages.
  - for *alloc*: initialization of the new pages by invocation of the page fault handler and then initializing the pages with zeros in physical memory.
- for correctness proofs:
  - annotate the flow chart. Just pick the right variants of the simulation relations for each place in the flow chart.
  - regaining C abstraction after 'process save'. Like for a goto; reconstruct the program rest using the successor function *succ*.
  - liveness of user step: at most two page faults per user step. Because of the FIFO strategy, handling the second page fault does not swap out the page swapped in at the first page fault.
  - recall how the user visible external interrupt signals were constructed.

Formally verified versions of results closely related to the material of this text have been produced in the Verisoft project [IdRP08]. In particular the formal verification of a page fault handler (which uses the formally verified disk driver from [AH08]) is reported in [ASS08], and the formal verification of a version of CVM (using the verified page fault handler) is reported in [AHL$^+$09]^3. Formal verification of abstract kernels together with the CVM layer, i.e., of $C0$ portions and of assembly portions of the kernel, is reported in [DSS09, Dör10].

The formal verification of SeL4, a kernel of industrial complexity, is reported in [KAE$^+$10], [SMK13]. So far the SeL4 project does not make use of a 'C+A' semantics yet; thus 'only' the C portions of the kernel can be completely verified. This is not a major problem: a quick way to remedy this would obviously be to treat SeL4 as an abstract kernel for CVM.

Encouraged by these results the formal verification of *HyperV*, a hypervisor with roughly 100,000 lines of code distributed by Microsoft together with Windows 7, was attempted from 2008 to 2010 in the Verisoft-XT project [CDH$^+$09]. The formal verification of a small sequential hypervisor programmed in 'C + macroassembly + assembly' is reported in [SS12, PSS12]. The full formal verification of the Microsoft product did not succeed in the time frame of the Verisoft-XT project. A main reason was that the theory, which is developed in this text for single-core kernels, turned out to be much harder to generalize to the multi-processor case than was originally expected. Development of this theory continued after 2010. A recent survey of the theory is given in [CPS13]. In this paper the remaining gaps in the theory are identified. At the time of writing the doctoral theses dealing with these gaps are nearing completion.

15.9 Exercises

Exercises for the last chapter of a lecture are slightly absurd: exercises always trail the material in the lecture. Thus the exercises of the last chapter never make it to the exercise sheet. We add them here anyway. After all there is nothing wrong with doing absurd things as long as they are harmless and fun.

1. Modify the processor hardware s.t. in untranslated mode accesses to addresses a with $a.px \geq P$ create a page table length exception (*ptle*) interrupt. You can assume that $P = 2^{\pi}$ is a power of two.
2. Pick some sections of the text and work out detailed correctness proofs instruction by instruction. Make sure to pick sections which force you to use all computational models developed so far and the specification of the disk drivers.
3. In the correctness argument for 'process save' we quietly assumed that the implementation was not interrupted. Why can we assume this
   a) for internal interrupts?
   b) for external interrupts?

---

^{3} In the formally verified version execution of the abstract kernel after 'process save' was resumed with *body*(*main*) as the program rest. Back in 2008 this simplified things somewhat.

4. Maintaining simulation relations.
   a) Which requirement guarantees that the abstract kernel does not break the invariants?
   b) Change the game by allowing the abstract kernel to use inline assembly. How could you destroy invariants for the process control blocks or the simulation of user memory while running such an abstract kernel?
5. Kernel with timer. Attach a timer to the physical processor and connect its interrupt signal to $eev[1]$ of the physical processor.
   a) Specify and implement a CVM primitive which allows the abstract kernel to access the timer.
   b) Assume the hardware is clocked at 1 GHz. Write a scheduler which runs user processes round robin (process $u + 1$ mod p after process u) and each process for at least 1 second.
   c) Why are disk driver runs not interrupted by the timer?
6. Handling heap overflow. In Sect. 12.3.2 the compiled code generates a system call if the execution of a *new* statement would produce a heap overflow. We are now in a position to handle this.
   a) Modify the compiled code such that it produces a system call with $gpr(1_5) = 1_{32}$ and $gpr(2_5) = mb$, where $\langle mb \rangle$ is the number of missing bytes. Pass the number of missing bytes as a parameter in $gpr[2]$. Pass the address where the computation is resumed after the handler has returned as a second parameter in register $gpr[3]$. Attention: *sysc* is a continue interrupt, thus you cannot use the exception *pc* that was saved.
   b) Assume you have a garbage collector. Specify a handler which calls the garbage collector. If it can collect enough garbage, it returns to the interrupted compiled user program, otherwise it does not return to the interrupted program.
7. Making the kernel interruptible. This is often desirable, but with the theory developed so far we cannot deal with this.
   a) For nested interrupts of the kernel we need a stack of process control blocks for the kernel, one for each level of nesting. Could we manipulate this within the abstract kernel?
   Hint: do our CVM primitives allow us to access the process control blocks of i) the users, ii) the kernel?
   b) Have we defined what happens *at the C0 level* if a compiled C0 instruction is interrupted?
   c) In order to avoid this difficulty we might be tempted to run inline assembly in the abstract kernel, while it is interruptible. Now CVM simulation by the concrete kernel would not be guaranteed any more. Why?

References

AA94.   American National Standards Institute and Accredited Standards Committee X3 for Information Technology. *ANSI X3.221-1994: AT Attachment Interface for Disk Drives*. American National Standards Institute, 1430 Broadway, New York, NY 10018, USA, 1994.

AH08.   Eyad Alkassar and Mark A. Hillebrand. Formal Functional Verification of Device Drivers. In Natarajan Shankar and Jim Woodcock, editors, *2nd IFIP Working Conference on Verified Software: Theories, Tools, and Experiments (VSTTE'08)*, Toronto, Canada, volume 5295 of *LNCS*, pages 225–239. Springer, 2008.

AHL+09.   Eyad Alkassar, Mark A. Hillebrand, Dirk Leinenbach, Norbert Schirmer, Artem Starostin, and Alexandra Tsyban. Balancing the Load. *J. Autom. Reasoning*, 42(2-4):389–454, 2009.

ASS08.   Eyad Alkassar, Norbert Schirmer, and Artem Starostin. Formal Pervasive Verification of a Paging Mechanism. In C. R. Ramakrishnan and Jakob Rehof, editors, *Proc. Tools and Algorithms for the Construction and Analysis of Systems (TACAS'08)*, Budapest, Hungary, volume 4963 of *LNCS*, pages 109–123. Springer, 2008.

Bau14.   Christoph Baumann. *Ownership-Based Order Reduction and Simulation in Shared-Memory Concurrent Computer Systems*. PhD thesis, Saarland University, Saarbrücken, 2014. http://www-wjp.cs.uni-saarland.de/publikationen/Ba14.pdf.

Bec87.   Bernd Becker. An Easily Testable Optimal-time VLSI-multiplier. *Acta Inf.*, 24(4):363–380, August 1987.

BJ01.   Christoph Berg and Christian Jacobi. Formal Verification of the VAMP Floating Point Unit. In Tiziana Margaria and Thomas F. Melham, editors, *Proc. 11th Advanced Research Working Conference on Correct Hardware Design and Verification Methods (CHARME'01)*, Livingston, Scotland, UK, volume 2144 of *LNCS*, pages 325–339. Springer, 2001.

BJK+03.   Sven Beyer, Christian Jacobi, Daniel Kroening, Dirk Leinenbach, and Wolfgang J. Paul. Instantiating Uninterpreted Functional Units and Memory System: Functional Verification of the VAMP. In Daniel Geist and Enrico Tronci, editors, *Correct Hardware Design and Verification Methods, Proc. 12th IFIP WG 10.5 Advanced Research Working Conference (CHARME'03)*, L'Aquila, Italy, volume 2860 of *LNCS*, pages 51–65. Springer, 2003.

© Springer International Publishing Switzerland 2016
W.J. Paul et al., *System Architecture*, DOI 10.1007/978-3-319-43065-2

504 References

BJMY89. William R. Bevier, Warren A. Hunt Jr., J Strother Moore, and William D. Young. An Approach to Systems Verification. *J. Autom. Reasoning*, 5(4):411–428, 1989.

BP05. Dominique Borrione and Wolfgang J. Paul, editors. *Proc. CHARME'05*, Saarbrücken, Germany, volume 3725 of *LNCS*. Springer, 2005.

CDH$^+$09. Ernie Cohen, Markus Dahlweid, Mark Hillebrand, Dirk Leinenbach, Michał Moskal, Thomas Santen, Wolfram Schulte, and Stephan Tobies. VCC: A Practical System for Verifying Concurrent C. In Stefan Berghofer, Tobias Nipkow, Christian Urban, and Markus Wenzel, editors, *Proc. TPHOLs'09*, Munich, Germany, volume 5674 of *LNCS*, pages 23–42. Springer, 2009.

CPS13. Ernie Cohen, Wolfgang Paul, and Sabine Schmaltz. Theory of Multi Core Hypervisor Verification. In Peter van Emde Boas, Frans C. A. Groen, Giuseppe F. Italiano, Jerzy Nawrocki, and Harald Sack, editors, *Proc. SOFSEM'13: Theory and Practice of Computer Science*, Spindleruv Mlyn, Czech Republic, volume 7741 of *LNCS*, pages 1–27. Springer, 2013.

CS10. Ernie Cohen and Bert Schirmer. From Total Store Order to Sequential Consistency: A Practical Reduction Theorem. In Matt Kaufmann and Lawrence C. Paulson, editors, *Proc. Interactive Theorem Proving, First International Conference (ITP'10)*, Edinburgh, UK, volume 6172 of *LNCS*, pages 403–418. Springer, 2010.

DHP05. Iakov Dalinger, Mark A. Hillebrand, and Wolfgang J. Paul. On the Verification of Memory Management Mechanisms. In Borrione and Paul [BP05], pages 301–316.

Dör10. Jan Dörrenbächer. *Formal Specification and Verification of a Microkernel*. PhD thesis, Saarland University, Saarbrücken, 2010. http://www-wjp.cs.uni-saarland.de/publikationen/JD10.pdf.

DSS09. Matthias Daum, Norbert W. Schirmer, and Mareike Schmidt. Implementation Correctness of a Real-Time Operating System. In Dang Van Hung and Padmanabhan Krishnan, editors, *Proc. SEFM'09*, Hanoi, Vietnam, pages 23–32. IEEE Computer Society, 2009.

HKS08. Ralf Huuck, Gerwin Klein, and Bastian Schlich, editors. *Proc. 3rd intl Workshop on System Software Verification (SSV'08)*, Sydney, Australia, volume 217 of *ENTCS*. Elsevier, 2008.

HT09. Mark A. Hillebrand and Sergey Tverdyshev. Formal Verification of Gate-Level Computer Systems. In Anna E. Frid, Andrey Morozov, Andrey Rybalchenko, and Klaus W. Wagner, editors, *Computer Science - Theory and Applications, Proc. 4th International Computer Science Symposium in Russia (CSR'09)*, Novosibirsk, Russia, volume 5675 of *LNCS*, pages 322–333. Springer, 2009.

HW73. C. A. R. Hoare and Niklaus Wirth. An Axiomatic Definition of the Programming Language PASCAL. *Acta Inf.*, 2:335–355, 1973.

IdRP08. T. In der Rieden and W. J. Paul. Beweisen als Ingenieurwissenschaft: Verbundprojekt Verisoft (2003–2007). In B. Reuse and R. Vollmar, editors, *Informatikforschung in Deutschland*, pages 321–326. Springer, 2008.

IdRT08. T. In der Rieden and A. Tsyban. CVM - A Verified Framework for Microkernel Programmers. In Huuck et al. [HKS08], pages 151–168.

ISO11. ISO. *ISO/IEC 9899:2011 Information technology — Programming languages — C.* International Organization for Standardization, Geneva, Switzerland, December 2011.

JR05. Warren A. Hunt Jr. and Erik Reeber. Formalization of the DE2 Language. In Borrione and Paul [BP05], pages 20–34.

KAE+10.   Gerwin Klein, June Andronick, Kevin Elphinstone, Gernot Heiser, David Cock, Philip Derrin, Dhammika Elkaduwe, Kai Engelhardt, Rafal Kolanski, Michael Norrish, et al. seL4: formal verification of an operating-system kernel. *Communications of the ACM*, 53(6):107–115, 2010.

KMP14.   M. Kovalev, S.M. Müller, and W.J. Paul. *A Pipelined Multi-core MIPS Machine: Hardware Implementation and Correctness Proof*, volume 9000 of *LNCS*. Springer, 2014.

KO63.   Anatoly A. Karatsuba and Y. Ofman. Multiplication of Multidigit Numbers on Automata. *Soviet Physics Doklady*, 7:595–596, 1963.

KP95.   Jörg Keller and Wolfgang J. Paul. *Hardware Design*. Teubner-Texte zur Informatik. Teubner, 1995.

Krö01.   Daniel Kröning. *Formal Verification of Pipelined Microprocessors*. PhD thesis, Saarland University, 2001. http://www-wjp.cs.uni-saarland.de/publikationen/Kr01.pdf.

Lam79.   Leslie Lamport. How to Make a Multiprocessor Computer That Correctly Executes Multiprocess Programs. *IEEE Trans. Computers*, 28(9):690–691, 1979.

Lan30.   Edmund Landau. *Grundlagen der Analysis*. Akademische Verlagsgesellschaft, 1930.

Ler09.   Xavier Leroy. Formal verification of a realistic compiler. *Communications of the ACM*, 52(7):107–115, 2009.

LMW89.   Jacques Loeckx, Kurt Mehlhorn, and Reinhard Wilhelm. *Foundations of Programming Languages*. John Wiley, 1989.

LP08.   D. Leinenbach and E. Petrova. Pervasive Compiler Verification – From Verified Programs to Verified Systems. In Huuck et al. [HKS08], pages 23–40.

LPP05.   D. Leinenbach, W. Paul, and E. Petrova. Towards the Formal Verification of a C0 Compiler: Code Generation and Implementation Correctness. In Bernhard K. Aichernig and Bernhard Beckert, editors, *Proc. SEFM'05*, Koblenz, Germany, pages 2–12. IEEE Computer Society, 2005.

MIP01.   MIPS Technologies, Inc. *MIPS32 Architecture For Programmers – Volume 2*, March 2001.

MP98.   Silvia M. Müller and Wolfgang J. Paul. On the correctness of hardware scheduling mechanisms for out-of-order execution. *Journal of Circuits, Systems, and Computers*, 8(02):301–314, 1998.

MP00.   Silvia M. Müller and Wolfgang J. Paul. *Computer Architecture, Complexity and Correctness*. Springer, 2000.

NK14.   Tobias Nipkow and Gerwin Klein. *Concrete Semantics - With Isabelle/HOL*. Springer, 2014.

Pau78.   Wolfgang J. Paul. *Komplexitätstheorie*. Leitfäden der angewandten Mathematik und Mechanik; Vol. 39. Teubner, 1978.

Pau13.   Wolfgang J. Paul. Computer Architecture 2 (unpublished lecture notes). Saarland University, 2013.

PH90.   David A. Patterson and John L. Hennessy. *Computer Architecture: A Quantitative Approach*. Morgan Kaufmann, 1990.

PSS12.   Wolfgang Paul, Sabine Schmaltz, and Andrey Shadrin. Completing the Automated Verification of a Small Hypervisor - Assembler Code Verification. In George Eleftherakis, Mike Hinchey, and Mike Holcombe, editors, *Proc. SEFM'12*, Thessaloniki, Greece, volume 7504 of *LNCS*, pages 188–202. Springer, 2012.

Rog67.   H. Rogers. *Theory of recursive functions and effective computability*. McGraw-Hill series in higher mathematics. McGraw-Hill, 1967.

SH02.   Jun Sawada and Warren A Hunt. Verification of FM9801: An out-of-order micro-
            processor model with speculative execution, exceptions, and program-modifying
            capability. *Formal Methods in System Design*, 20(2):187–222, 2002.
SMK13.   Thomas Sewell, Magnus Myreen, and Gerwin Klein. Translation Validation for a
            Verified OS Kernel. In Hans-Juergen Boehm and Cormac Flanagan, editors, *Proc.
            ACM-SIGPLAN Conference on Programming Language Design and Implemen-
            tation (PLDI'13)*, Seattle, Washington, USA, pages 471–481. ACM, 2013.
SS12.   Sabine Schmaltz and Andrey Shadrin. Integrated Semantics of Intermediate-
            Language C and Macro-Assembler for Pervasive Formal Verification of Oper-
            ating Systems and Hypervisors from VerisoftXT. In Rajeev Joshi, Peter Müller,
            and Andreas Podelski, editors, *4th International Conference on Verified Software:
            Theories, Tools, and Experiments (VSTTE'12)*, Philadelphia, USA, volume 7152
            of *LNCS*, pages 18–33. Springer, 2012.
SU70.   Ravi Sethi and J. D. Ullman. The Generation of Optimal Code for Arithmetic
            Expressions. *J. ACM*, 17(4):715–728, October 1970.
SVZN+13.   Jaroslav Sevcik, Viktor Vafeiadis, Francesco Zappa Nardelli, Suresh Jagannathan,
            and Peter Sewell. CompCertTSO: A Verified Compiler for Relaxed-Memory
            Concurrency. *Journal of the ACM*, 60(3):22, 2013.
VBC+15.   Viktor Vafeiadis, Thibaut Balabonski, Soham Chakraborty, Robin Morisset, and
            Francesco Zappa Nardelli. Common Compiler Optimisations are Invalid in the
            C11 Memory Model and what we can do about it. In Sriram K. Rajamani and
            David Walker, editors, *Proc. 42nd Symposium on Principles of Programming
            Languages (POPL'15)*, Mumbai, India, pages 209–220. ACM, 2015.
VN13.   Viktor Vafeiadis and Chinmay Narayan. Relaxed separation logic: a program
            logic for C11 concurrency. In Antony L. Hosking, Patrick Th. Eugster, and
            Cristina V. Lopes, editors, *Proc. ACM SIGPLAN International Conference on Ob-
            ject Oriented Programming Systems Languages & Applications, (OOPSLA'13)*,
            Indianapolis, IN, USA, pages 867–884. ACM, 2013.
Wal64.   C.S. Wallace. A suggestion for a fast multiplier. *Electronic Computers, IEEE
            Transactions on*, EC-13(1):14–17, February 1964.

Index

A operand, *see* sequential processor
ABI, *see* application binary interface
abstract kernel, 434
adder, 81
  carry-chain, 82
  carry-look-ahead, 88
  CSA, 83
addition, 9
  algorithm, *see* school method
address range
  of translated code, 255
  of variables, 259
address translation, 384
alignment, 113, 120, 123
ALU, *see* arithmetic logic unit
application binary interface, 498
arithmetic logic unit, 94
arithmetic sum, 25
arithmetic unit, 89
  negative bit, 90
  negative signal, 92
  overflow bit, 90
  overflow signal, 94
ASCII symbols, 182
assembler syntax, 145
asymptotic growth, 17
AU, *see* arithmetic unit

B operand, *see* sequential processor
base address, *see* C0 variable
base pointer, 255
BCE, *see* branch condition evaluation
big O notation, *see* asymptotic growth

binary numbers, 33
  addition, 36
  subtraction, 39
binary representation, 35
bit operations, 20
bit strings, 20
  as binary numbers, 33
  as two's complement numbers, 37
  decomposition, 35
bit vectors, *see* bit strings
Boolean equations, 41
  identities, 41, 42
  solving equations, 41, 44
Boolean expression, 39
  as a circuit, 54
  evaluation, 40
  syntax, 169
Boolean values, 17
boot loader, 472
border word, *see* derivation tree
branch condition evaluation, 98
  in sequential processor, 133
buffer, *see* hard disk ports
bytes, 20
  byte-addressable memory, 112
  consecutive bytes in memory, 113

C abstraction, 333
C address, *see* sequential processor
C0
  configuration, 199
  initial, 209
  well-formed, 334

© Springer International Publishing Switzerland 2016
W.J. Paul et al., *System Architecture*, DOI 10.1007/978-3-319-43065-2

Printed in the United States
By Bookmasters